# Statistical Power Analysis

*Statistical Power Analysis* explains the key concepts in statistical power analysis and illustrates their application in both tests of traditional null hypotheses (that treatments or interventions have no effect in the population) and in tests of the minimum-effect hypotheses (that the population effects of treatments or interventions are so small that they can be safely treated as unimportant). It provides readers with the tools to understand and perform power analyses for virtually all the statistical methods used in social and behavioral sciences.

Brett Myors and Kevin Murphy apply the latest approaches of power analysis to both null hypothesis and minimum-effect testing using the same basic unified model. This book starts with a review of the key concepts that underly statistical power. It goes on to show how to perform and interpret power analyses and the ways to use them to diagnose and plan research. We discuss the uses of power analysis in correlation and regression, in the analysis of experimental data, and in multilevel studies. This edition includes new material and new power software. The programs used for power analysis in this book have been rewritten in **R**, a language that is widely used and freely available. The authors include **R** codes for all programs, and we have also provided a web-based app that allows users who are not comfortable with **R** to perform a wide range of analyses using any computer or device that provides access to the web.

*Statistical Power Analysis* helps readers design studies, diagnose existing studies, and understand why hypothesis tests come out the way they do. The fifth edition includes updates to all chapters to accommodate the most current scholarship, as well as recalculations of all examples. This book is intended for graduate students and faculty in the behavioral and social sciences; researchers in other fields will find the concepts and methods laid out here valuable and applicable to studies in many domains.

**Kevin Murphy** is a professor emeritus, University of Limerick, and an organizational psychologist. He is an author and editor of over 13 books and over 200 articles and chapters in areas ranging from data analysis and research design to performance appraisal and performance management.

**Brett Myors** received his PhD in psychology from the University of New South Wales, with a postdoctoral appointment at Colorado State University. He served as director of organisational psychology at Griffith University and has published methodological research in several leading journals. He currently resides in the United Kingdom.

# Statistical Power Analysis
A Simple and General Model for Traditional and Modern Hypothesis Tests

Fifth Edition

**Kevin Murphy and Brett Myors**

Taylor & Francis Group

NEW YORK AND LONDON

Cover image: © Kevin Murphy and Brett Myors

First published 2023
by Routledge
4 Park Square, Milton Park, Abingdon, Oxon OX14 4RN

and by Routledge
605 Third Avenue, New York, NY 10158

*Routledge is an imprint of the Taylor & Francis Group, an informa business*

© 2023 Kevin Murphy and Brett Myors

The right of Kevin Murphy and Brett Myors to be identified as authors of this work has been asserted in accordance with sections 77 and 78 of the Copyright, Designs and Patents Act 1988.

All rights reserved. No part of this book may be reprinted or reproduced or utilised in any form or by any electronic, mechanical, or other means, now known or hereafter invented, including photocopying and recording, or in any information storage or retrieval system, without permission in writing from the publishers.

*Trademark notice*: Product or corporate names may be trademarks or registered trademarks, and are used only for identification and explanation without intent to infringe.

*British Library Cataloguing-in-Publication Data*
A catalogue record for this book is available from the British Library

ISBN: 978-1-032-28301-2 (hbk)
ISBN: 978-1-032-28300-5 (pbk)
ISBN: 978-1-003-29622-5 (ebk)

DOI: 10.4324/9781003296225

Typeset in Bembo
by MPS Limited, Dehradun

# Contents

*Acknowledgments* x
*Preface* xi

**1 The Power of Statistical Tests** 1
   *The Structure of Statistical Tests 2*
      *Null Hypotheses vs. Nil Hypotheses 4*
      *Understanding Conditional Probability 6*
   *The Mechanics of Power Analysis 9*
      *Understanding Sampling Distributions 9*
      *d vs. delta vs. g 16*
   *Statistical Power of Research in the Social and Behavioral Sciences 17*
      *Power and the Replication Crisis 18*
   *Using Power Analysis 20*
      *The Meaning of Statistical Significance 20*
   *Hypothesis Tests vs. Confidence Intervals 23*
      *Accuracy in Parameter Estimation 25*
   *What Can You Learn from a Null Hypothesis Test? 26*
   *Notes 27*

**2 A Simple and General Model for Power Analysis** 28
   *The General Linear Model, the F Statistic, and Effect Size 30*
      *Effect Size 30*
      *Understanding Linear Models 32*
   *The F Distribution and Power 33*
      *Confidence Intervals for PV and d 35*
   *Using the Noncentral F Distribution to Assess Power 37*
   *Translating Common Statistics and ES Measures into F 38*
      *Worked Example – Hierarchical Regression 39*
      *Worked Examples Using the d Statistic 40*

vi  Contents

    *Defining Large, Medium, and Small Effects 42*
    *Nonparametric and Robust Statistics 43*
    *From F to Power Analysis 44*
    *Analytic and Tabular Methods of Power Analysis 44*
    *Using the One-Stop F Table 46*
    *Simple and General Software for Power Analysis 47*
        *R code for Power Analysis for Traditional and Modern Hypothesis Tests 47*
    *A Web-Based App for Power Analysis 49*
    *Notes 51*

**3 Power Analyses for Minimum-Effect Tests**     52
    *Nil Hypothesis Testing 52*
    *The Nil Hypothesis is Almost Always Wrong 54*
    *Implications of the Conclusion That the Nil Hypothesis Is Almost Always Wrong 56*
        *Polar Bear Traps: Why Type I Error Control is a Bad Investment 57*
    *The Nil may not be True, but it is Often Fairly Accurate 58*
    *Minimum-Effect tests as Alternatives to Traditional Null Hypothesis Tests 59*
        *Sometimes a Point Hypothesis is also a Range Hypothesis 60*
        *How do you Know the Effect Size? 62*
    *Testing the Hypothesis that Treatment Effects are Negligible 63*
    *Using the One-Stop Tables to Assess Power to Test Minimum-Effect Hypotheses 67*
        *A Worked Example of Minimum-Effect Testing 69*
    *Type I Errors in Minimum-Effect Tests 71*
    *Notes 72*

**4 Using Power Analyses**     73
    *Estimating the Effect Size 74*
        *Using the One-Stop Tables and the R Code/Shiny Web app to Perform Power Analyses 75*
        *Worked Example: Calculating F-equivalents and Power 77*
    *Four Applications of Statistical Power Analysis 79*
    *Calculating Power 79*
    *Determining Sample Sizes 81*
        *A Few Simple Approximations for Determining Sample Size Needed 83*

*Determining the Sensitivity of Studies 84*
*Determining Appropriate Decision Criteria 85*
   *Finding a Sensible Alpha 88*
*Post-Hoc Power Analysis Should be Avoided 89*
*Notes 89*

## 5  Correlation and Regression — 90

*The Perils of Working with Large Samples 90*
*Multiple Regression 93*
   *Testing Minimum-Effect Hypotheses in Multiple*
      *Regression 94*
*Power in Testing for Moderators 97*
   *Power Analysis for Moderators 99*
*Implications of Low Power in Tests for Moderators 100*
*If You Understand Regression, You Will Understand (Almost)*
   *Everything 101*
*Notes 102*

## 6  $t$-Tests and the One-Way Analysis of Variance — 103

*The t-Test 103*
   *The t Distribution versus the Normal Distribution 105*
*Independent Groups t Test 105*
*One-Tailed versus Two-Tailed Tests 108*
   *Re-analysis of Smoking Reduction Treatments: One-Tailed*
      *Tests 109*
*Repeated Measures or Dependent t-Test 109*
*The Analysis of Variance 111*
   *Retrieving Effect Size Information from F Ratios 113*
*Which Means Differ? 114*
*Designing a One-way ANOVA Study 117*
*Notes 118*

## 7  Multifactor ANOVA Designs — 119

*The Factorial Analysis of Variance 119*
   *Calculating PV from F and df in Multi-Factor ANOVA:*
      *Worked Example 124*
*Factorial ANOVA from Means and Standard Deviations 126*
   *Reconstructing ANOVA results from descriptive statistics: A*
      *Worked Example 127*
   *Eta squared vs. partial eta squared 130*

General Design Principles for Multifactor ANOVA 131
Fixed, Mixed, and Random Models 133
Note 134

8 **Studies with Multiple Observations for Each Subject: Repeated-Measures and Multivariate Analyses** — 135
Randomized Block ANOVA: An Introduction to Repeated Measures Designs 135
Independent Groups versus Repeated Measures 137
Complexities in Estimating Power in Repeated-Measures Designs 141
Mixed Designs: Split-Plot Factorial ANOVA 142
Estimating Power for a Split Plot Factorial ANOVA 145
Power for Within-Subject versus Between-Subject Factors 145
Split-Plot Designs with Multiple Repeated-Measures Factors 146
The Multivariate Analysis of Variance 147

9 **Power Analysis for Multilevel Studies** — 149
What do Multilevel Analyses Tell You? 150
The Multilevel Equation 152
Are Multilevel Models Necessary? – The Intraclass Correlation 155
An Illustration of Multilevel Analysis 156
Remember, It's All Regression 159
Effect Sizes in Multilevel Analysis 160
R code for obtaining $R^2$ and pseudo-$R^2$ estimates 161
Power for What? 162
Using Changes in Model Fit as a Basis for Power Analysis in Multilevel Modeling 164
R code for calculating critical Chi-squared values and power for minimum-effect comparisons of models 165
Sample Size – Some General Guidance 168
Notes 169

10 **The Implications of Power Analyses** — 170
Tests of the Traditional Null Hypothesis 170
Tests of Minimum-Effect Hypotheses 172
Type I Errors in Minimum-Effect Tests Revisited 174
Statistical Power and the Replication Crisis 177
Power Analysis: Benefits, Costs, and Implications for Hypothesis Testing 179

*Direct Benefits of Power Analysis 179*
    *Is HARKing a Serious Problem? 181*
*Indirect Benefits of Power Analysis 182*
*Costs Associated With Power Analysis 183*
*Implications of Power Analysis: Can Power be too High? 184*
*Note 185*

| | |
|---|---:|
| *Appendix A Translating Common Statistics into F-Equivalent and PV Values* | 186 |
| *Appendix B One-Stop F Table* | 187 |
| *Appendix C One-Stop PV Table* | 204 |
| *Appendix D $df_{err}$ Needed for Power of .80 for Nil and Minimum-Effect Hypothesis Tests* | 221 |
| *References* | 227 |
| *Index* | 234 |

# Acknowledgments

We gratefully acknowledge the contributions of Allen Wolach, co-author of several previous volumes, for his contributions to the development of the ideas, examples, and software that provided valuable in developing this current edition.

# Preface

One of the most common statistical procedures in the behavioral and social sciences is to test the hypothesis that treatments or interventions have no effect, or that the correlation between two variables is equal to zero, etc. – i.e., tests of the null hypothesis. Researchers have long been concerned with the possibility that they will reject the null hypothesis when it is in fact correct (i.e., make a Type I error), and an extensive body of research and data-analytic methods exists to help understand and control these errors. Unfortunately, less attention has been devoted to the possibility that researchers will fail to reject the null hypothesis, when in fact treatments, interventions, etc. have some real effect (i.e., make a Type II error). Statistical tests that fail to detect the real effects of treatments or interventions might substantially impede the progress of scientific research.

The statistical power of a test is the probability that it will lead you to reject the null hypothesis when that hypothesis is in fact wrong. Because most statistical tests are done in contexts where treatments have at least some effect (although it might be minuscule), power often translates into the probability that your test will you lead to a correct conclusion about the null hypothesis. Viewed in this light, it is obvious why researchers have become interested in the topic of statistical power, and in methods of assessing and increasing the power of their tests.

In response to criticisms of traditional null hypothesis testing, several researchers have developed methods for testing what we refer to as "minimum-effect" hypotheses – i.e., the hypothesis that the effect of treatments, interventions, etc. – exceeds some specific minimal level. For example, you might test the hypothesis that the effect of treatments is trivially small (e.g., that treatments account for less than 1% of the variance in outcomes) rather than the hypothesis that treatments have no effect whatsoever. While the difference between hypothesizing that treatment effects are, at best, trivially small might not seem all that different than hypothesizing that treatments have no effect at all, this shift from traditional null hypothesis testing to testing the effects that treatment effects are so small as to be trivial revolutionizes many aspects of null hypothesis testing and it solves many of the most intractable problems with traditional tests.

Ours is the first book to discuss in detail the application of power analysis to both traditional null hypothesis tests and to minimum-effect tests. We show how the same basic model applies to both types of testing and illustrate applications of power analysis to both traditional null hypothesis tests (i.e., tests of the hypothesis that treatments have no effect) and to minimum-effect tests (i.e., tests of the hypothesis that the effects of treatments exceeds some minimal level). This book presents a simple and general model for statistical power analysis that is based on the widely used $F$ statistic. A wide variety of statistics used in the social and behavioral sciences can be thought of as special applications of the "general linear model" (e.g., $t$-tests, analysis of variance and covariance, correlation, multiple regression), and the $F$ statistic can be used in testing hypotheses about virtually any of these specialized applications. The model for power analysis laid out here is quite simple, and it illustrates how these analyses work and how they can be applied to problems of study design, to evaluating others' research, and even to problems such as choosing the appropriate criterion for defining "statistically significant" outcomes.

Most of the analyses presented in this book can be carried out using a single table, the One-Stop $F$ Table, presented in Appendix B. Appendix C presents a comparable table that expresses statistical results in terms of the percentage of variance explained ($PV$) rather than the $F$ statistic. These two tables make it easy to move back and forth between assessments of statistical significance and assessments of the strength of various effects in a study. Virtually all the power analyses discussed in this book can be performed using simple **R** programs (**R** code is provided). For users who are not familiar with or comfortable with **R**, we provide a web-based app (see link below) that performs the analyses discussed in this book without any need to enter or run **R** code.

This book is intended for a wide audience, including advanced students and researchers in the social and behavioral sciences, education, health sciences, and business. Presentations are kept simple and nontechnical whenever possible. Although most of the examples in this book come from social and behavioral sciences, general principles explained in this book should be useful to researchers in diverse disciplines.

## Changes in the New Edition

Previous editions of this book have included calculators and computer programs for performing power analysis for traditional and modern hypothesis tests, but these programs were sometimes difficult to use and maintain. We have completely rewritten the programs used to perform power analyses, generated the tables shown in Appendices B–D, and determine sample sizes needed to achieve adequate power using the computer language **R**. **R** is freely available and widely used in the scientific community, and it provides a simple and robust platform for power analysis. We provide the **R** code for all programs, which allows users to examine, modify, and customize these programs. For users who are not comfortable with **R**, we provide a web app (https://murphy0921.shinyapps.io/ShinyPower/) that allows you to estimate power and determine sample sizes needed to obtain specific levels of power without obtaining or running **R**.

In this edition, we introduce a discussion of power analysis for multilevel analysis. In addition, we have provided many new worked examples and discussions of analyses. Like our previous edition, we include **Boxed Material** in each of the chapters that provide insights into the meaning and implications of power analysis. We have added new material and expanded discussions to all our chapters, and we believe this new material makes it easier for readers to understand and apply the material covered in this book.

Kevin Murphy and Brett Myors

# 1 The Power of Statistical Tests

In the social and behavioral sciences, statistics serve two general purposes. First, they can be used to describe what happened in a particular study (descriptive statistics). Second, they can be used to help draw conclusions about what those results mean in some broader context (inferential statistics). The main question in inferential statistics is whether a result, finding, or observation from a study reflects some meaningful phenomenon in the population from which that study was drawn. For example, if 100 college sophomores are surveyed and it is determined that most of them prefer pizza to hot dogs, does this mean that people in general (or college students in general) also prefer pizza? If a medical treatment yields improvements in 6 out of 10 patients, does this mean that it is an effective treatment that should be approved for general use? The goal of inferential statistics is to determine what sorts of inferences and generalizations can be made based on data of this sort and to assess the strength of evidence and the degree of confidence one can have in these inferences.

The process of drawing inferences about populations from samples is a risky one, and a great deal has been written about the causes and cures for errors in statistical inference. Statistical power analysis (Cohen, 1988; Kraemer & Thiemann, 1987; Lipsey, 1990) falls under this general heading. Studies with too little statistical power can lead to erroneous conclusions about the meaning of the results of a particular study. In the example cited above, the fact that a medical treatment worked for 6 out of 10 patients is probably insufficient evidence that it is truly safe and effective, and if you have nothing more than this study to rely on, you might conclude that the treatment has not been proven effective. Does this mean that you should abandon the treatment, or that it is unlikely to work in a broader population? The conclusion that the treatment has not been shown to be effective may say as much about the low level of statistical power in your study as about the value of the treatment.

In this chapter, we will describe the rationale for and applications of statistical power analysis. In most of our examples, we describe or apply power analysis in studies that assess the effect of some treatment or intervention (e.g., psychotherapy, reading instruction, performance incentives)

DOI: 10.4324/9781003296225-1

by comparing outcomes for those who have received the treatment to outcomes of those who have not (non-treatment or control group). However, power analysis can be applied to virtually all statistical tests, and the same simple and general model can be applied to virtually all of the statistical analyses you are likely to encounter in the social and behavioral sciences.

## The Structure of Statistical Tests

To understand statistical power, you must first understand the ideas that underlie statistical hypothesis testing. Suppose 100 children are randomly divided in two groups. Fifty children receive a new method of reading instruction, and their performance on reading tests is on average six points higher (on a 100-point test) than the other 50 children who received standard methods of instruction. Does this mean that the new method is truly better? A six-point difference *might* mean that the new method is better, but it is also possible that there is no real difference between the two methods, and that this observed difference is the result of the sort of random fluctuation you might expect when you use the results from a single sample to draw inferences about the effects of these two methods of instruction in the population.

One of the most basic ideas in statistical analysis is that results obtained in a sample do not necessarily reflect the state of affairs in the population from which that sample was drawn. For example, the fact that scores averaged six points higher in this group of children does not necessarily mean that scores will be six points higher in the population, or that the same six-point difference would be found in another study examining a new group of students. Because samples do not (in general) perfectly represent the populations from which they were drawn, you should expect some instability in the results obtained from each sample. This instability is usually referred to as "sampling error". The presence of sampling error is what makes drawing inferences about populations from samples difficult. One of the key goals of statistical theory is to estimate the amount of sampling error that is likely to be present in different statistical procedures and tests and thereby gaining some idea about the amount of risk involved in using a particular procedure.

Statistical significance tests can be thought of as decision aids. That is, these tests can help you reach conclusions about whether the findings of your study are likely to represent real population effects, or whether they fall within the range of outcomes that might be produced by random sampling error. For example, there are two possible interpretations of the findings in this study of reading instruction:

1 the difference between average scores from the two programs is so small that it might reasonably represent nothing more than sampling error
   vs.
2 the difference between average scores from the two programs is so large that it cannot be reasonably explained in terms of sampling error

The most common statistical procedure in the social and behavioral sciences is to pit a null hypothesis ($H_0$) against an alternative ($H_1$). In this example, the null and alternative hypotheses might take the forms:

$H_0$ – Reading instruction has no effect. It doesn't matter how you teach children to read because in the population, there is no difference in the average scores of children receiving either method of instruction
vs.
$H_1$ – Reading instruction has an effect. It does matter how you teach children to read because in the population; there *is* a difference in the average scores of children receiving different methods of instruction

Although null hypotheses usually refer to "no difference" or "no effect", it is important to understand that there is nothing magic about the hypothesis that the difference between two groups is zero. It might be perfectly reasonable to evaluate the following set of possibilities:

$H_0$ – In the population, the difference in the average scores of those receiving these two methods of reading instruction is six points
vs.
$H_1$ – In the population, the difference in the average scores of those receiving these two methods of reading instruction is *not* six points

Another possible set of hypotheses is:

$H_0$ – In the population, the new method of reading instruction is *not* better than the old method; the new method might even be worse
vs.
$H_1$ – In the population, the new method of reading instruction *is* better than the old method

This last set of hypotheses leads to what is often called a "one-tailed" statistical test, in which the researcher not only asserts that there is a real difference between these two methods but also describes the direction or the nature of this difference (i.e., that the new method is not just different from the old one, it is also better). We will discuss one-tailed tests in several sections of this book, but in most cases will focus on the more widely-used two-tailed tests that compare the null hypothesis that nothing happened with the alternative hypothesis that something happened; unless we specifically note otherwise, the traditional null hypothesis tests discussed in this book will be assumed to be two-tailed. However, the minimum effect tests we introduce in Chapter 2 and discuss extensively throughout the book have all the advantages and few of the drawbacks of traditional one-tailed tests of the null hypothesis.

> **Null Hypotheses vs. Nil Hypotheses**
>
> The most common structure for tests of statistical significance is to pit the null hypothesis that treatments have no effect, or that there is no difference between groups, or that there is no correlation between two variables against the alternative hypotheses that there is *some* treatment effect. In fact, this structure is so common that most people assume that the "null hypothesis" is essentially a statement that there is no difference between groups, no treatment effect, no correlation between variables, etc. This is not true. The null hypothesis is simply the hypothesis you actually test, and if you reject the null, you are left with the alternative. That is, if you reject the hypothesis that the effect of an intervention of treatment is "X", you are left to conclude that the alternative hypotheses that the effect of treatments is "not-X" must be true. If you test and reject the hypothesis that treatments have no effect, you are left with the conclusion that they must have some effect. If you test and reject the hypothesis that a particular diet will lead to a 20% weight loss, you are left with the conclusion that the diet will *not* lead to a 20% weight loss (if might have no effect, if might have a smaller effect, it might even have a larger effect).
>
> Following Cohen's (1994) suggestion, we think it is useful to distinguish between the null hypothesis in general and its very special and very common form, the "nil hypothesis" – i.e., the hypothesis that treatments, interventions, etc. have no effect whatsoever. The nil hypothesis is common because it is very easy to test and because it leaves you with a fairly simple and concrete alternative. If you reject the nil hypothesis that nothing happened, the alternative hypothesis you should accept is that something happened. However, as we will show in this chapter and in the chapters that follow, there are often important advantages to testing null hypotheses that are broader than the traditional nil hypothesis.

Most treatments of power analysis focus on the statistical power of tests of the nil hypothesis – i.e., tests of the hypothesis that treatments or interventions have no effect whatsoever. However, there many advantages to posing and testing substantive hypotheses about the size of treatment effects (Murphy & Myors, 1999). For example, it is easy to test the hypothesis that the effects of treatments are negligibly small (e.g., they account for 1% or less of the variance in outcomes, or that the standardized mean difference is .10 or less). If you test and reject this hypothesis, you are left with the alternative hypothesis that the effect of treatments is *not* negligibly small, but rather large enough to deserve at least some attention. The methods of power analysis described in this book are easily extended to such minimum-effect tests and are not limited to traditional tests of the null hypothesis that treatments have no effect.

## *What Determines the Outcomes of Statistical Tests?*

Four outcomes are possible when you use the results obtained in a particular sample to draw inferences about a population; these outcomes are shown in Figure 1.1.

As Figure 1.1 shows, there are two ways to make errors when testing hypotheses. First, the treatment (e.g., a new method of instruction) might have no real effect in the population, but the results in your sample might lead you to believe that it does have some effect. If the results of this study lead you to incorrectly conclude that the new method of instruction does work better than the current method, when in fact there were no differences, you would be making a *Type I* error (sometimes called an *alpha* error). Type I errors might lead you to waste time and resources by pursuing what are essentially dead ends, and researchers have traditionally gone to great lengths to avoid Type I errors.

There is extensive literature dealing with methods of estimating and minimizing the occurrence of Type I errors (e.g., Keselman, Miller & Holland, 2011; Zwick & Marascuilo, 1984). The probability of making a Type I error is in part a function of the standard or decision criterion used in testing your hypothesis (often referred to as alpha, or $\alpha$). A very lenient standard (e.g., if there is *any* difference between the two samples, you will conclude that there is also a difference in the population) might lead to

|  | What is True in the Population? | |
|---|---|---|
|  | Treatments Have No Effect | Treatments Have An Effect |
| **Conclusion Reached in a Study** — No Effect | Correct Conclusion ($p = 1-\alpha$) | Type II Error ($p = \beta$) |
| **Conclusion Reached in a Study** — Treatment Effect | Type I Error ($p = \alpha$) | Correct Conclusion ($p = 1-\beta$) |

*Figure 1.1* Outcomes of Statistical Tests.

more frequent Type I errors, whereas a more stringent standard might lead to fewer Type I errors.[1]

The second type of error (referred to as *Type II* error, or a *beta* error) is also common in statistical hypothesis testing (Cohen, 1994; Sedlmeier & Gigerenzer, 1989). A Type II error occurs when you conclude in favor of $H_0$, when in fact $H_1$ is true. For example, if you conclude that there are no real differences in the outcomes of these two methods of instruction, when in fact one really is better than the other in the population, you have made a Type II error.

Statistical power analysis is concerned with Type II errors (i.e., the probability of making a Type II error is $\beta$, power = $1 - \beta$). Another way of saying this is to note that power is the (conditional probability) probability that you will *avoid* a Type II error. Studies with high levels of statistical power will rarely fail to detect the effects of treatments. If we assume that most treatments have at least some effect, the statistical power of a study often translates into the probability that the study will lead to the correct conclusion – i.e., that it will detect the effects of treatments.

---

### Understanding Conditional Probability

Conditional probability is an important concept in power analysis, and it can be a difficult to understand. An example might help. Suppose a doctor sees 100 patients, and 5 of them have prostate cancer. You bump into one of this doctor's patients in the street. The simple probability that this patient has prostate cancer is:

Probability(cancer) = # who have cancer/total # of patients = 5/100 = .05

Assessments of conditional probability always require some additional information. Suppose that all patients are given routine screening exams for prostate cancer, and 20 patients receive high scores (indicating higher cancer risk). Suppose further that 4 of these 20 patients have cancer. Assessments of conditional probability ask about the likelihood of cancer, given that we know a person has received a high score on the cancer screening test. There are only 20 people like this, and an assessment of the conditional probability of cancer, given a high score on this test is defined by

Probability(cancer|high score on screening test) = 4/20 = .20

In diagnostic testing, this type of conditional probability is referred to as the *sensitivity* of the test. The analysis above suggests that the test is not very sensitive. Sixteen of the 20 people who will get a cancer scare because of this test will *not* have cancer. Sensitivity can be contrasted with specificity, which is defined by a comparison between those who do not have cancer with those who do not receive high scores on the screening test. *Specificity* is also a type of conditional probability, defined here as

Probability(no cancer|low score on screening test) = 79/80 = .98

In this context, the test is very specific. People who receive a low score on the test rarely have cancer. The best diagnostic tests have high sensitivity (i.e., they almost always detect the condition) and high specificity (i.e., they rarely give healthy patients a false cancer scare).

The key difference between simple and conditional probability is the idea of a condition. That is, specificity is the likelihood that you are cancer-free given that you receive a low score on the screening test. The idea of a condition also illustrates why we have such a low opinion of traditional null hypothesis testing, which evaluates the probability that (for example) two sample means will be found to be significantly different given the assumption that there is no difference in the population. We think there are ample reasons to question the belief that there are absolutely no differences between the populations from which just about any two samples are drawn, which means that the central assumption behind traditional null hypothesis tests is often questionable.

## *Effects of Sensitivity, Effect Size, and Decision Criteria on Power*

The power of a statistical test is a function of its sensitivity, the size of the effect in the population, and the standards or criteria used to test statistical hypotheses. Tests have higher levels of statistical power when:

1  *Studies are highly sensitive.* Researchers can increase sensitivity by using better measures or using study designs that allow them to control for unwanted sources of variability in your data (for the moment, we will define sensitivity in terms of the degree to which sampling error introduces imprecision into the results of a study; a fuller definition will be presented later in this chapter). The simplest method of increasing the sensitivity of a study is to increase its sample size ($N$) or the number of observations.[2] As $N$ increases, statistical estimates become more precise and the power of statistical tests increases.
2  *Effect sizes (ES) are large.* Different treatments have different effects. It is easiest to detect the effect of treatments if that effect is large (e.g., when

treatment outcomes are very different or when treatments account for a substantial proportion of variance in outcomes; we will discuss specific measures of effect size later in this chapter and in the chapters that follow). When treatments have very small effects, these effects can be difficult to reliably detect. As ES values increase, power increases.

3 *Criteria for statistical significance are lenient.* Researchers decide about the standards that are required to reject $H_0$. It is easier to reject $H_0$ when the significance criterion, or alpha ($\alpha$) level, is .05 than when it is .01 or .001. As the standard for determining significance becomes more lenient, power increases.

Power is highest when all three of these conditions are met (i.e., sensitive study, large effect, a lenient criterion for rejecting the null hypothesis). In practice, sample size (which affects sensitivity) is probably the more important determinant of power. Effect sizes in the social and behavioral sciences tend to be small or moderate (if the effect of a treatment is so large that it can be seen by the naked eye, even in small samples, there may be little reason to test for it statistically), and researchers are often unwilling to abandon the traditional criteria for statistical significance that are accepted in their field (usually, alpha levels of .05 or sometimes .01; Cowles & Davis, 1982). Thus, effect sizes and decision criteria tend to be similar across a wide range of studies. In contrast, sample sizes vary considerably, and they directly impact levels of power.

With a sufficiently large $N$, virtually any test statistic will be "significantly" different from zero, and virtually any nil hypothesis can be rejected. Large $N$ makes statistical tests highly sensitive, and virtually any specific point hypothesis (e.g., the difference between two treatments is zero, the difference between two reading programs is six points) can be rejected if the study is sufficiently sensitive. For example, suppose you are testing a new medicine that will result in a .0000001% increase in success rates for treating cancer. This increase is larger than zero, and if researchers include enough subjects in a study evaluating this treatment, they will almost certainly conclude that the new treatment is statistically different from existing treatments. On the other hand, if very small samples are used to evaluate a treatment that has a real and substantial effect, statistical power might be so low that you incorrectly conclude that the new treatment is not different from existing treatments.

Studies can have very low levels of power (i.e., are likely to make Type II errors) when: (1) they use small samples, (2) when the effect being studied is a small one, and/or (3) stringent criteria are used to define a "significant" result. The worst case occurs when a researcher uses a small sample to study a treatment that has a very small effect, *and* he or she uses a very strict standard for rejecting the null hypothesis. Under those conditions, Type II errors may be the norm. To put it simply, studies that use small samples and stringent criteria for statistical significance to examine treatments that have small effects will almost always lead to the wrong conclusion about those treatments (i.e., to the conclusion that treatments have no effect whatsoever).

## The Mechanics of Power Analysis

When a sample is drawn from a population, the exact value of any statistic (e.g., the mean, difference between two group means) is uncertain, and that uncertainty is reflected by a statistical distribution. Suppose, for example, that you evaluate a treatment that you expect has no real effect (e.g., you use astrology to advise people about career choices) by comparing outcomes in groups who receive this treatment to outcomes in groups who do not receive it (control groups). You will *not* always find that treatment and control groups have exactly the same scores, even if the treatment has no real effect. Rather, there is some range of values that can be expected for any test statistic in a study like this, and the standards used to determine statistical significance are based on this range or distribution of values. In traditional null hypothesis testing, a test statistic is "statistically significant" at the .05 level if its actual value is outside of the range of values you would observe 95% of the time in studies where the treatment had no real effect. If the test statistic is outside of this range, the usual inference is that the treatment *did* have some real effect.

For example, suppose that 62 people are randomly assigned to treatment and control groups, and the $t$ statistic is used to compare the means of the two groups. If the treatment has no effect whatsoever, the $t$ statistic should usually be near zero and will have a value less than or equal to approximately 2.00 in 95% of the time. If the $t$ statistic obtained in a study is larger than 2.00, you can safely infer that treatments are very likely to have some effect; if there was no real treatment effect, values above 2.00 would be a very rare event.

---

### Understanding Sampling Distributions

In the example above, 62 people are randomly assigned to groups that either receive astrology-based career advice or who do not receive such advice. Even though you might expect that the treatment has no real effect, you would probably not expect that the difference between the average level of career success of these two groups will always be exactly zero. Sometimes the astrology group might do better and sometimes it might do worse.

Suppose you repeated this experiment 1,000 times and noted the difference between the average level of career success in the two groups. The distribution of scores would look something like Figure 1.2 below:

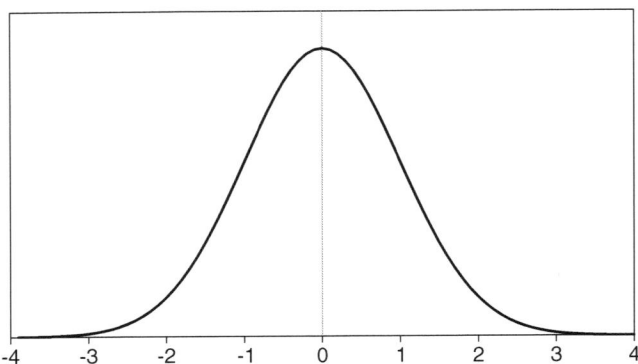

*Figure 1.2* Distribution of Group Differences Expected If Astrology Has No Effect.

This distribution is referred to as a *sampling distribution*, and it illustrates the extent to which differences between the means of these two groups might be expected to vary as a result of chance or sampling error. Most of the time, the differences between these groups should be near zero, because we expect that advice based on astrology has no real systematic effect. The variance of this distribution illustrates the range of differences in outcomes you might expect if your hypothesis that astrology has no effect is true. In this case, about 95% of the time, you expect the difference between the astrology and the no-astrology groups to be about two points or less. If you find a bigger difference between groups (suppose the average success score for the astrology group is five points higher than the average for the no-astrology group), you should reject the null hypothesis that the career advice has no systematic effect.

You might ask why anyone in their right mind would repeat this study 1,000 times. Luckily, statistical theory allows us to estimate sampling distributions based on a few simple statistics. Virtually all the statistical tests discussed in this book are conducted by comparing the value of some test statistic to its sampling distribution, so understanding the idea of a sampling distribution is essential to understanding hypothesis testing and statistical power.

As the example above suggests, if treatments have no effect whatsoever in the population, you should not expect to always find a difference of precisely zero between samples of those who receive the treatment and those who do not. Rather, there is some range of values that might be found for any test statistic in a sample (e.g., in the example cited earlier, you expect the value of the difference in the two means to be near zero, but you also know it might range from about −2.00 to +2.00). The same is true if treatments have a real effect. For example, if a researcher expects that the mean in a treatment group that receives career advice based on valid measures of work interests will be 10 points higher than the mean in a control group (e.g., because this is the size of the difference in the population), that researcher should also expect some variability around that figure. Sometimes the difference between two samples might be 9 points, and sometimes it might be 11 or 12. The key to power analysis is estimating the range of values one might reasonably expect for some test statistic if the real effect of treatments is small, or medium, or large.

Figure 1.3 illustrates the key ideas in statistical power analysis.

Suppose you use a $t$-test to determine whether the difference in average reading test scores of 500 pupils randomly assigned to two different types of reading instruction are statistically significant. You do not make any specific prediction about which reading program will be better, and therefore test the two-tailed hypothesis that the two programs lead to systematically different outcomes. To be "statistically significant", the value of this test statistic must be 1.96 or larger. As Figure 1.3 suggests, the likelihood you will reject the null hypothesis that there is no difference between the two groups depends substantially on whether the true effect of treatments is small or large.

If the null hypothesis that there is no real effect was true, you would expect to find values of 1.96 or higher for this test statistic in 5 tests out of every 100 performed (i.e., $\alpha = .05$). This is illustrated in Section 1 of Figure 1.3. Section 2 of Figure 1.3 illustrates the distribution of test statistic values you might expect if treatments had a small effect on the dependent variable. You might notice that the distribution of test statistics you would expect to find in studies of a treatment with this sort of effect has shifted a bit, and in this case 25% of the values you might expect to find are greater than or equal to 1.96. That is if you run a study under the scenario illustrated in Section 2 of this figure (i.e., treatments have a small effect), the probability you will reject the null hypothesis is .25. Section 3 of Figure 1.3 illustrates the distribution of values you might expect if the true effect of treatments is large5. In this distribution, 40% of the values are 2.00 or greater, and the probability you will reject the null hypothesis is .40. ***The power of a statistical test is the proportion of the distribution of test statistics expected for a study like this that is above the critical value used to establish statistical significance.***

The qualifier "for a study like this" is important because the distribution of test statistics you should reasonably expect in a particular study depends on both the population effect size and the sample size. If the power of a study is .80, that is the same thing as saying that if you draw a distribution of the test statistic values you expect to find based on the population effect size and the sample size, 80% of these will be equal to or larger than the critical value needed to reject the null hypothesis.

## 12   The Power of Statistical Tests

**1. Distribution Expected if Treatments Have No Effect**

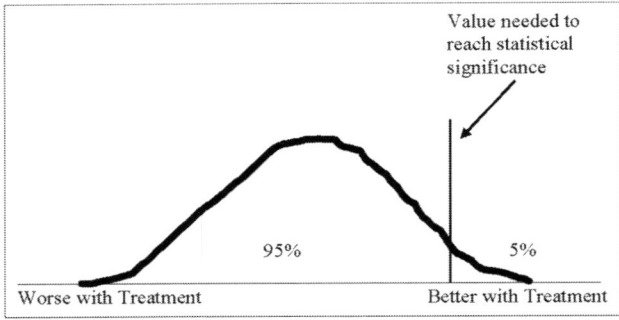

**2. Distribution Expected if Treatments Have Small Effect**

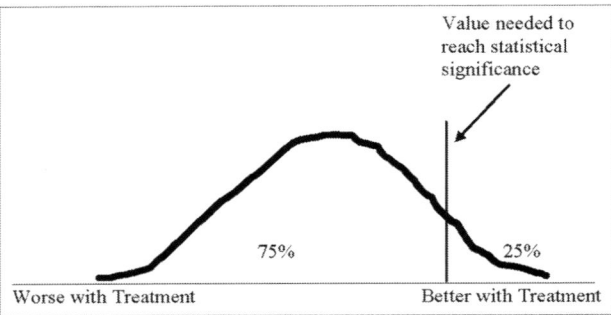

**3. Distribution Expected if Treatments Have Larger Effect**

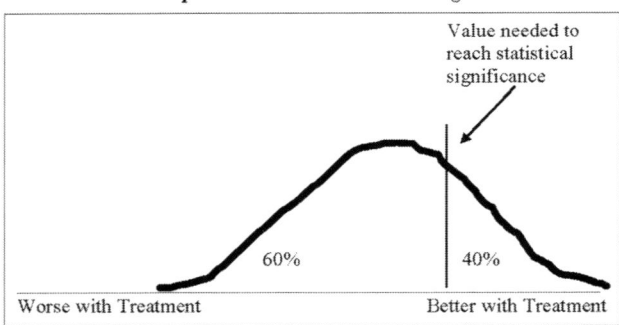

*Figure 1.3* Essentials of Power Analysis.

No matter what hypothesis you are testing, or what statistic you are using to test that hypothesis, power analysis always involves three basic steps, which are listed in Table 1.1. First, a criterion or critical value for "statistical significance" must be established. For example, the tables found in the back of virtually any statistics textbook can be used to determine such critical values for testing the traditional null hypothesis. If the test statistic a researcher

*Table 1.1* The Three Steps to Determining Statistical Power

1. Establish a criterion or critical value for statistical significance
   * What is the hypothesis that is being tested? (e.g., traditional null hypothesis, minimum-effect tests)
   * What level of confidence is desired? (e.g., a = .05 vs. a = .01)
   * What is the critical value for your test statistic [these critical values are determined on the basis of the degrees of freedom (df) for the test and the desired confidence level]
2. Estimate the effect size (ES)
   * Are treatments expected to have large, medium, or small effects?
   * What is the range of values researchers expect to find for the test statistic, given this ES?
3. Determine where the critical value lies in relation to the distribution of test statistics expected if the null hypothesis is true (i.e., the sampling distribution). The power of a statistical test is the proportion of the distribution of test statistics expected for a study (based on the sample size and the estimated ES) that is above the critical value used to establish statistical significance

computes exceeds this critical value the researcher will reject the null hypothesis. However, these tables are not the only basis for setting such a criterion. Suppose you wanted to test the hypothesis that the effects of treatments are so small that they can safely be ignored. This might involve specifying some range of effects that would be designated as "negligible", and then determining the critical value of a statistic needed to reject this hypothesis. Chapter 2 shows how such tests are performed, and lays out the implications of such hypothesis testing strategies for statistical power analysis.

Power analysis required researchers to make their best guess of the size of effect that treatments are likely to have on the dependent variable(s); methods of estimating effect sizes are discussed later in this chapter. As we noted earlier, if there are good reasons to believe that treatments have a very large effect, it should be quite easy to reject the null hypothesis. On the other hand, if the true effects of treatments are small and subtle, it might be very hard to reject the hypothesis that treatments have no real effect.

Once you have estimated ES, it is also possible to use that estimate to describe the distribution of test statistics that should be expected. We will describe this process in more detail in Chapter 2, but a simple example will show you what we mean.

Suppose you are using a $t$-test to assess the difference in the mean scores of those receiving two different treatments. If there was no real difference between the treatments, you would expect to find $t$ values near zero most of the time, and you can use statistical theory to tell how much these $t$ values might depart from zero as a result of sampling error. The $t$ tables in most statistics textbooks tell you how much variability you might expect with samples of different sizes, and once the mean (here, zero) and the standard deviation of this distribution are known, it is easy to estimate what proportion of the distribution falls above or below any critical value. If there is a large difference between the treatments (e.g., the dependent variable has a mean of 500 and a standard deviation of 100, and the mean for one treatment is usually 80 points higher than the mean for another), large $t$ values should be expected most of the time.

The final step in power analysis is a comparison between the values obtained in the first two steps. For example, if you determine that a $t$ value of 1.96 is needed to reject the null hypothesis and also determine that because the treatments being studied have very large effects, you are likely to find $t$ values of 1.96 or greater 90% of the time, the power of this test — i.e., power is .90.

### *Sensitivity and Power*

Sensitivity refers to the precision with which a statistical test distinguishes between true treatment effects and differences in scores that are the result of sampling error. As noted above, the sensitivity of statistical tests is largely a function of the sample size. Large samples provide very precise estimates of population parameters, whereas small samples produce results than can be unstable and untrustworthy. For example, if 6 children in 10 do better with a new reading curriculum than with the old one, this might reflect nothing more than simple sampling error. If 600 out of 1,000 children do better with the new curriculum, this is powerful and convincing evidence that there are real differences between the new curriculum and the old one.

In a study with low sensitivity, there is considerable uncertainty about statistical outcomes. As a result, it might be possible to find a large treatment effect in a sample, even though there is no true treatment effect in the population. This translates into both substantial variability in study outcomes and the need for relatively demanding tests of "statistical significance". If outcomes can vary substantially from study to study, researchers need to observe relatively large effects to be confident that they represent true treatment effects and not mere sampling error. As a result, it is often difficult to reject the hypothesis that there is no true effect when small samples are used, and many Type II errors should be expected.

In a highly sensitive study, there is very little uncertainty or random variation in study outcomes, and virtually any difference between treatment and control groups is likely to be accepted as an indication that the treatment has an effect in the population.

### *Effect Size and Power*

Effect size is a key concept in statistical power analysis (Cohen, 1988; Rosenthal, 1991; Tatsuoka, 1993a). At the simplest level, effect size measures provide an index of how much impact treatments have on the dependent variable; if $H_0$ states that treatments have no impact whatsoever, the effect size can be thought of as an index of just how wrong the null hypothesis is.

One of the most common ES measures is the standardized mean difference, $d$, defined as $d = (M_t - M_c)/SD$, where $M_t$ and $M_c$ are the treatment and control group means, respectively, and SD is the pooled standard deviation. By expressing the difference in group means in standard deviation units, the $d$ statistic provides a simple metric that allows one to compare treatment effects from different studies, areas, or research, etc., without having to keep track of the units of measurement used in different studies or areas of research. For example, Lipsey and Wilson (1993) cataloged the effects of a wide range of psychological, educational, and behavioral treatments, all expressed in terms of $d$. Examples of interventions in these areas that have relatively small, moderately large, and large effects on specific sets of outcomes are presented in Table 1.2.

For example, worksite smoking cessation/reduction programs have a relatively small effect on quit rates ($d = .21$). The effects of class size on achievement or of juvenile delinquency programs on delinquency outcomes are similarly small. Concretely, a $d$ value of .20 means that the difference between the average score of those who receive the treatment and those who do not is only 20% as large as the standard deviation of the outcome measure within each of the treatment groups. This standard deviation measures the variability in outcomes, independent of treatments, so $d = .20$ indicates that the average effect of treatments is only 1/5th as large as the variability in outcomes among people who receive the same treatments. In contrast, interventions such as psychotherapy, meditation and relaxation, or positive reinforcement in the classroom have relatively large effects on outcomes such as functioning levels, blood pressure, and learning ($d$ values range from .85 to 1.17).

*Table 1.2* Examples of Effect Sizes Reported in Lipsey and Wilson (1993) Review

|  | Dependent Variable | d |
|---|---|---|
| *Small Effects (d = .20)* | | |
| Treatment programs for juvenile delinquents | delinquency outcomes | .17 |
| Worksite smoking cessation/reduction programs | quit rates | .21 |
| Small vs. large class size, all grade levels | achievement measures | .20 |
| *Medium Effects (d = .50)* | | |
| Behavior therapy vs. placebo controls | various outcomes | .51 |
| Chronic disease patient education | compliance and health | .52 |
| Enrichment programs for gifted children | cognitive, creativity, affective outcomes | .55 |
| *Large Effects (d = .80)* | | |
| Psychotherapy | various outcomes | .85 |
| Meditation and relaxation techniques | blood pressure | .93 |
| Positive reinforcement in the classroom | learning | 1.17 |

> **d vs. delta vs. g**
>
> Several effect size measures represent variations on the same theme as the $d$ statistic described here. $d$ represents the difference between means, expressed in standard deviation units. Which standard deviation you should use to describe this difference, however, is not always a simple question. Glass introduced a statistic delta ($\Delta$) which divides the difference in means by the standard deviation in the control group – the rationale being that treatments or interventions can change both the mean and the standard deviation (Glass, McGaw & Smith, 1981). Cohen's $d$ uses the average of the SDs in each group as a standard for comparison. Hedges (1981) introduced a statistic $g$, which corrects for biases in $d$ when small samples are used.
>
> When samples are very small, these different statistics may yield different values, but as $N$ increases, they tend to be more and more similar. More to the point, they all express the same concept, that the differences between means are best understood in comparison to a measure of the variability of scores – i.e., the standard deviation.

It is important to keep in mind that "small", "medium", or "large" effect refers to the size of the effect, but not necessarily to its importance. For example, a new security screening procedure might lead to a small change in rates of detecting threats, but if this change translates into hundreds of lives saved at a small cost, the effect might be judged to be both important and worth paying attention to.

When the true treatment effect is very small it might be hard to detect this effect in successive samples. For example, aspirin can be useful in reducing heart attacks, but the effects are relatively small ($d$ = .068; See, however, Rosenthal, 1993). As a result, studies of 20 or 30 patients taking aspirin or a placebo will not consistently detect the true and life-saving effects of this drug. Large-sample studies, however, provide compelling evidence of the consistent effect of aspirin on heart attacks. On the other hand, if the effect is relatively large, it is easy to detect, even with a relatively small sample. For example, cognitive ability has a strong influence on performance in school ($d$ is about 1.10), and the effects of individual differences in cognitive ability are readily noticeable even in small samples of students.

### *Decision Criteria and Power*

Finally, the standard or decision criteria used in hypothesis testing have a critical impact on statistical power. The standards that are used to test statistical hypotheses are usually set with a goal of minimizing Type I errors;

alpha levels are usually set at .05, .01, or some other similarly low level, reflecting a strong bias against treating study outcomes that might be due to nothing more than sampling error as meaningful (Cowles & Davis, 1982). Setting a more lenient standard makes it easier to reject the null hypothesis, and while this can lead to Type I errors in those rare cases where the null is true, anything that makes it easier to reject the null hypothesis also increases the statistical power of your study.

As Figure 1.1 shows, there is always a trade-off between Type I and Type II errors. If you make it very difficult to reject the null hypothesis, you will minimize Type I errors (incorrect rejections), but you will also increase the number of Type II errors. That is, if you rarely reject the null, you will often incorrectly dismiss sample results as mere sampling error, when they may in fact indicate the true effects of treatments. Numerous authors have noted that procedures to control or minimize Type I errors can substantially reduce statistical power, and may cause more problems (i.e., Type II errors) than they solve (Cohen, 1994; Sedlmeier & Gigerenzer, 1989).

*Power Analysis and the General Linear Model*

In the chapters that follow, we will describe a simple and general model for statistical power analysis. This model is based on the widely used $F$ statistic. This statistic and variations on the $d$ are used to test a wide range of statistical hypotheses in the context of the general linear model (Cohen & Cohen, 1983; Horton, 1978; Tatsuoka, 1993b). The general linear model provides the basis for correlation, multiple regression, analysis of variance, discriminant analysis, and all of the variations of these techniques. The general linear model subsumes a large proportion of the statistics that are widely used in the behavioral and social sciences, and by tying statistical power analysis to this model, we will show how the same simple set of techniques can be applied to an extraordinary range of statistical analyses.

## Statistical Power of Research in the Social and Behavioral Sciences

Research in the social and behavioral sciences often shows shockingly low levels of power. Starting with Cohen's (1962) review of research published in *Journal of Abnormal and Social Psychology*, studies in psychology, education, communication, journalism, and other related fields have routinely documented power in the range of .20 - .50 for detecting small to medium treatment effects (Sedlmeier & Gigerenzer, 1989). Power is similarly low across several domains of biomedical research (Dumas-Mallet, Button, Boraud, Gonon & Mufanò, 2017). Despite decades of warnings about the consequences of low levels of statistical power in the behavioral and social sciences, the level of power encountered in published studies is lower than .50 (Mone, Mueller & Mauland, 1996). In other words, it is typical for studies in these areas to have less than a 50% chance of rejecting the null hypothesis. If you believe that the null hypothesis is virtually always wrong (i.e., that treatments have at least *some* effect, even if it is a very small one), this means that at least half of all studies in the social and behavioral sciences (perhaps as many as 80%) are likely to reach the wrong conclusion by making a Type II error when testing the null hypothesis.

These figures are even more startling and discouraging when you realize that these reviews have examined the statistical power of *published research*. Given the strong biases against publishing methodologically suspect studies or studies reporting null results, it is likely that the studies that survive the editorial review process are better than the norm, that they show stronger effects than similar unpublished studies, and that the statistical power of unpublished studies is even lower than the power of published studies.

### Power and the Replication Crisis

A widely publicized study in *Science* (Open Science Collaboration, 2015) carefully replicated 100 studies published in top psychology journals. According to criteria that included both the size of the effect and whether the results reported in each study were statistically significant, they judged that only 39% of the published results held up in replication. Their team of 270 contributing authors took considerable care to select a representative set of studies from these journals and to reproduce as faithfully as possible the original studies, often working with the original study authors to duplicate original materials, study conditions and subject populations, and the frequent failure to replicate is discouraging.

Scientists in many disciplines are well aware of the difficulty in reproducing published results (Ioannidis, 2005; McNutt, 2014; Pashler & Wagenmakers, 2012). Based on analyses that take into account the low statistical power of studies across many disciplines, Ioannidis (2005) concluded that more than half of all published studies might produce untrustworthy results. There are many reasons for the failure to replicate, but the low power of studies in so many fields is almost certainly a major cause.

Publication is often difficult if your study fails to reject the null hypothesis. If you run studies with low levels of power, the only time you will reject $H_0$ is if you find an unusually strong effect. The studies the Open Science Collaboration (2015) replicated often used shockingly small samples (the median $N$ was 54), and these studies would never have come close to rejecting $H_0$ unless they found unusually strong effects. Unusual results just don't replicate well, and it is a good bet that replication would be much more likely if the original samples had used large rather than small samples.

Studies that do not reject the null hypothesis are often regarded by researchers as failures. The levels of power reported above suggest that "failure", defined in these terms, is quite common. If a treatment effect is small, and a study is designed with a power level of .20 (which is depressingly common), researchers are four times as likely to fail (i.e., fail to reject the null) as to succeed. Power of .50 suggests that the outcome of a study is basically like the flip of a coin. A researcher whose study has power of .50 is just as likely to fail to reject the null hypothesis as he or she is to succeed. It is likely that much of the apparent inconsistency in research findings is due to nothing more than inadequate power (Schmidt, 1992). If 100 studies are conducted, each with power of .50, about half of them will reject the null, and about half will not. Given the stark implications of low power, it is important to consider *why* research in the social and behavioral sciences is so often conducted in a way in which failure is more likely than success.

The most obvious explanation for a low level of power in the social and behavioral sciences is the belief that social scientists tend to study treatments, interventions, etc. that have small and unreliable effects. Until recently, this explanation was widely accepted, but the widespread use of meta-analysis in integrating scientific literature suggests that this is not necessarily the case. There is now ample evidence from literally hundreds of analyses of thousands of individual studies that the treatments, interventions, and the like studied by behavioral and social scientists have substantial and meaningful effects (Haase, Waechter & Solomon, 1982; Hunter & Hirsh, 1987; Lipsey, 1990; Lipsey & Wilson, 1993; Schmitt, Gooding, Noe & Kirsch, 1984); these effects are of a similar magnitude as many of the effects reported in the physical sciences (Hedges, 1987).

A second possibility is that the decision criteria used to define "statistical significance" are too stringent. We will argue in several of the chapters that follow that the researchers who use statistical analyses to make sense of data are often too concerned with Type I errors and insufficiently concerned with statistical power. However, the use of overly stringent decision criteria is probably not the best explanation for low levels of statistical power.

The best explanation for the low levels of power observed in many areas of research is many studies use samples that are much too small to provide accurate and credible results. Researchers routinely use samples of 20, 50, or 75 observations to make inferences about population parameters. When sample results are unreliable, it is necessary to set some strict standard to distinguish real treatment effects from fluctuations in the data that are due to simple sampling error, and studies with these small samples often fail to reject null hypotheses, even when the population treatment effect is fairly large.

On the other hand, very large samples will allow you to reject the null hypothesis even when it is very nearly true – i.e., when the effect of treatments is very small. The effects of sample size on statistical power are so profound that it is tempting to conclude that a significance test is little more than a roundabout measure of how large the sample is. If the sample is sufficiently small, you will virtually never reject the null hypothesis. If the sample is sufficiently large, you will virtually always reject the null hypothesis.

## Using Power Analysis

Statistical power analysis can be used for both planning and diagnosis. Power analysis is frequently used in designing research studies. The results of power analysis can help in determining how large your sample should be, or in deciding what criterion should be used to define "statistical significance". Power analysis can also be used as a diagnostic tool, to determine whether a specific study has adequate power for specific purposes, or to identify the sort of effects that can be reliably detected in that study.

Because power is a function of the sensitivity of your study (which is essentially a function of $N$), the size of the effect in the population (ES), and the decision criterion that is used to determine statistical significance, we can solve for any of the four values (i.e, power, $N$, ES, $\alpha$), given the other three. However, none of these values is necessarily known in advance, although some values may be set by convention. The criterion for statistical significance (i.e., $\alpha$) is often set at .05 or .01 by convention, but there is nothing sacred about these values. As we will note later, one important use of power analysis is in making decisions about what criteria *should* be used to describe a result as "significant".

---

### The Meaning of Statistical Significance

Suppose a study leads to the conclusion that "there is a statistically significant correlation between the personality trait of conscientiousness and job performance". What does "statistically significant" mean?

"Statistically significant" clearly does *not* mean that this correlation is large, meaningful, or important (although it might be all of these). If the sample size is large, a correlation that is quite small will still be "statistically significant". For example, if $N = 20,000$, a correlation of $r = .02$ will be significantly ($\alpha = .05$) different from zero. The term "statistically significant" can be thought of as shorthand for the following statement:

> "In this particular study, there is sufficient evidence to allow the researcher to reliably distinguish (with a level of confidence defined by the alpha level) between the observed correlation of .02 and a correlation of zero"

In other words, the term "statistically significant" does not describe the result of a study, but rather describes the sort of result this particular study can reliably detect. The same correlation will be statistically significant in some studies (e.g., those that use a large $N$ or a lenient alpha) and not significant in others. In the end, "statistically significant" usually says more about the design of the study than about the results. Studies that are designed with high levels of statistical power will, by definition, usually produce significant results. Studies that are designed with low levels of power will not yield significant results. A significant test usually tells you more about the study design than about the substantive phenomenon being studied.

The ES depends on the treatment, phenomenon, or variable you are studying, and is usually not known in advance. Sample size is rarely set in advance, and $N$ often depends on some combination of luck and resources on the part of the investigator. Actual power levels are rarely known, and it can be difficult to obtain sensible advice about how much power you should have. It is important to understand how each of the parameters involved is determined when conducting a power analysis.

### *Determining the Effect Size*

There is a built-in dilemma in power analysis. To determine the statistical power of a study, ES must be known. But if you already knew the exact strength of the effect the particular treatment, intervention, etc., you would not need to do the study! The whole point of doing a study is to find out what effect the treatment has, and the true ES in the population is unlikely to ever be known.

Statistical power analyses are always based on *estimates* of ES. In many areas of study, there is a substantial body of theory and empirical research that will provide a well-grounded estimate of ES. For example, there are hundreds of studies of the validity of cognitive ability tests as predictors of job performance (Hunter & Hirsch, 1987; Schmidt, 1992), and this literature suggests that the relationship between test scores and performance is consistently strong (corrected correlations of about .30–.50 are common). Even where there is not extensive literature available, researchers can often use their experience with similar studies to realistically estimate effect sizes.

When there is no good basis for estimating effect sizes, power analyses can still be carried out by making a conservative estimate. A study that has adequate power to reliably detect small effects (e.g., a $d$ of .20, or a correlation of .10) will also have adequate power to detect larger effects. On the other hand, if researchers design studies with the assumption that effects will be large, they might have insufficient power to detect small but important effects. Earlier, we noted that the effects of taking aspirin on heart attacks are relatively small, but that there is still a substantial payoff for taking the drug. If the initial research that led to the use of aspirin for this purpose had been conducted using small samples, the researchers would have had little chance of detecting the life-saving effect of aspirin.

### *Determining the Desired Level of Power*

In determining desired levels of power, researchers must weigh the risks of running studies without adequate power against the resources needed to attain high levels of power. Researchers can always achieve high levels of power by using very large samples, but the time and expense required may not always justify the effort.

There are no hard-and-fast rules about how much power is enough, but there does seem to be consensus about two things. First, if at all possible, power should be above .50. When power drops below .50, a study is more likely to fail (i.e., it is unlikely to reject the null hypothesis) than succeed. It is hard to justify designing studies in which failure is the most likely outcome. Second, power of .80 or above is usually judged to be adequate. The .80 convention is arbitrary (in the same way that significance criteria of .05 or .01 are arbitrary), but it seems to be widely accepted, and it can be rationally defended.

Power of .80 means that success (rejecting the null) is four times as likely as a failure. It can be argued that some number other than four might represent a more acceptable level of risk (e.g., if power = .90, success is nine times as likely as a failure), but it is often prohibitively difficult to achieve power much above .80. For example, to have a power of .80 in detecting a small treatment effect (where the difference between treatment and control groups is $d = .20$), a sample of about 775 subjects is needed. If power of .95 is desired, a sample of about 1,300 subjects will be needed. Most power analyses specify .80 as the desired level of power to be achieved, and this convention seems to be widely accepted.

### *Applying Power Analysis*

There are four ways to use power analysis: (1) in determining the sample size needed to achieve desired levels of power, (2) in determining the level of power in a study that is planned, (3) in determining the size of the effect that can be reliably detected by a particular study, and (4) in determining sensible criteria for "statistical significance". The chapters that follow will lay out the actual steps in doing a power analysis, but it is useful at this point to get a preview of the four potential applications of this method. Power analysis can be used in:

1 *Determining sample size* – given a particular ES, significance criterion and a desired level of power, it is easy to solve for the sample size

needed. For example, if researchers think the correlation between a new test and performance on the job is .30, and they want to have at least an 80% chance of rejecting the null hypothesis (with a significance criterion of .05), they need a sample of about 80 cases. When planning a study, researchers should routinely use power analysis to help make sensible decisions about the number of subjects needed.

2   *Determining power levels* – if $N$, ES, and the criterion for statistical significance are known, researchers can use power analysis to determine the level of power for that study. For example, if the difference between treatment and control groups is small (e.g., $d = .20$), there are 50 subjects in each group, and the significance criterion is $\alpha = .01$, power will be only .05! Researchers should certainly expect that this study will fail to reject the null, and they might decide to change the design of this study considerably (e.g., use larger samples, more lenient criteria).

3   *Determine ES levels* – researchers can also determine what sort of effect could be reliably detected, given $N$, the desired level of power, and $\alpha$. In the example above, a study with 50 subjects in both the treatment and control groups would have power of .80 to detect a very large effect (approximately $d = .65$) with a .01 significance criterion, or a large effect ($d = .50$) with a .05 significance criterion.

4   *Determine criteria for statistical significance* – given a specific effect, sample size, and power level, it is possible to determine the significance criterion. For example, if you expect a correlation coefficient to be .30, $N = 67$, and you want power to equal or exceed .80, you will need to use a significance criterion of $\alpha = .10$ rather than the more common .05 or .01.

Some researchers conduct post-hoc power analyses. That is, they conduct a study and then use the results of their study (the outcome of significance tests, the observed effect size) to calculate the power of the study they have just done. In earlier editions of this book, we had discussed ways of conducting these analyses and had provided post-hoc power calculations in the software we provided. We have become convinced that these analyses are not a good idea. First, as Hoenig and Heisey (2001) observe, once a study is completed, all of the information provided by a post-hoc power analysis is already provided by the exact $p$ value. More to the point, when your study is done, you know whether you have rejected the null hypothesis. Computing the probability that you would reject the null does not provide new information and does not help you determine whether this conclusion would replicate (e.g., would a similar study also reject the null).

## Hypothesis Tests vs. Confidence Intervals

Null hypothesis testing has been criticized on several grounds (e.g., Schmidt, 1996), but perhaps the most persuasive critique is that null hypothesis tests provide so little information. It is widely recognized that the use of confidence intervals and other methods of portraying levels of uncertainty about the outcomes of statistical procedures have many advantages over simple null hypothesis tests (Wilkinson et al., 1999).

Suppose a study is performed that examines the correlation between scores on an ability test and measures of performance in training. The authors find a correlation of $r = .30$, and on the basis of a null hypothesis test, decide that this value is significantly (e.g., at the .05 level) different from zero. That test tells them something, but it does not tell them whether the finding that $r = .30$ represents a good or a poor estimate of the relationship between ability and training performance. A confidence interval would provide that sort of information.

Staying with this example, suppose researchers estimate the amount of variability expected in correlations from studies like this and conclude that a 95% confidence interval ranges from .05 to .55. This confidence interval would tell researchers exactly what they learned from the significance test – i.e., that they could be quite sure the correlation between ability and training performance was not zero. A confidence interval would also tell them that $r = .30$ might not be a good estimate of the correlation between ability and performance; the confidence interval suggests that this correlation could be much larger or much smaller than .30. Another researcher doing a similar study using a larger sample might find a much smaller confidence interval, indicating a good deal more certainty about the generalizability of sample results.

As the previous paragraph implies, most of the statements that can be made about statistical power also apply to confidence intervals. That is, if you design a study with low power, you will also find that it produces wide confidence intervals (i.e., that there is considerable uncertainty about the meaning of sample results). If you design studies to be sensitive and powerful, these studies will yield smaller confidence intervals. Although the focus of this book is on hypothesis tests, it is important to keep in mind that the same facets of the research design ($N$, the alpha level) that cause power to go up or down also cause confidence intervals to shrink or grow. A powerful study will not always yield precise results (e.g., power can be high in a poorly designed study that examines a treatment that has very strong effects), but in most instances, whatever researchers do to increase power will also lead to smaller confidence intervals and more precision in sample statistics.

By almost any criterion, statistical analyses that include confidence intervals are preferable to analyses that rely on null hypothesis tests alone. Unfortunately, the process of changing researchers' regarding data analysis is very slow and uncertain, and too few researchers incorporate confidence intervals when they report the results of statistical analyses (Fidler, Thomason, Cumming, Finch & Leeman, 2004).

## *Do We* **Really** *Need Null Hypotheses Tests or Power Analysis?*

You can make a good case that confidence intervals give you all the information a null hypothesis test provides and more. This leads to the question of whether we would be better off doing away with tests of null hypotheses altogether. Although we find confidence intervals very useful, we believe that it is not a good idea to drop significance testing altogether, and that it is better to reform this method so that the tests provide useful information. There are two reasons for this recommendation.

First, the recommendation that researchers rely on confidence intervals rather than significance tests have been around for at least 50 years, and there is no indication that the majority of researchers will finally adopt it. Like it or not, significance tests are here to stay. Second, there are methods for planning studies to achieve specific levels of accuracy or confidence in results (see the boxed section describing methods for specifying accuracy in parameter estimation), but their uptake is still sporadic at best. Power analysis is a more familiar method, albeit an underused one, and power analyses allow you to attack questions not so easily handled when working with confidence intervals or similar methods, such as an assessment of the size of the effect you will need to find in a study to have genuine confidence that your results are not trivial.

The real question is not confidence intervals vs. significance testing, but rather how to best combine them. When reporting results, the best practice is to report: (1) the outcome of a significance test and the statistics that support that test, such as $F$; (2) a good measure of effect size, such as $PV$; and (3) confidence intervals around the key findings, such as the difference between means, or around your estimate of effect size. There are few situations, if any, in which you should simply report the outcome of a significance test (e.g., by putting stars in a table to indicate significant findings) and leave it at that. It is almost always better to report effect size measures and/or confidence intervals along with significance and to report all three whenever possible.

---

### Accuracy in Parameter Estimation

Statistical power analysis is concerned to do with null hypothesis testing, and with the right null hypothesis (e.g., one that specifies a range of values rather than a single point) this approach can provide some real value. In particular, it can help you make good choices regarding sample size. An alternative perspective for determining the sample sizes needed in a particular study is to do so in terms of the desired level of accuracy in parameter estimation (AIPE). For example, if the focus of your study is on the squared multiple correlation between academic achievement and five measures of school quality, you could use power analysis to determine the sample size needed to reject the traditional null hypothesis that this correlation is exactly zero, or the sample size needed to reject the null hypothesis that this correlation is so small as to be trivial in its magnitude. The AIPE approach focuses on the level of accuracy desired in estimating the relationship between school quality and academic achievement. For example, if you believed that school quality accounted for 20% of the variance in achievement and you wanted to achieve a confidence interval sufficiently small to be relatively certain that the population R squared was between .10 and .30 in value, equations developed by Kelly (2008) show that you would need a sample of $N = 239$.

Kelly and his colleagues have developed methods and software for determining the sample size needed to reach desired levels of accuracy in evaluating the standardized difference between two means (d – See Kelly & Rausch, 2006), squared multiple correlations (Kelly, 2008), and various parameters in structural equation modeling (Kelly & Lai, 2011; Lai & Kelly, 2012; Maxwell, Kelly & Rausch, 2008). As with power analysis, the biggest problem with the AIPE approach is that it requires you to have at least a general idea of the expected size of the effect in the population. As in power analysis, the best strategy for planning sample sizes when there is not a good estimate of the effect size in the population, the best thing to do is to plan as if the population effect is a small one.

## What Can You Learn from a Null Hypothesis Test?

Tests of the traditional nil hypothesis, that the difference between two treatments is exactly zero can be informative, but unless they are accompanied with some sort of effect size measures and unless the power of these tests is known, it can be difficult to draw any sensible conclusions from this type of test. Consider the example of a study that compares two alternative treatments for colon cancer (e.g., radiation vs. chemotherapy). The study fails to reject the null hypothesis. If you do not know anything about the power of this test or the likely size of the difference in the effects of these two treatments in the population, you are left with two alternatives.

1   The study has too little power to detect a real difference between these two treatments
2   The difference between these treatments is either exactly zero (something that is extremely unlikely) or it is so small that it is probably not meaningful

In this book, we are skeptical about the value of most null hypothesis tests because of the difficulty in determining which alternative of more likely to be true. In particular, we will show you why the failure to reject the null hypothesis in most studies is more likely to say something about the design of the study (i.e., the study did not have sufficient power) than about the phenomenon being studied. The methods of framing and testing minimum-effect hypotheses introduced in Chapter 2 can be applied to reframe the questions that are asked in a null hypothesis test in a way that will make the outcomes of significance tests both more informative and more easily understood.

## Notes

1 It is important to note that Type I errors can only occur when the null hypothesis is true. If the null hypothesis is that there is *no* true treatment effect (a nil hypothesis) this will rarely be the case. As a result, Type I errors are probably quite rare in published tests of the traditional null hypothesis, and efforts to control these errors at the expense of making more Type II errors might be ill-advised (Murphy, 1990).
2 In some studies, it is possible to obtain multiple observations from each study participant. In general, studies of this sort in which $N$ represents the number of observations will have more statistical power than studies involving the same number of data points in which only one observation is obtained from each participant. This phenomenon is discussed in more detail in Chapter 8.

# 2 A Simple and General Model for Power Analysis

This chapter develops a simple approach to statistical power analysis that is based on the widely used $F$ statistic. The $F$ statistic (or some transformation of $F$) is used to test statistical hypotheses in the general linear model (Horton, 1978; Tatsuoka, 1993b), a model that includes all of the variations of correlation and regression analysis (including multiple regression), analysis of variance and covariance (ANOVA and ANCOVA), $t$-tests for differences in group means, tests of the hypothesis that the effect of treatments takes on a specific value, or a value different from zero. Most of the statistical tests that are used in the social and behavioral sciences can be treated as special cases of this general linear model.

Analyses based on the $F$ statistic are not the only approach to statistical power analysis. For example, in the most comprehensive work on power analysis Cohen (1988) constructed power tables for a wide range of statistics and statistical applications, using separate ES measures and power calculations for each class of statistics. Kramer and Thiemann (1987) derived a general model for statistical power analysis based on the intraclass correlation coefficient, and developed methods for evaluating the power of a wide range of test statistics using a single general table based on the intraclass correlation coefficient ($\rho$). Lipsey (1990) used the $t$-test as a basis for estimating the statistical power of several statistical tests. Nevertheless, an analytic model based on the $F$ statistic has two substantial advantages. First, it can be applied to a wide array of statistical analyses, even those (e.g., correlation coefficients) that do not appear to rely on an $F$ distribution. Second, this model can be easily adapted to test a much wider range of hypotheses than the traditional nil hypothesis that the difference between two groups or two treatments, or that the relationship between two variables is exactly zero.

The idea of using the $F$ distribution as the basis for a general system of statistical power analysis is hardly an original one; Pearson and Hartley (1951) proposed a similar model over 70 years ago. It is useful, however, to explain the rationale for choosing the $F$ distribution in some detail, because the family of statistics based on $F$ have several characteristics that help to take the mystery out of power analysis.

DOI: 10.4324/9781003296225-2

Basing a model for statistical power analysis on the $F$ statistic provides a nice balance between applicability and familiarity. First, the $F$ statistic is familiar to most researchers. This chapter and the one that follows show how to transform a wide range of test statistics and measures into $F$ statistics, and how to use those $F$ values in statistical power analysis. Because such a wide range of statistics can be transformed into $F$ values, structuring power analysis around the $F$ distribution allows one to cover a great deal of ground with a single set of tables.

Second, the approach to power analysis developed in this chapter is flexible. Unlike other presentations of power analysis, we do not limit ourselves to tests of the traditional null hypothesis (i.e., the hypothesis that treatments have no effect whatsoever). Traditional null hypothesis tests have been roundly criticized (Amrhein, Greenland & McShane, 2019; Cohen, 1994; Meehl, 1978; Morrison & Henkel, 1970; Wasserstein & Lazar, 2016; Wasserstein, Schirm, & Lazar, 2019), and there is a need to move beyond such limited tests. Our discussions of power analysis consider several methods of statistical hypothesis testing and show how power analysis can be easily extended beyond the traditional null hypothesis test. In particular, we show how the model developed here can be used to evaluate the power of "minimum-effect" hypothesis tests – i.e., tests of the hypothesis that the effects of treatments are either trivially small or exceed some minimum level.

In the last 25 years, researchers have devoted considerable attention to alternatives to traditional null hypothesis tests (e.g., Murphy & Myors, 1999; Rouanet, 1996; Serlin and Lapsley, 1985, 1993), particularly on tests of the hypothesis that the effect of treatments falls within or outside of some range of values. For example, Murphy and Myors (1999) discuss alternatives to tests of the traditional null hypothesis that involve specifying some range of effects that would be regarded as negligibly small and then testing the hypothesis that the effect of treatments either falls within this range ($H_0$) or falls above this range ($H_1$).

The $F$ statistic is particularly well suited to tests of the hypothesis that effects fall within some range that can be reasonably described as "negligible" vs. falling above that range. The $F$ statistic ranges in value from zero to infinity, with larger values indicating stronger effects. As we will show in sections that follow, this property of the $F$ statistic makes it easy to adapt familiar testing procedures to evaluate the hypothesis that effects exceed some minimum level, rather than simply evaluating the possibility that treatments have no effect.

Finally, the $F$ distribution explicitly incorporates one of the key ideas of statistical power analysis – i.e., that the range of values that might reasonably be expected for a variety of test statistics depends in part on the size of the effect in the population. As we note below, the concept of ES is reflected very nicely in one of the three parameters that determine the distribution of the $F$ statistic (i.e., the noncentrality parameter).

## The General Linear Model, the $F$ Statistic, and Effect Size

Before exploring the $F$ distribution and its use in power analysis, it is useful to describe the key ideas in applying the general linear model as a method of structuring statistical analyses, show how the $F$ statistic is used in testing hypotheses according to this model, and describe a very general index of whether treatments have large or small effects.

Suppose 200 children are randomly assigned to one of two methods of reading instruction. Each child receives instruction that is either accompanied by audiovisual aids (computer software that "reads" to the child while showing pictures on a screen) or given without the aids. At the end of the semester, each child's reading performance is measured.

One way to structure research on the possible effects of reading instruction is to construct a mathematical model to explain why some children read well and others read poorly. This model might take a simple additive form:

$$y_{ijk} = a_i + b_j + ab_{ij} + e_{ijk} \qquad (2.1)$$

where: $y_{ijk}$ = the score for child $k$, who received instruction method $i$ and audiovisual aid $j$

$a_i$ = the effect of the method of reading instruction
$b_j$ = the effect of audiovisual aids
$ab_{ij}$ = the effect of the interaction between instruction and audiovisual aids
$e_{ijk}$ = the part of the child's score that cannot be explained by the treatments received

When a linear model is used to analyze a study of this sort, researchers can ask several sorts of questions. First, it makes sense to ask whether the effect of a particular treatment or combination of treatments is large enough to rule out sampling error as an explanation for why people receiving one treatment obtain higher scores than people not receiving it. As we explain below, the $F$ statistic is well suited for this purpose.

---

### Effect Size

As Preacher and Kelly (2011) note, the term "effect size" is widely used in the social and behavioral sciences, but its meaning is not always clear or consistent. An "effect size" might refer to a sample result (i.e., the sample $d$ is often referred to as an effect size measure), to a population parameter ($d$ in the population is unlikely to be the same size as the sample $d$), or it might refer to a verbal description of a range of effects (e.g., a small effect). They offer a very useful general definition of the term, noting that "Effect size is defined as a quantitative reflection of the magnitude of some phenomenon that is used for the purpose of addressing a question of interest" (Kelly & Preacher, 2012, p. 140). That is, effect size is not a statistic or a specific parameter, it is any quantitative measure that reflects the magnitude of a phenomenon.

Preacher and Kelly (2011) note that effect sizes can be discussed at three different levels of analysis. First, there are the dimensions and units that define the way we discuss effect size. For example, in this book, we often use the percentage of variance in one variable explained by some other variable or set of variables as a working definition of effect size. Second, effect size can be discussed in terms of some specific index (e.g., $d$). Third, they might be discussed in terms of some specific value of that index (e.g., $d = .20$ is a widely accepted definition of a small difference between two means).

Preacher and Kelly (2011) lay out several criteria for good effect size measures:

- Effect size values should be scaled in a way that makes sense to the reader.
- Effect size values should be accompanied by confidence intervals.
- Your estimate of the population effect size value should be independent of sample size.
- Estimates of effect sizes values should be unbiased (their expected values should equal the corresponding population values), consistent (they should converge to the corresponding population value as sample size increases), and efficient (they should have minimal variance among competing measures).

In this book, we will use the percentage of variance (PV) in some variable or outcome that is explained as a general measure of effect size. This is not the only possible effect size measure, and it may not even be the best in some specific circumstances, but it meets the criteria laid out by Preacher and Kelly (2011), and it provides a common language for researchers in different areas to use when discussing their results, even when the statistical methods used (e.g., ANOVA vs. multiple regression) appear to differ. Some alternatives, like $d$, $g$, or McGraw and Wong's (1992) CLE (common language effect size) have many attractive features but are limited in their applicability (e.g., they apply to comparisons between two groups). PV is as close as possible to a universal effect size measure.

Second, it makes sense to ask whether the effects of treatments are relatively large or relatively small. There are a variety of statistics that might be used in answering this question, but one very general approach is to estimate the percentage of variance in scores ($PV$) that is explained by various effects included in the model (Valentine & Cooper, 2003). We use this measure extensively in this book, in part because it can be easily linked with the widely used $F$ statistic. Regardless of the specific approach taken in statistical testing under the general linear model (e.g., analysis of variance, multiple regression, $t$-tests), the goal of the model is always to explain variance in the dependent variable (e.g., to help understand why some children obtained higher scores than others).

Linear models like the one shown above divide the total variance in scores into variance that can be explained by methods and treatment effects (i.e., the combined effects of instruction audiovisual aids) and variance that cannot be explained in terms of the treatments received by subjects. The percentage of variance ($PV$) associated with each effect in a linear model provides one very general measure of whether treatment effects are large or small (i.e., whether they account for a lot of the variance in the dependent variable or only a little). As noted above, the value of $PV$ is closely linked to $F$.

There are several specific statistics that describe $PV$, notably eta squared and $R^2$ (the two statistics are typically encountered in the contexts of the analysis of variance and multiple regression, respectively). We prefer to use the more general term $PV$ because it describes a general index of the effects of treatments or interventions and is not limited to any specific statistic or statistical approach. As we will show later, estimates of $PV$ are extremely useful in structuring statistical power analyses for virtually any of the specific applications of the general linear model.

### Understanding Linear Models

Linear models combine simplicity, elegance, and robustness, but for people who are the least bit math-phobic, they can seem intimidating. Consider the model illustrated in Equation 2.1 (presented earlier in this chapter). The mathematical form of this model is:

$$y_{ijk} = a_i + b_j + ab_{ij} + e_{ijk}$$

This model is easier to understand if we re-write it as a story or an explanation. This equation says that to understand why some children perform better than others in reading ($y_{ijk}$), it is important to consider three things. First, the type of instruction ($a_i$) matters. Second, it makes a difference whether or not the child gets audiovisual aids ($b_j$).

Third, these two variables might interact; the effects of methods of instruction might be different for children who receive audiovisual aids than for those who do not ($ab_{ij}$). Finally, there are all sorts of other variables that might be important, but the effects of these variables were not measured and cannot be estimated in this experiment. The combined effects of all of the other things that affect performance introduce some error ($e_{ijk}$) into the explanation of why some children end up performing better than others.

Linear models are easiest to understand if you think of them as an answer to the question "Why do the scores people receive on the dependent variable (Y) vary?" All linear models answer that question by identifying some systematic sources of variance (here $a_i + b_j + ab_{ij}$); whatever cannot be explained in terms of these systematic sources is explained in terms of error ($e_{ijk}$).

## The *F* Distribution and Power

If you take the ratio of any two independent estimates of the variance in a population (e.g., $s^2_1$ and $s^2_2$), this ratio is distributed as *F*, where:

$$F = s_1^2/s_2^2 \qquad (2.2)$$

The most familiar example of the *F* statistic is in the analysis of variance, where $s_1^2$ is a measure of variability between groups and $s_2^2$ is a measure of variability within groups. The distribution of this *F* statistic depends on the degrees of freedom for the numerator ($s_1^2$) and the denominator ($s_2^2$); these degrees of freedom are a function of the number of pieces of information used to create $s_1^2$ (e.g., the number of groups) and $s_2^2$ (e.g., the number of subjects within each group). The *F* ratio can be used to test a wide range of statistical hypotheses (e.g., testing for the equality of means or variances). In the general linear model, the *F* statistic is used to test the null hypothesis (e.g., that means are the same across treatments), by comparing a measure of the variability in scores due to the treatments to a measure of the variability in scores you might expect as a result of sampling error. In its most general form, the *F* test in general linear models is:

$$F = \text{variability due to treatments}/\text{variability due to error} \qquad (2.3)$$

The distribution of the statistic *F* is complex, and it depends in part on the degrees of freedom of the hypothesis or effect being tested ($df_{hyp}$) and the degrees of freedom for the estimate of error used in the test ($df_{err}$). If treatments have no real effect, the expected value of *F* is very close to 1.0 [$E(F) = df_{err}/(df_{err} - 2)$]. That is, if the traditional null hypothesis is true the $F$ ratios you find in studies will usually be approximately 1.0. Sometimes, the $F$ ratio might be a bit larger than 1.0 and sometimes it might be a bit smaller; depending on the degrees of freedom ($df_{hyp}$ and $df_{err}$), the $F$ values that are expected if the null hypothesis is true might cluster closely around 1.00, or they might vary considerably. The $F$ tables shown in most statistics textbooks provide a sense of how much $F$ values might vary as a function of sampling error, given various combinations of $df_{hyp}$ and $df_{err}$.

The $F$ and Chi-squared distributions are closely related. The ratio of two Chi-squared variables, each divided by its degrees of freedom, is distributed as $F$, and both distributions are special cases of a more general form (the gamma distribution). Similarly, the $t$ distribution is closely related to the $F$ distribution; when the means of two groups are compared, the value if $t$ is identical to the square root of $F$.

### The F Statistic and Effect Size Measures

The method of power analysis that this book describes is built around the $F$ distribution. There are three reasons for using $F$ as the centerpiece of our methods. First, as we noted earlier, most of the statistics used in the social and behavioral sciences can be thought of as specialized applications of the general linear model, and therefore can be easily re-expressed in terms of $F$ statistics. This means that you do not need different power tables for $F$, $t$, $r$, etc. Second, as we describe below, the well-known noncentral $F$ distribution provides a simple method for evaluating power for a much wider range of hypotheses than the traditional nil hypothesis. Thus, if you want to test the hypotheses that differences between means are large enough to be worth paying attention to (rather than the hypothesis that they are exactly zero), the noncentral $F$ makes this easy. Finally, it is easy to translate $F$ statistics into effect size measures, in particular the percentage of variance in your dependent variable that is explained by the variable or variables that represent your independent variables. This effect size measure is easily applicable to any analysis involving correlation, regression, $t$-tests, analyses of variance and covariance, Chi-squared statistics, and many more. The ability to express so many methods of analysis in terms of the $F$ statistic and to express the size of the effect any of these analyses involves is so useful and so fundamental to understanding how power analysis works and how it can be creatively applied that our first appendix (which is the one we think you will consult most often) is devoted to laying out the simple equations that allow one to translate statistical tests involving correlation, regression, $t$-tests, analyses of covariance, Chi-squared statistics, and the like into their equivalent $F$ values *and* to determine the proportion of the variance in your dependent variable that is explained in any of these analyses. Appendix A provides equations for expressing a wide range of statistics in terms of $F$ and for evaluating the proportion of variance that is explained in the dependent variable.

## Confidence Intervals for *PV* and *d*

The effect size measures *d* and *PV* are both informative, but they only tell you what you found in a particular study. It is even more useful to draw confidence intervals around *d* or *PV*, which will tell you how much uncertainty there is in your effect size estimate. For example, suppose you find a very large effect in your study (e.g., $d = 1.20$). You might interpret this very differently if the confidence interval around *d* is small, meaning that you can put some faith in the large effect than if the confidence interval ranged from .20 to 2.20. This large interval would suggest that you could not be very confident in the effect size estimate coming from your study.

Suppose you do a study comparing the means in two groups ($n_1 = 40$, $n_2 = 50$) and find $d = 40$, a small to moderately large difference. You can get a very good approximation of the 95% confidence interval around this *d* using the formula:

$$95\%\text{CI for } d = d \pm 1.96 * \sqrt{\frac{2}{\bar{n}}\left(1 + \frac{\frac{d^2 * \bar{n}}{2}}{2N}\right)}$$

where

$\bar{n}$ is the harmonicmean of n1 and n2 — i.e., $\bar{n} = (2(n1 * n2))/n1 + n2)$.

In this study, the 95% confidence interval ranges from −.01 to .81, suggesting that you cannot have much confidence in the .40 estimate.

Suppose in another study, with $n = 150$, you use four variables to predict Y and find an $R^2$ (i.e., *PV*) value of .20. Cohen, Cohen, Aiken, and West (2003) note that the standard error of $R^2$ is given by:

$$SE_R^2 = \sqrt{\frac{4R^2(1-R^2)^2(n-k-1)^2}{(n^2-1)(n+3)}}$$

In this study, the standard error of $R^2 = .055$ and the 95% confidence interval for *PV* ranges from .090 to .311, suggesting you can be confident that the value of *PV* is in the moderately large to large range.

As an alternative to using these formulas, you can easily obtain the confidence intervals around *d* and *PV* using mathematical functions built into the programming language **R**.[1] We discuss the use of **R** for power analysis in a section to follow.

### The Noncentral F

Most familiar statistical tests are based on the central $F$ distribution – i.e., the distribution of $F$ statistics you expected when the traditional nil hypothesis is true. However, interventions or treatments normally have at least some effect and the distribution of $F$ values that you actually would expect in any particular study is likely to take the form of a *singly noncentral F distribution*. The power of a statistical test is defined by the proportion of that singly noncentral $F$ distribution that exceeds the critical value used to define "statistical significance".

The shape and range of values in this noncentral $F$ distribution is a function of three parameters: (1) $df_{hyp}$ (2) $df_{err}$ and (3) the "noncentrality" parameter ($\lambda$). One way to think of the noncentrality parameter is that it is a function of just how wrong the traditional null hypothesis is in the population. That is, the larger the difference between treatments in the populations, the larger the value of the noncentrality parameter. When $\lambda = 0$, the traditional null hypothesis is true and the noncentral $F$ is identical to the central $F$ that is tabled in most statistics texts.

The exact value of the noncentrality parameter is a function of both ES and the sensitivity of the statistical test (which is largely a function of the number of observations, $N$). For example, in a study where $n$ subjects are randomly assigned to each of four treatment conditions, $\lambda = [n\Sigma(\mu_j - \mu)^2] / \sigma_e^2$, where $\mu_j$ and $\mu$ represent the population mean in treatment group j and the population mean over all four treatments, and $\sigma_e^2$ represents the variance in scores due to sampling error. Horton (1978) noted that in many applications of the general linear model:

$$\lambda_{est} = SS_{hyp}/MS_{err} \tag{2.4}$$

where $\lambda_{est}$ represents an estimate of the noncentrality parameter, $SS_{hyp}$ represents the sum of squares for the effect of interest, and $MS_{err}$ represents the mean square error term used to test hypotheses about that effect. Using $PV$ to designate the proportion of the total variance in the dependent variable explained by treatments (which means that $1 - PV$ refers to the proportion not explained), the noncentrality parameter can be estimated using the following equation:

$$\lambda_{est} = df_{err}[PV/(1 - PV)] \tag{2.5}$$

Equations 2.4 and 2.5 provide a practical method for estimating the value of the noncentrality parameter in many of the most common applications of the general linear model.[2] A more general form of Equation 2.5 that is useful form complex analyses of variance is:

$$\lambda_{est} = [N-p] * [PV/(1 - PV)] \tag{2.6}$$

where:

N = number of subjects
p = number of terms in the linear model

The noncentrality parameter reflects the positive shift of the $F$ distribution as the size of the effect in the population increases (Horton, 1978). For example, if $N$ subjects are randomly assigned to one of k treatments, the mean of the noncentral $F$ distribution is approximately $[((N-k)(k+\lambda-1))/((k-1)(N-3))]$ as compared to an approximate mean of 1.0 for the central $F$ distribution. More concretely, assume that 100 subjects are assigned to one of four treatments. If the null hypothesis is true, the expected value of $F$ is approximately 1.0, and the value of $F$ needed to reject the null hypothesis ($\alpha = .05$) is 2.70. However, if the effect of treatments in the population is large (e.g., $PV = .25$), you should expect to find $F$ values substantially larger than 1.0 most of the time; given this effect size, the mean of the noncentral $F$ distribution is approximately 11.3, and more than 99% of the values in this distribution exceed 2.70. In other words, the power of this $F$ test, given the large effect size of $PV = .25$, is greater than .99. If the population effect is this large, a statistical test comparing k groups and using a sample of $N = 100$ is virtually certain to correctly reject the null hypothesis.

The larger the effect, the larger the noncentrality parameter, and the larger the expected value of $F$. The larger the $F$, the more likely it is that $H_0$ will be rejected. Therefore, all other things being equal, the more noncentrality (i.e., the larger the effect or the larger the $N$) the higher the power.

## Using the Noncentral F Distribution to Assess Power

Chapter 1 laid out the three steps in conducting a statistical power analysis, determining the critical value for significance, estimating ES, and estimating the proportion of test statistics likely to exceed that critical value. Applying these three steps here, it follows that power analysis involves:

1. *Deciding the value of* F *that is needed to reject* $H_0$. As we will see later in this chapter, this depends in part on the specific hypothesis being tested.
2. *Estimating the ES and the degree of noncentrality.* Estimates of *PV* can be used to estimate the noncentrality parameter of the *F* distribution.
3. *Estimating the proportion of the noncentral* F *that lies above the critical* F *from step 1*

In the chapters that follow, we present a simple method of conducting power analysis that is based on the noncentral *F* distribution. This method can be used with a wide variety of statistics and can be used for testing both nil hypotheses and null hypotheses that are based on specifications of negligible versus important effects. Methods of approximating the singly noncentral F distribution are presented in subsequent chapters, and methods of estimating more complex noncentral F distributions are included in the software that is distributed with this book. Appendix B presents a table of F values obtained by estimating the noncentral F distribution over a range of $df_{hyp}$, $df_{err}$, and effect size values. Appendix C presents a parallel table that presents the percentage of variance explained ($PV$) that corresponds with each of the noncentral F values shown in Appendix B.

The tables presented in Appendices B and C save you the difficulty of estimating noncentral F values, and more importantly, of directly computing power estimates for each statistical test. This table can be used to test both traditional and minimum-effect null hypotheses, and to estimate the statistical power of tests of both types of hypotheses.

## Translating Common Statistics and ES Measures into F

The model developed here is expressed in terms of the F statistic, which is commonly reported in studies that employ analysis of variance or multiple regression. However, many studies report results in terms of statistics other than the F value. It is useful to have at hand formulas for translating common statistics and effect size. Appendix A presents these formulas.

Suppose a study compared the effectiveness of two smoking cessation programs, using a sample of 122 adults, who were randomly assigned to treatments. The researchers used an independent-groups $t$-test to compare scores in these two treatments and reported $t$ value of 2.48. This $t$ value would be equivalent to an F value of 6.15, with 1 and 120 degrees of freedom. The tabled value for F with 1 and 120 degrees of freedom ($\alpha = .05$) is 3.91, and the researchers would be justified in rejecting the null hypothesis.

Suppose another study ($N = 103$) reported a squared multiple correlation of $R^2 = .25$ between a set of four vocational interest tests and an occupational choice measure. Applying the formula shown in Appendix A, this $R^2$ value would yield an F value of 8.33, with 4 and 100 degrees of freedom. The critical value of F needed to reject the traditional null hypothesis ($\alpha = .05$) is 2.46; the reported $R^2$ is significantly different from zero.

Suppose that in another study, hierarchical regression is used to determine the incremental contribution of several new predictor variables over and above the set of predictor variables already in an equation. For example, in a study with $N = 250$, two spatial ability tests were used to predict performance as an aircraft pilot; scores on these tests explained 14% of the variability in pilots' performance (i.e., $R^2 = .14$). Four tests measuring other cognitive abilities were added to the predictor battery, and this set of four tests explained an additional 15% of the variance in performance (i.e., when all six tests are used to predict performance, $R^2 = .29$). The F statistic that corresponds to this increase in $R^2$ is $F(4, 243) = 12.83$.

## Worked Example – Hierarchical Regression

In Appendix A, we presented a formula that can be used to calculate the F-equivalent of the increase in $R^2$ reported in a study that used hierarchical regression. The formula is:

$$F(df_{hyp}, df_{err}) = \frac{(R_F^2 - R_R^2)/df_{hyp}}{(1 - R_F^2)/df_{err}}$$

In this study ($N = 250$), the researchers started with two predictors and reported $R^2 = .15$, and then added four more predictors and reported $R^2 = .29$. It follows that:

---

$df_{hyp} = 4$  This is the number of predictors that are added to the original set of two predictors. The null hypothesis is that adding these four predictors leads to no real increase in $R^2$.
$df_{err} = 250 - 6 - 1 = 243$
$R^2_F = .29$  The full model containing all six tests explains 29% of the variance in performance.
$R^2_R = .14$  The restricted model that contains only the first two tests explains 14% of the variance.
$F(4,243) = [(.29 - .14)/4]/[(1 - .29)/243]$
$F(4,243) = .037/.000292 = 12.83$

---

Most $F$ tables will not report critical values for 4 and 243 degrees of freedom, but if you interpolate between the critical values of $F$ ($\alpha = .05$) for 4 and 200 degrees of freedom and 4 and 300 degrees of freedom, you will find that $F$ values of 2.41 or larger will allow you to reject the null hypothesis. With $F(4, 243) = 12.83$, you can easily reject the null hypothesis and conclude that adding these four predictors does lead to a change in $R^2$.

As Appendix A shows, $\chi^2$ values can be translated in F-equivalents. For example, if a researcher found a $\chi^2$ value of 24.56 with 6 degrees of freedom, the equivalent $F$ value would be 4.09 (i.e., 24.56/6), with $df_{hyp} = 6$ and $df_{err}$ being infinite. Because the $F$ table asymptotes as $df_{err}$ grow larger, $df_{err} = 10,000$ (which is included in the $F$ table listed in Appendix C) represents an excellent approximation to infinite degrees of freedom for the error term. The critical value of $F$ ($\alpha = .05$) needed to reject the null hypothesis in this study is 2.10, so once again, the null hypothesis will be rejected.

Appendix A also includes the ES measure d. This statistic is not commonly used in hypothesis testing per se, but it is widely used in describing the strength of effects, particularly when the scores of those receiving a treatment are compared to scores in a control group. This statistic can also be easily transformed into its F-equivalent using the formulas shown in Appendix A.

Finally, a note concerning terminology. In the section above, and in several sections that follow, we use the term "F-equivalent". We use this term to be explicit in recognizing that even when the results of a statistical test in the general linear model are reported in terms of some statistic other than F (e.g., r, t, d), it is, nevertheless, usually possible to transform these statistics into the F value that is equivalent in meaning.

---

### Worked Examples Using the d Statistic

The d statistic can be used to describe the strength of an effect in a single study. For example, if a researcher was comparing two treatments and reported d = .50, this would indicate that the difference between treatment means was one-half as large as the standard deviation within each group. An even more common use of d is in meta-analyses, in which the results of several studies are summarized to estimate how large an effect is in the population.

Suppose, for example, that previous research suggests that the effect size d should be about .25. This ES measure would be extremely useful in conducting power analyses. Given this value of d, a study in which 102 subjects were randomly assigned to one of these two treatments would be expected to yield:

$$F(1, \text{df}_{err}) = \frac{d^2 \text{df}_{err}}{4}$$

where:

$d = .25$
$\text{df}_{err} = N - 2 = 102 - 2 = 100$
and
$F(1,100) = [.25^2 * 100]/4$
$F(1,100) = [.0625 * 100]/4 = 1.56$

If you check an F table, you will find that this F is not close to the value needed to reject a null hypothesis (for df = 1,100, you need an F value of 3.93 to reject the null hypothesis). That is given the effect size expected here, a sample of N = 102 will not provide very much statistical power.

The calculations above are based on the independent-groups $t$, in which participants are randomly assigned to one of two treatments. Suppose you used a repeated measures design (e.g., one in which scores of 102 subjects on a pretest and a posttest were compared, with a correlation of .60 between these scores). If you expect that $d = .25$, the formula for transforming repeated-measures $t$ yields:

$$F(1, df_{err}) = \frac{d^2 df_{err}}{\sqrt[4]{1 - r_{ab}}}$$

where:

$d = .25$
$df_{err} = N - 2 = 102 - 2 = 100$
$r_{ab} = .60$
and:
$F(1,100) = [.25^2 * 100]/[4 * .632]$ [the square root of $(1 - .60)$ is .632]
$F(1,100) = [.0625 * 100]/2.529 = 2.47$

The critical value for $F$ ($\alpha = .05$) when the degrees of freedom are 1 and 100 is 3.93, suggesting that the studies described above would not have sufficient power to allow you to reject the traditional null hypothesis.

As you may have noticed, in the examples above we included the sample size. The reason for this is that the value and the interpretation of the $F$ statistic depend in part on the size of the sample (in particular, on the degrees of freedom for the error term, or $df_{err}$). In the preceding paragraph, a $d$ value of .25 in a sample of 102 would yield $F(1,100) = 1.56$. In a study that randomly assigned 375 participants to treatments, the same $d$ would translate into $F(1,373) = 5.82$, which would be statistically significant. This reflects the fact that the same difference between means is easier to statistically detect when the sample (and therefore $df_{err}$) is large than when the sample is small. Small samples produce unstable and unreliable results, and in a small sample, it can be hard to distinguish between true treatment effects and simple sampling error.

## Transforming from F to PV

Appendix A shows how to transform commonly used statistics and effect size estimates into their equivalent $F$ values. It can also often be used transform $F$ values into ES measures. For example, suppose you randomly assigned participants into four different treatments and used analysis of variance to analyze the data. You reported a significant $F$ value [$F(3, 60) = 2.80$], but this does not provide information about the strength of the effect. Formula 2.6 allows you to obtain an estimate of the proportion of variance in the dependent variable explained by the linear model (i.e., $PV$), given the value of $F$:

$$PV = (df_{hyp} * F)/[(df_{hyp} * F) + df_{err}] \qquad (2.7)$$

which yields:

$$PV = (3 * 2.80)/[(3 * 2.80) + 60] \qquad (2.8)$$

$$PV = 8.4/68.4 = .122$$

In other words, the $F$ value reported in this study allows you to determine that treatments accounted for 12% of the variance in outcomes.

Formula 2.6 cannot be used in complex, multifactor analyses of variance, because the $F$ statistic for any particular effect in a complex ANOVA model and its degrees of freedom do not contain all of the information needed to estimate $PV$. In Chapters 7 and 8, we will discuss the application of power analyses to these more complex designs and will show how the information presented in significance tests can be used to estimate ES.

## Defining Large, Medium, and Small Effects

Cohen's books and papers on statistical power analyses (e.g., Cohen, 1988) have suggested several conventions for describing treatment effects as "small", "medium", or "large". These conventions are based on surveys of the literature, and seem to be widely accepted, at least as approximations. Table 2.1 presents conventional values for describing large, medium, and small effects, expressing these effects in terms of many widely used statistics.

For example, a small effect might be described as one that accounts for about 1% of the variance in outcomes, or one where the treatment mean is about one-fifth of a standard deviation higher in the treatment group than in the control group, or as one where the probability that a randomly selected member of the treatment group will have a higher score than a randomly selected member of the control group is about .56.

These are, of course, not the only suggestions for describing effects as small, medium, or large. For example, Ferguson (2009) describes a correlation of .20 or a squared correlation ($PV$) of .04 as the smallest value that is likely to be practically significant. Our preference is to use the more conservative threshold suggested by Cohen (i.e., $PV = .01$ or lower) to

Table 2.1 Some Conventions for Defining Effect Sizes

|  | PV | r | d | $f^2$ | Probability of a higher score in treatment group |
|---|---|---|---|---|---|
| Small effects | .01 | .10 | .20 | .02 | .56 |
| Medium effects | .10 | .30 | .50 | .15 | .64 |
| large effects | .25 | .50 | .80 | .35 | .71 |

From: Cohen (1988), Grissom (1994), Ferguson (2009).
Note: Cohen's $f^2 = R^2/(1 - R^2) = \eta^2/(1 - \eta^2) = PV/(1 - PV)$, where $\eta^2 = SS_{treatments}/SS_{total}$.

describe effects that are likely to be too small to matter in most real-world settings.

The values in Table 2.1 are approximations and nothing more. A few minutes with a calculator makes it clear that they are not all exactly equivalent (e.g., if you square an *r* value of .30, you get an estimate of $PV =$ .09, not $PV = .10$). Although they are not exact or completely consistent, the values in Table 2.1 are nevertheless very useful. These conventions provide a starting point for statistical power analysis, and they provide a sensible basis for comparing the results in any one study with a more general set of conventions.

## Nonparametric and Robust Statistics

The decision to anchor our model for statistical power analysis to the *F* distribution is driven primarily by the widespread use of statistical tests based on that distribution. The *F* statistic can be used to test virtually any hypothesis that falls under the broad umbrella of the "general linear model". There are, however, some important statistics that do not fall under this umbrella and these are not easily transformed into a form that is compatible with the *F* distribution.

For example, several of robust or "trimmed" statistics have been developed in which outliers are removed from observed distributions prior to estimating standard errors and test statistics (Wilcox, 1992; Yuen, 1974). Trimming outliers from data can sometimes substantially reduce the effects of sampling error, and trimmed statistics can have more power than their normal-theory equivalents (Wilcox, 1992). The power tables developed in this book are not fully appropriate for trimmed statistics and can substantially underestimate the power of these statistics when applied in small samples.

A second family of statistics that is not easily accommodated using the model developed here are that group of statistics referred to as "nonparametric" or distribution-free statistics. Nonparametric statistics do not require a priori assumptions about distributional forms and tend to use little information about the observed distribution of data in constructing statistical tests. The conventional wisdom has long been that nonparametric tests have less power than their parametric equivalents (Siegel, 1956), but this is not always the case. Nonparametric tests can have more power than their parametric equivalents under a variety of circumstances, especially when conducting tests using distributions with heavy tails (i.e., more extreme scores than would be expected in a normal distribution; Zimmerman & Zumbo, 1993). The methods developed here do not provide accurate estimates of the power of robust or nonparametric statistics.

## From $F$ to Power Analysis

Earlier in this chapter, we noted that statistical power analysis involves a three-step process. First, you must determine the value of $F$ that is needed to reject the null hypothesis – i.e., the critical value of $F$. Virtually any statistics text is likely to include a table of critical $F$ values that can be used to test the traditional null hypothesis. The great advantage of framing statistical power analysis in terms of the noncentral $F$ distribution is that this approach makes it easy to test many alternatives to the nil hypothesis. As we will show in Chapter 3 this approach to statistical power analysis makes it easy to evaluate the power of tests of the hypothesis that treatments have effects that are not only greater than zero but are also sufficiently large that they are substantively meaningful. Like tests of the traditional null hypothesis, these minimum-effect tests start with the identification of critical values for the $F$ statistic.

Once the critical value of $F$ is established for any test, the assessment of statistical power is relatively easy. As we noted in Chapter 1, *the power of a statistical test is the proportion of the distribution of test statistics expected for a particular study that is above the critical value used to establish statistical significance.* If you determine that the critical value for the $F$ statistic that will be used to test hypotheses in your study is equal to 6.50, all you need to do to conduct a power analysis is to determine the noncentral $F$ distribution that corresponds with the design of your study (the design of your study determines $df_{hyp}$ and $df_{err}$) and the ES you expect in that study, which determines the degree of noncentrality of the $F$ distribution. If the critical value of $F$ is equal to 6.50, power will be defined as the proportion of this noncentral $F$ distribution that is equal to or greater than 6.50.

Several methods can be used to carry out statistical power analyses. For the mathematically inclined, it is always possible to analytically estimate the noncentral $F$ distribution; we will discuss analytic methods below. However, most users of power analysis are likely to want a simpler set of methods. We will discuss the use of power tables and power calculators in the section below.

## Analytic and Tabular Methods of Power Analysis

Analytic methods of power analysis are the most flexible and the most exact, but also the most difficult to implement. Power tables or graphs provide good approximations of the statistical power of studies under a wide range of conditions. Our preference is to work with tables, in part the type of graphs needed to plot a variable (i.e., power) as a function of three other variables (i.e., $N$, $\alpha$, and ES) strike us as complicated and difficult to use. Software and calculators that can be used to estimate statistical power combine the flexibility and precision of analytic methods with the ease of use of power tables.

*Analytic Methods*

The most general method for evaluating statistical power involves estimating the noncentral $F$ distribution that corresponds with the design of your study and the ES you expect. While this analytic method is both precise and flexible, it is also relatively cumbersome and time consuming. That is, the direct computation of statistical power involves: (1) determining some standard for statistical or practical significance (2) estimating the noncentral $F$ distribution that corresponds to the statistic and study being analyzed, and (3) determining the proportion of that noncentral $F$ distribution that lies above the standard. For readers interested in analytic approaches to estimate this distribution, there are web-based calculators and **R** functions that calculate key features of noncentral $F$ distributions.[3] A more user-friendly approach is to develop tables or software that contain the essential information needed to estimate statistical power. We include tables and software that can be used for power analyses for a wide range of analyses that are commonly used in the behavioral and social sciences.

*Power Tables*

Several excellent books present extensive tables describing the statistical power of numerous tests; Cohen (1988) is the most complete source currently available. The approach presented here is simpler (although it provides a bit less information), but considerably more compact. Unlike Cohen (1988), who developed tables for many different statistics, our approach involves translating different statistics into their $F$-equivalents. This allows us to present virtually all of the information needed to do significance tests and power analyses for statistical tests in the general linear model in a single table.

Appendix B contains a table we call the "One-Stop $F$ Table". This is called a "one-stop" table because each cell contains the information you need for: (1) conducting traditional significance tests, (2) conducting power analyses at various key levels of power, (3) testing the hypothesis that the effect in a study exceeds various criteria used to define negligibly small or small to moderate effects, and (4) estimating power for these "minimum-effect" tests. Minimum-effects tests are explained in detail in Chapter 3; at this point, we will focus on the use of the One-Stop $F$ Table for testing the traditional null hypothesis and for evaluating the power of these tests.

## Using the One-Stop F Table

Each cell in the One-Stop $F$ Table contains 12 pieces of information. The first four values in each cell are used for testing significance and estimating power for traditional null hypothesis tests. The next eight values in each cell are used for testing significance and estimating power when testing the hypothesis that treatment effects are negligible, using two different operational definitions of a "negligible" effect (i.e., treatments account for 1% or less of the variance, or that they account for 5% or less of the variance). In this chapter, we will focus on the first four pieces of information presented for each combination of $df_{hyp}$ and $df_{err}$.

This table presents the critical value of $F$ for testing the traditional null hypothesis, using $\alpha$ values of .05 and .01, respectively. We label these "nil .05" and "nil .01". For example, consider a study in which 54 subjects are randomly assigned to one of four treatments, and the analysis of variance is used to analyze the data. This study will have $df_{hyp}$ and $df_{err}$ values of 3 and 50, respectively. The critical values of $F$ for testing the nil hypothesis will be 2.79 when $\alpha$ equals .05; the critical value of $F$ for testing the nil hypothesis will 4.20 when $\alpha$ equals .01. In other words, to reject the nil hypothesis ($\alpha = .05$), the value of the mean square for treatments will have to be at least 2.79 times as large as the value of the mean square for error (i.e., $F = MS_{treatments}/MS_{error}$).

The next two values in each cell are $F$-equivalents of the effect size values needed to obtain power of .50 and power of .80 (we label these "pow .50" and "pow .80"), given an $\alpha$ level of .05 and the specified $df_{hyp}$ and $df_{err}$. The values of "pow .50" and "pow .80" for 3 and 50 degrees of freedom are in the table are 1.99 and 3.88, respectively. That is, a study designed with 3 and 50 degrees of freedom will have power of .50 for detecting an effect that is equivalent to $F = 1.99$. It will have power of .80 for detecting an effect that is equivalent to $F = 3.88$. $F$-equivalents are handy for creating tables, but to interpret these values, it is necessary to translate them into ES measures.

Because subjects are randomly assigned to treatments and the simple analysis of variance is used to analyze the data, we can use Formula 2.6, presented earlier in this chapter, to transform these $F$ values into equivalents $F$ values (i.e., $PV = (df_{hyp} * F)/[(df_{hyp} * F) + df_{err}])$. Applying this formula, you will find that in a study with $df_{hyp}$ and $df_{err}$ values of 3 and 50, you will need a moderately large ES to achieve power of .50. Translating the F value of 2.16 into its equivalent $PV$, you will find:

$$PV = (3 * 1.99)/[(3 * 1.99) + 50] = 5.97/55.97 = .106$$

That is, to achieve a power of .50 with $df_{hyp}$ and $df_{err}$ values of 3 and 50, you will need to be studying treatments that account for at least 10% of the variance in outcomes. Applying the same formula to the $F$ needed to achieve power of .80 (i.e., $F = 3.88$), you will find that you need an effect size of $PV = .188$. In other words, to achieve power of .80 with this study, you will need to be studying a truly large effect, on which treatments account for about 19% or more of the variance in outcomes.

Suppose that based on your knowledge of the scientific literature, you expect the treatments being studied to account for about 15% of the variance in outcomes. This ES falls between $PV = .10$ and $PV = .19$, which implies that power of this study will fall somewhere between .50 and .80. You could use the **R** code or the Shiny Web app described below to get a more precise estimate of power for this specific study.

## Simple and General Software for Power Analysis

Free power analysis software that is relatively simple to use is readily available. For example, we often recommend G∗Power [4] to our students. It is flexible and informative, but it has one limitation. It only allows you to test the nil hypothesis – the hypothesis that treatments have exactly zero effect, or that the parameter (e.g., the correlation between X and Y) you are testing is exactly zero in the population.

We have developed a simpler and more flexible approach that allows you to test the hypothesis that treatments have no effect whatsoever, but also to test the hypothesis that treatments have an effect that is so small that it is trivial (e.g., that treatments account for 1% of the variance or less). In previous editions of this book, we had included a software developed using Visual Basic that was sometimes hard to apply (e.g., it worked on Windows, but not other operating systems), and more important not fully transparent (i.e., we provided software but no code). In this edition, we have rewritten our software in **R**, and provide both **R** code and a Shiny Web app that allows users who are not familiar with **R** to implement this code by simply going to a website and entering the information needed to perform a power analysis.

**R** is a programming language that has emerged as the default method for data visualization and data analysis in many scientific fields. It is free[5] and many tools are available to help users implement **R**. **R** includes programs, referred to as "packages" for carrying out virtually any type of data analysis and for creating useful data visualizations. The **R** code for the power analysis method described in this book is included in the boxed section below. We also show the **R** code that can be used to generate the tables shown in Appendices B, C, and D at the end of each table.

---

### R Code for Power Analysis for Traditional and Modern Hypothesis Tests

**R** code is shown below. A description of what each of the values you need to enter (shown in boldface below) means, the output of the program, and what it tells you is provided in the section that follows, which describes the implementation of this software as a Shiny Web app. **R** users can paste the code below into an **R** project and conduct power analyses for a wide range of analyses. Readers who are not familiar or comfortable with **R** can use our web-based app to carry out these same analyses.

```
# Begin R Code
# enter df, alpha, the the df for the hypothesis you are testing and for error
alpha=.05
dfhyp=5
dferr=100
# enter the minimum effect (percentage of variance explained) that represents the upper bound
# for a trivially small effect. This is often .01 – i.e., that any effect that explains 1% or less
# of the variance is described as having a trivially small effect
# to test the traditional null hypothesis (nil hypothesis), enter .00
mineffect=.00
# enter the assumed effect size (in terms of percentage of variance explained) in the population. # A common convention is that
# a small effect would correspond to one percent of the variance explained, a moderate effect
# would correspond to 10% of the vairance explained and a large effect, 25% of the variance
# explained
assumedeff=.05
# calculate power
noncen=dferr/((1/mineffect)-1)
critF<-qf((1-alpha),dfhyp,dferr,noncen)
reqiv<-(dfhyp*critF)/((dfhyp*critF)+ dferr)
noncen1=dferr/((1/assumedeff)-1)
omp<-pf(critF,dfhyp,dferr,noncen1)
power=1-omp
#calculate power as N varies
narray<-array(1:20000,dim=c(10000,2))
dferror=0
for (i in 1:10000){
dferror=i
noncen2=dferror/((1/mineffect)-1)
critFn<-qf((1-alpha),dfhyp,dferror,noncen2)
noncen3=dferror/((1/assumedeff)-1)
omp<-pf(critFn,dfhyp,dferror,noncen3)
powern=1-omp
narray[i,1] = i
narray[i,2] = powern
```

```
if(narray[i,2]>=.80){
break
}
powernum=narray[i,1]+ dfhyp + 1
}
# Display critical value of F and the power of this test
res1<-as.matrix(c(mineffect,assumedeff,alpha,dfhyp,dferr))
rownames(res1)<-c("Definition of minimum effect","Assumed effect
size","alpha","df for the hypothesis","df error")
colnames(res1)<-c("Parameter Values")
res2<-as.matrix(c(critF,reqiv,power,powernum))
rownames(res2)<-c("Critical value of F for Min Effect Test",
"PV equivalent of critical F","Power","N needed for power of .80")
colnames(res2)<-c("Power Analysis Results")
options(scipen= 3)
res1
res2
#End R Code
```

## A Web-Based App for Power Analysis

The **R** code shown in the boxed section is handy, but suppose you are not an **R** user. The simplest way to apply the program shown in the boxed section **"R Code for Power Analysis for Traditional and Modern Hypothesis Tests"** is to use our Shiny Web app, which can be accessed at https://murphy0921.shinyapps.io/ShinyPower/.

Use of this app does not require any knowledge of **R** and it does not even require that you install **R** on your computer. If you go to this website, you will see a screen like the one illustrated in Figure 2.1.

To calculate power for any test, you need to enter a few pieces of information. The web page will always have values entered as placeholders, and all you need to do is check them and modify where needed to fit your own study.

First, you also need to enter the degrees of freedom for the hypothesis being tested and for error. Suppose, for example, that there are 106 subjects randomly assigned to one of six treatments. This yields degrees of freedom of 5 and 100 for the hypothesis being tested ($df_{hyp}$) and for error ($df_{err}$).

Next, you need to enter the alpha ($\alpha$) level, usually .05.

Next, you need to describe what type of test you are performing. If you are testing the traditional null (nil) hypothesis that treatments have no effect whatsoever, enter .00 in the "Minimum effect size" box. If you want to test the minimum-effect hypothesis that treatments account for 1% of the variance or less, enter .01.

## Power Analysis for Traditional and Modern Hypothesis Tests

Specify df, alpha, minimum effect (effect so small it is regarded as trivial) and assumed effect size in the population

degrees of freedom for the hypothesis:
1

degrees of freedom for error:
10

alpha level:
0.05

Minimum effect size - Percentage of variance (PV) explained that defines a trivially small effect:
0.01

assumed effect size (PV) in the population:
0.05

```
Critical value of F        PV equivalent of critical F
           5.4600                              0.3532
            Power          N needed for power of .80
           0.0867                            380.0000
```

*Figure 2.1* Shiny Web App for Power Analysis for Traditional and Modern Hypothesis Tests.

Finally, enter the assumed affect size – i.e., the proportion of variance in the population you think is explained by treatments or the predictors you are studying. In this example, let's suppose you expect treatments to account for 5% of the variance in the population, and you want to test a nil hypothesis.

So, if you want to test the traditional null hypothesis, with $\alpha = .05$, $df_{hyp} = 5$, $df_{err} = 100$, and an assumed effect size in the population of .05 (i.e., treatments are assumed to account for 5% of the variance in the population), the web page will immediately calculate four quantities: (1) how large $F$ needs to be to reject the null hypothesis you are testing, (2) the $PV$ value that is equivalent to this $F$ – i.e., how large the observed effect will need to be to reject your null, (3) the power of this test, and (4) the $N$ needed to achieve power of .80. If you enter the values laid out above, you will find:

| | |
|---|---|
| Critical value of $F$ for Min Effect Test | 2.305 |
| $PV$ equivalent of critical $F$ | 0.103 |
| Power | 0.361 |
| $N$ needed for power of .80 | 255.000 |

That is, you will reject $H_0$ if the observed value of $F$ is 2.30 or bigger and/or if the variance accounted for is .103 or larger in your study. The power of this study design is low (.36), and if you want to achieve power of .80 in this study, you will need an $N$ of 255.

Suppose you wanted to test a different hypothesis, that the effects of treatments are trivially small, defined as treatments that explain 1% of the variance or less. If you change the value in the "Minimum effect size" box from .00 to 01, you will find:

| | |
|---|---:|
| Critical value of F for Min Effect Test | 2.748 |
| PV equivalent of critical F | 0.121 |
| Power | 0.247 |
| N needed for power of .80 | 453.000 |

As we will explain in Chapter 3, minimum-effect hypotheses are more challenging to test than nil hypotheses (e.g., they require larger samples to achieve comparable power), but these tests are often much more informative.

## Notes

1. Install the **R** package MBESS and the confidence interval for $d$ is given by **ci.smd (ncp = 1.88, n.1 = 40, n.2 = 50, conf.level = 0.95)**, where ncp is equal to the value of the $t$ statistic used to test the significance of differences between means, which in this case equals 1.88. The confidence interval for $PV$ is given by **ci.pvaf (F.value = 7.2, df.1 = 4, df.2 = 145, N = 150, conf.level = .95)**, where the F.value is the $F$ that corresponds to $R^2 = .20$, given by $F = \left(\frac{R^2}{1-R^2}\right) * \left(\frac{df_{err}}{df_{hyp}}\right)$

2. Equations 2.4 and 2.5 are based on simple linear models, in which there is only one effect being tested, and the variance in scores is assumed to be due to either the effects of treatments or to error (e.g., this is the model that underlies the $t$-test or the one-way analysis of variance). In more complex linear models, $df_{err}$ does not necessarily refer to the degrees of freedom associated with variability in scores of individuals who receive the same treatment (within-cell variability in the one-way ANOVA model), and Equation 2.6, a more general form of Equation 2.5 ($\lambda_{est} = [(N-k) * (PV/(1-PV))]$, where $N$ represents the number of observations and $k$ represents the total number of terms in the linear model) is needed. When $N$ is large, Equation 2.5 yields very similar results to those of Equation 2.6.

3. See https://keisan.casio.com/exec/system/1180573165. The **R** function qf(.95, df1 = 2, df2 = 120, ncp = 5) gives you the 95th percentile of the noncentral $F$ distribution 2 and 120 degrees of freedom and a noncentrality paramater value of 5.0.

4. You can download this software at https://www.psychologie.hhu.de/arbeitsgruppen/allgemeine-psychologie-und-arbeitspsychologie/gpower.

5. You can download the latest version of **R** at https://www.r-project.org. You can download **R Studio**, an application that makes it easier to use **R** and to organize your work in **R** at https://www.rstudio.com. Both are free, open-source programs.

# 3 Power Analyses for Minimum-Effect Tests

The traditional null hypothesis is that treatments, interventions, etc. have no effect; in Chapters 1 and 2, we used the term "nil hypothesis" to describe this version of $H_o$. The nil hypothesis is so common and so widely used that most researchers assume that the hypothesis that treatments have no effect, or that the correlation between two variables is zero *is* the null hypothesis. This is wrong. The null hypothesis is simply the specific hypotheses that is being tested (and that might be nullified by the data), and there are an infinite number of null hypotheses researchers might test. One researcher comparing two treatments might test the hypothesis that there is no difference between the mean scores of people who receive different treatments. A different researcher might test the hypothesis that one treatment yields scores that are, on average, five points higher than those obtained using another treatment. Yet another researcher might test the hypothesis that treatments have a very large effect, accounting for at least 25% of the variance in outcomes. These are all null hypotheses. Knowing that there are so many null hypotheses that might be tested, it is useful to understand why one special form – i.e., the nil hypothesis, is the one that is tested in most statistical analyses.

## Nil Hypothesis Testing

There are two advantages to testing the nil hypothesis: (1) it is easy to test this hypothesis; tests of this hypothesis represent the standard method presented in statistics textbooks, data analysis packages, etc., and the derivation of test statistics designed to evaluate the nil hypothesis is often comparatively simple, and (2) if a researcher rejects the hypothesis that treatments have no effect, he or she is left with the alternative that treatments have *some* effect. If $H_0$ states that nothing happened as a result of treatments, the alternative hypothesis ($H_1$) is that *something* happened as a result of treatments. In contrast, tests of some alternatives to the nil hypothesis can lead to confusing results.

Suppose a researcher tests and rejects that hypothesis that the difference between treatment means is 5.0. The alternative hypothesis ($H_1$) is that the difference in treatment means is *not* 5.0, and this includes quite a wide range of possibilities. Treatments might lead to differences larger than 5.0. They might have no effect whatsoever. They might have some effect but lead to differences smaller than 5.0. They might even have the opposite effect than the researchers expected (i.e., a new treatment might lead to worse outcomes than the old one). All a researcher learns by rejecting this null hypothesis is that the difference is not 5.0.

As we noted in Chapters 1 and 2, nil hypothesis testing has been widely criticized (Amrhein et al., 2019; Cohen, 1994; Meehl, 1978; Morrison & Henkel, 1970; Murphy, 1990; Schmidt, 1992, 1996; Wasserstein et al., 2019. For discussions of the advantages of this approach, see Chow, 1988; Cortina & Dunlap, 1997; Hagen, 1997). The most general criticism of the traditional approach to null hypothesis testing is that very few researchers believe that the nil hypothesis is likely to be correct. That is, it is rare to encounter serious treatments or interventions that have no effect whatsoever, which is the traditional null hypothesis. In a later section, we will discuss in detail why the nil hypothesis is almost always false. Here, we will simply note that if the null hypothesis to be tested is known in advance to be false, or is very likely to be false, tests of that hypothesis have very little value (Murphy, 1990).

The second critique of nil hypothesis testing is that the outcomes of tests of the traditional null hypothesis are routinely misinterpreted. As we will show below, the outcomes of standard statistical tests probably say more about the power of your study than the phenomenon you are studying. If you design a study with sufficient power, you will almost always reject the nil hypothesis. If you design a study with insufficient power, you will usually fail to reject the nil hypothesis. These facts are well understood by most researchers, but it is still common to find that when researchers fail to reject the null hypothesis, they end up drawing conclusions about the treatments or about the relationships being studied. In particular researchers who fail to reject the null hypothesis all too often conclude that the treatments or interventions being studied did not work, or that the variables being studied are not correlated. Similarly, researchers who *do* reject the null hypothesis often conclude that there are meaningful treatment effects, or that the intervention being studied worked. It should be clear by now that the outcomes of nil hypothesis tests are not driven solely, or even mainly by the substantive phenomenon being studied. The outcomes of nil hypothesis tests usually tell researchers more about the study than they do about the substantive phenomenon being studied. The role of $N$ is so dominant in determining the outcome of a nil hypothesis test that these tests are often little more than an indication of how many subjects showed up for a study.

A third criticism of the traditional null hypothesis is that showing that a result is "significant" at the .05 level does not necessarily imply that it is important or that it is likely to be replicated in a future study (Cohen, 1994). This significance test merely shows that the results reported in a study probably would not have been found if the true effect of treatments was zero.

Unfortunately, researchers routinely misinterpret the results of significance tests (Cohen, 1994; Cowles, 1989; Greenwald, 1993). This is entirely understandable; most dictionary definitions of "significant" include synonyms such as "important" or "weighty". However, these tests do not directly assess the size or importance of treatment effects, nor do they assess the likelihood that future studies will find similar effects.

Tests of the traditional nil hypothesis are more likely to tell you about the sensitivity of your study than about the phenomenon being studied. With large samples, statistical tests of the traditional nil hypothesis become so sensitive that they can detect the slightest difference between a sample result and the specific value that characterized the null hypothesis, even if this difference is negligibly small. With small samples, on the other hand, it is difficult to establish that *anything* has a statistically significant effect. The best way to get an appreciation of the limitations of traditional null hypothesis tests is to scan the tables in any power analysis book (Cohen, 1988). What you will find is that, regardless of the true strength of the effect, the likelihood of rejecting the traditional null hypothesis is very small when samples are small, and the likelihood of rejecting this hypothesis is extremely high when samples are large. There is a clear need for approaches to significance testing that tell researchers more about the phenomenon being studied than about the size of their samples. Alternatives to traditional nil hypothesis tests are described later in this chapter.

## The Nil Hypothesis Is Almost Always Wrong

In our view, the most serious problem with nil hypothesis testing is that we can be virtually certain that the null hypothesis, in its literal form, is wrong (Cohen, 1994; Kirk, 1996). That is, regardless of the treatments, interventions, correlations, etc. being studied, it is very unlikely that the true population effect (the difference between two means, the correlation between two variables) is *precisely* zero. Treatment effects might be small (they often are), but the likelihood that they are exactly zero is so small that we can often dismiss it without any type of statistical test. We lay out four reasons for believing that the nil hypothesis is almost always wrong below.

### *The Nil Is a Point Hypothesis*

The main reason that the nil hypothesis is almost always wrong is it that is a point hypothesis. That is, the hypothesis being tested is that the effect of treatments is exactly zero, even to the millionth decimal place or beyond. The traditional nil hypothesis represents a convenient abstraction, similar to the mythical "friction-less plane" encountered by freshmen in solving physics problems, in which potentially small effects are treated as zero for the sake of simplicity. Many real-world phenomena seem to mirror the traditional nil hypothesis, the most obvious being flipping a coin (see Fick, 1995, for examples that seem to show how the traditional null could be correct). However, even in studies that involved repeated flips of a fair coin, the hypothesis is that there is no difference whatsoever in the probability of getting a head or a tail is simply not true. It is impossible to mill a coin that is so precisely balanced that the likelihood of getting heads or tails is exactly equivalent; imbalance at the ten-billionth of the ounce will lead you to favor heads or tails if you flip the coin a sufficient number of times.

### The Nil Hypothesis Is Impossibly Precise

There is nothing special about the hypothesis that the difference between two treatments is zero. All point hypotheses (e.g., that the difference between two treatments is 5.0) suffer from the same problem. In the abstract, they might be true, but in reality there is no way to ever demonstrate that they *are* true. Because the nil hypothesis is infinitely precise, none of the real-world phenomena it is designed to test can possibly be assessed at that level of precision. Therefore, if you showed that there was no difference in two treatments, even at the billionth decimal place, the possibility that some difference might emerge with an even finer-grained analysis could never be completely dismissed.

### The Domain of Alternative Hypotheses Is Infinitely Large

A third argument for the conclusion that the nil hypothesis is almost always wrong is best conveyed using a spatial analogy. Suppose you used a standard football field to represent all the outcomes that could possibly occur when you compare the means of two populations, each of whom received different treatments. One possibility is that there is no difference. Another possibility is that one treatment mean is .01 units higher than the other. Another possibility is that one treatment mean is .02 units higher than the other, and so on. Divide the football field so that $H_0$ represents the one outcome (no difference) that corresponds with the nil hypothesis and $H_1$ represents all the other outcomes. How much space do you think you would devote to $H_0$ versus $H_1$? The most optimistic proponent of nil hypothesis testing would only devote a tiny patch of dirt to $H_0$. Virtually the entire football field would have to be devoted to $H_1$ because there is an effectively infinite number of outcomes that could happen in this experiment. $H_0$ represents one of these outcomes, and $H_1$ represents all the rest. In spatial terms, the solution space that represents $H_o$ is essentially infinitely small, whereas the solution space that represents $H_1$ is infinitely large.

### Nil Treatment Effects Exist Only in the World of Abstractions

The argument against the traditional nil hypothesis is not only a philosophical one; there are also abundant data to suggest that treatments in the social and behavioral sciences virtually always have at least some effect (Lipsey & Wilson, 1993; Murphy & Myors, 1999). It may not be possible to devise a real treatment that has no effect whatsoever; if it is possible, it would take real effort. For example, suppose you try to come up with a treatment for the common cold that has no effect whatsoever. Virtually any type of medication, dietary change, or changes in behavior are probably out of bounds. Even if the treatment itself has no effect, there is always the possibility of placebo effects or Hawthorn effects (people who think you are doing something to make them feel better often report feeling a bit better, even if the thing you do is crazy).

To be sure, there is no shortage of crackpot interventions, junk science, and treatments that have no meaningful effect (e.g., wearing magnetized bracelets as a way of treating cancer). However, if your goal is to evaluate serious treatments devised by someone who had any reason to believe that they might work, that the likelihood that treatments will have no effect whatsoever is so low that tests of the nil hypothesis is almost certainly pointless (Murphy, 1990).

## Implications of the Conclusion That the Nil Hypothesis Is Almost Always Wrong

The belief that the traditional nil hypothesis is almost always wrong has important implications for thinking about Type I and Type II errors. We would argue that tests of the traditional nil hypothesis make sense *only* if you believe that the nil hypothesis is often correct, and we know of no researchers who believe this. To understand why assumptions about the likelihood that the nil hypothesis is true are so important, consider Figure 3.1, which illustrates two ways of making errors in a nil hypothesis test. First, it is possible that $H_0$ is true, and that researchers will wrongly conclude that treatments

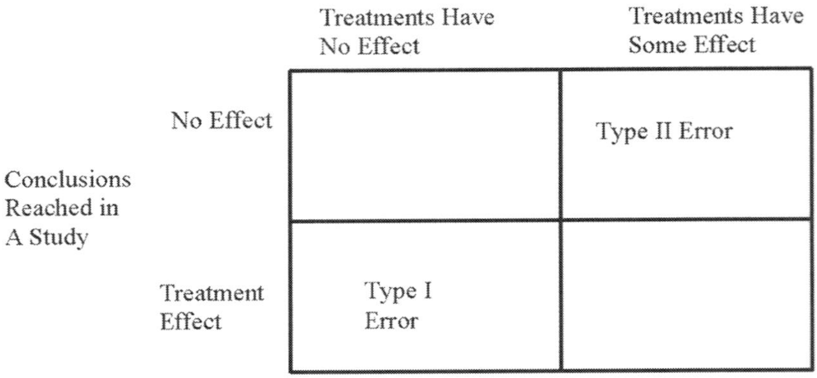

|  | Treatments Have No Effect | Treatments Have Some Effect |
|---|---|---|
| No Effect |  | Type II Error |
| Treatment Effect | Type I Error |  |

Conclusions Reached in A Study

*Figure 3.1* Two Types of Errors in Statistical Tests.

have some effect – i.e., a Type I error. There is a large and robust literature detailing methods for controlling Type I errors (e.g., Wilkinson et al., 1999; Zwick & Marascuilo, 1984). As Figure 3.1 makes clear, the likelihood of making a Type I error depends first and foremost on whether the nil hypothesis that treatments have no effect is true (Murphy, 1990). If you believe that the nil hypothesis is virtually never true, it follows that Type I errors are virtually impossible – these errors can occur if and only if $H_o$ is true. Unless you believe that the nil hypothesis is right in many studies, it is unlikely that you will ever have much reason to be concerned about Type I errors.

---

**Polar Bear Traps: Why Type I Error Control Is a Bad Investment**

A simple analogy helps to drive home the importance of the low likelihood that the nil hypothesis is true. Suppose you are a homeowner, and a salesman comes to the door selling polar bear traps. He makes a compelling case that a polar bear attack would be very unpleasant and that these are good traps. Would you buy traps? We would not. The likelihood that a polar bear will attack is so small that the traps strike us as a waste of money.

You should think of procedures designed to control or reduce Type I errors as polar bear traps. They give you a means of controlling or limiting something that you know is highly unlikely in the first place. More important, these traps cost something. Virtually anything you do to control Type I errors will increase the likelihood of Type II errors. For example, the common rationale for using a stringent alpha level, such as $\alpha = .01$ rather than a less demanding criterion (e.g., $\alpha = .05$) in defining statistical significance is that the stricter alpha will lead to fewer Type I errors. If you think that Type I errors are virtually impossible in the first place, you are unlikely to believe a more stringent alpha is worthwhile. In general, steps you take to control Type I errors usually lead to reduced levels of power. If you think about Type I error control as a polar bear trap, you are unlikely to see it as a wise investment.

---

In Figure 3.1, we presented the traditional fourfold table for describing decision errors when testing nil hypotheses. Although this type of figure is widely used, it is fundamentally misleading. A more accurate fourfold table would look something like the one shown in Figure 3.2.

First, even if you believe that the nil hypothesis might sometimes be true, there is no serious argument that it is just as likely that treatments have no effect as that they have some effect (albeit often a small one). If anything, the "Treatments have no effect" column should probably be even narrower than what is shown in this figure. Second, studies in many sciences show low levels of power,

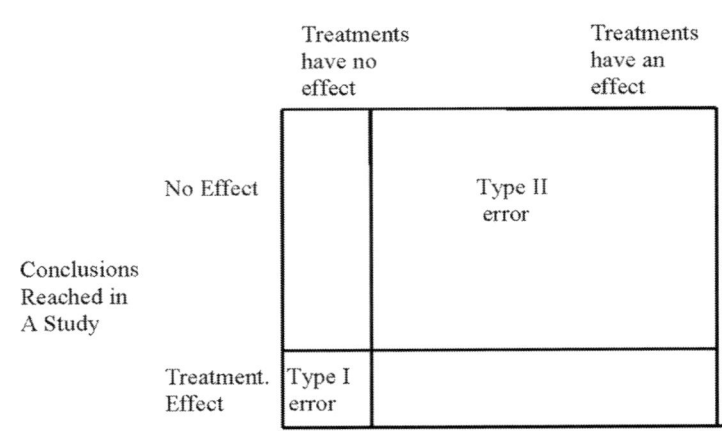

*Figure 3.2* A More Realistic Decision Table.

meaning that many are likely to reach the conclusion (probably an incorrect one) that treatments have no effect. If the hypothesis that treatments have no effect is usually wrong and we design studies that will often have a small chance of rejecting the null hypothesis, Type II errors are likely to be distressingly common. Researchers who worry about Type I errors and who tolerate lots of Type II errors have things backwards. If the nil hypothesis is almost always wrong, the only way you are likely to make an error in hypothesis testing is to fail to reject a nil hypothesis you know is almost certainly wrong.

## The Nil May Not Be True, but It Is Often Fairly Accurate

The conclusion that the nil hypothesis is almost always wrong is not the same as the conclusion that most treatments, interventions, etc. work well. On the contrary, it is reasonable to believe that many treatments and interventions have very small effects and that the statement that they had no effect at all is often close to the truth. As we noted in Chapters 1 and 2, most studies you read in the literature are likely to deal with treatments, interventions, etc. that have small (but perhaps important) effects. If the effects of treatments were large and obvious, there would not be much demand for further studies testing the nil hypothesis. Many of the most important and interesting studies you are likely to read will deal with new treatments, novel interventions, and innovative approaches, and many of these just won't work very well. As a point hypothesis, the nil hypothesis is almost always wrong, but as a general description of what researchers expect might happen, it is often a good approximation. What is needed is an approach that captures the worrisome possibility that many treatments, interventions, etc. might have very small effects, without all the problems that accompany tests of point hypotheses. A method that overcomes many of the limitations of traditional nil hypothesis tests is presented in the section that follows.

## Minimum-Effect Tests as Alternatives to Traditional Null Hypothesis Tests

The criticisms of nil hypothesis tests outlined above have led some critics to call for abandoning null hypothesis testing altogether (e.g., Schmidt, 1992, 1996). Rather than take this drastic step, we think it is better to reform the process. In our view, the most serious problem with most null hypotheses tests is that the specific hypothesis being tested, i.e., that these treatments have no effect whatsoever, is neither credible nor informative (Murphy, 1990). There are several alternatives to testing the "nil hypothesis", and all of these are a marked improvement over the standard procedure of testing the hypothesis that the effect of treatments is precisely zero.

Serlin and Lapsley (1993) show how researchers can test the hypothesis that the effect of treatments falls within or outside of some range of values that is "good enough" to establish that one treatment is meaningfully better than the other. Rouanet (1996) shows how Bayesian methods can be used to assert the importance or negligibility of treatment effects. Both methods allow researchers to directly test credible and meaningful hypotheses.

The method described in this book involves testing "minimum-effect" hypotheses (Murphy & Myors, 1999). The traditional nil hypothesis involves a choice between:

$H_0$ – treatments have no effect
vs.
$H_1$ – treatments have some effect

In contrast, minimum-effect hypotheses involve a choice between:

$H_0$ – treatments have an effect that is so small that it can be described as negligible
vs.
$H_1$ – treatments have an effect that is so large that they should be described as meaningful or important

Minimum-effect tests require researchers decide what sort of treatment effect is so small that it should be labeled as negligible. This decision may be a difficult one, but if a reasonable standard for defining negligibly small effects can be determined, it is quite easy to develop the appropriate tests, using the same noncentral-$F$-based model we have used in presenting power analysis.

The central assumption of minimum effect tests is that there is some range of possible ES values that are so small that they might as well be zero. For example, if there is a substantive reason to believe that a treatment effect that accounts for less than 1% of the variance is simply too small to care about, it does not matter whether the true treatment effect is $PV = 0.0$, $PV = 0.001$, $PV = 0.002$, etc. Anything that falls below $PV = 0.01$ will be regarded as too small to be meaningful or important, and the null hypothesis in this minimum-effect test is that the population treatment falls somewhere within this range. A researcher who can reject this null hypothesis is left with the alternative hypothesis that effects are large enough to be important.

Minimum-effect tests have many advantages over traditional nil hypothesis tests. First, both $H_0$ and $H_1$ are substantively meaningful and intrinsically interesting to most researchers. Second, the minimum-effect null hypothesis, that treatments have effects so small that they are negligible, is one that might be true. Third, these tests can be developed in ways that allow researchers to apply virtually all the procedures, standards, and metrics they have learned in the context of nil hypothesis testing to ask whether the effects of treatments are either negligibly small or large enough to be of interest.

### Minimum-Effect Tests Are Meaningful

There is much to be learned by conducting tests of the hypothesis that effects of treatments are negligibly small (e.g., they account for 1% or less of the variance in outcomes). In contrast to tests of the traditional null, tests of this sort are far from trivial. First, when conducting minimum-effects tests, researchers do not know in advance whether $H_0$ is right or wrong. Second, these tests involve questions that are of real substantive interest. Because these include some range of values rather than a single exact point under the "null umbrella", the results of these tests are not a foregone conclusion. Although it might be impossible to devise a treatment, intervention, etc. that had *no* effect whatsoever, there are any number of treatments whose effects fall somewhere between zero and whatever percentage of variance you designate as representing a negligibly small effect. The hypothesis that treatments might have a negligibly small effect is both real and meaningful and researchers can learn something important about their treatments by testing this hypothesis.

### Sometimes a Point Hypothesis Is Also a Range Hypothesis

The hypothesis that treatments account for 1% of the variance in the dependent variable is a point hypothesis, just like the nil hypothesis. We have been highly critical of tests of the nil hypothesis, and you might wonder why these criticisms of nil hypothesis tests do not apply to minimum-effect tests.

One of the benefits of building a system of power analysis around the $F$ statistic is that the $F$ distribution is unidirectional. That is, the lowest possible value of $F$ is zero, and the larger the value of $F$ the larger the value of $PV$. More important, an $F$ large enough to allow you to reject the hypothesis that treatments account for 1% of the variance with $\alpha = .05$ is also large enough to allow you to reject the hypothesis that treatments account for 0.5% of the variance or that treatments account for no variance at all with at least 95% confidence.

For example, suppose you are using three variables to predict Y in a sample of 145 subjects. In this study, $df_{hyp} = 2$ and $df_{error} = 142$. The values to $F$ and the observed $PV$ values needed to reject the null hypothesis ($\alpha = .05$) for various nulls are shown below.

| Hypothesis | Critical Value of F | PV |
|---|---|---|
| Treatments account for 1% of the variance in Y | 4.92 | .064 |
| Treatments account for 0.5% of the variance in Y | 4.05 | .054 |
| Treatments account for 0.1% of the variance in Y | 3.27 | .044 |
| Treatments account for none of the variance in Y (nil) | 3.04 | .041 |

If you find that $F_{(2,142)} = 6.17$, $R^2 = .08$, you could reject all of the hypotheses listed above with $\alpha = .05$ or lower. More generally, if your $F$ is large enough to reject the hypothesis that treatments account for any specific percentage of the variance with $\alpha = .05$, you can also reject any of the null hypotheses that fall within the range of 0% to that percentage of the variance explained with at least the same level of confidence. If your $F$ is large enough to reject a null hypothesis at the top of any range of values of $PV$, you can reject the entire range of possible null hypothesis form the nil (0) to the top of that range of $PV$ values.

## The Minimum-Effect Hypothesis Has a Reasonable Chance of Being True

In an earlier section of this chapter, we noted that when testing the traditional null hypothesis, a researcher can guarantee the outcome if the sample is sufficiently large. That is, with a sufficiently sensitive study, the statistical power of tests of the traditional null hypothesis will approach 1.00. The reason it is possible to reach power levels of essentially 1.00, regardless of the substantive question being examined in a study, is that traditional models of power analysis are based on the entirely realistic assumption that $H_o$ is virtually never true.

In contrast to the nil hypothesis, the minimum-effect null hypothesis is often a realistic possibility. That is, there certainly are treatments, interventions, etc.

that turn out to have truly small effects. Because there is a real possibility that the effect of treatments will turn out to be negligible, $H_0$ might really be true, Type I errors might really occur, and the whole structure for defining errors in significance testing illustrated in Figure 3.1 become relevant once again.

Because there is a very real possibility that the effects of treatments will turn out to be negligibly small, the upper bound of the power of minimum-effects tests will generally be lower than 1.00. That is, no matter how sensitive the study, researchers can never be certain in advance of the result of a minimum-effect test. We regard this as a good thing. One major criticism of tests of the traditional nil hypothesis is that they are pointless (Murphy, 1990). Because researchers know in advance that the traditional $H_0$ is almost certainly wrong, they are unlikely to learn anything new about $H_0$, regardless of the outcome of the significance test. Minimum-effect tests, on the other hand, can be informative, particularly if these tests are conducted with acceptable levels of statistical power. If a researcher designs a study that has a great deal of power and still fails to reject the hypothesis that the effects of treatments are negligible, this failure to reject the null hypothesis can be interpreted as strong evidence that the treatment effect truly *is* negligible. In contrast, failure to reject the nil hypothesis usually tells you that your study does not have enough power.

### How Do You Know the Effect Size?

You have probably noticed that there is some circularity in many discussions of power analysis. To do a good job in power analysis, you need to know the effect size in the population (e.g., how much of the variance in job performance can be explained by individual differences in general cognitive ability). However, if you *knew* this effect size, you probably would not have needed to do your study in the first place. More to the point, if you are quite certain about the value of the effect size in the population, it is unlikely that whatever you find in a study will lead you to change your mind about the size of the effect you are studying (Newman, Jacobs & Bartram, 2007).

In Chapter 1, we observed that in power analysis, precision is not always necessary or even useful. This is especially the case when there is not good basis for estimating the effect size in the population (such as a recent and comprehensive meta-analysis). In the absence of this type of evidence, what is a researcher to do?

Our advice is to be conservative in estimating population effect size values. If you design a study that has sufficient power to test a hypothesis when the population $PV = .02$, that study will have even more power for than your calculations suggest if the population ES is larger than $PV = .02$. On the other hand, over-estimating the size of the effect in a population could lead you to design and execute studies that have less power than you think they have. Starting with a conservative estimate of the ES covers just about all of the bases.

## Testing the Hypothesis That Treatment Effects Are Negligible

The best way to describe the process of testing a minimum-effect hypothesis is to compare it to the process used in testing the traditional nil hypothesis. The significance of the $F$ statistic is usually assessed by comparing the value of the $F$ obtained in a study to the value listed in a standard $F$ table. The tabled values presented in virtually every statistics text correspond to specific percentiles in the central $F$ distribution. For example, if $df_{hyp} = 2$ and $df_{err} = 100$, the tabled values of the $F$ statistic that is used in testing the nil hypothesis are 3.09 and 4.82, for $\alpha = .05$ and $\alpha = .01$, respectively. In other words, if the nil hypothesis is true and there are 2 and 100 degrees of freedom, you should expect to find $F$ values of 3.09 or lower 95% of the time, and values of 4.82 or lower 99% of the time. If the $F$ in a particular study is larger than these values, the researcher can reject the null hypothesis and conclude that treatments probably do have some effect.

Tests of minimum-effect hypotheses proceed in exactly the same way, the only difference being that they use a different set of tabled $F$ values (Murphy & Myors, 1999). The $F$ tables found in the back of most statistics texts are based on the central $F$ distribution, or the distribution of the $F$ statistic that would be expected if the traditional nil hypothesis were true. Tests of minimum-effect hypotheses are based on a noncentral $F$ distribution rather than the central $F$ that is used in testing the nil hypothesis.

For example, suppose a researcher decides that treatments that account for 1% or less of the variance in outcomes have a "negligible" effect. It is then possible to estimate a noncentrality parameter (based on $PV = .01$), and to estimate the corresponding noncentral $F$ distribution for testing the hypothesis that treatment effects are at best negligible. If $PV = .01$, $df_{hyp} = 2$, and $df_{err} = 100$, 95% of the values in this noncentral $F$ distribution will fall at or below 4.49 and 99% of the values in this distribution will fall at or below 6.76 (as we note below, $F$ values for testing minimum-effect hypotheses are listed in Appendix B). In other words, if the observed $F$ in this study is greater than 4.49, the researcher could be confident ($\alpha = .05$) in rejecting the hypothesis that treatments accounted for 1% or less of the variance. Later in this chapter, we will discuss standards that might be used in designating effects as "negligible".

You might wonder how a single critical $F$ value allows you to test for a whole range of null possibilities. As we noted in the boxed section

"Sometimes a Point Hypothesis is also a Range Hypothesis" one characteristic of the $F$ statistic is that it ranges from zero to infinity, with larger $F$ values indicating larger effects. Therefore, if you can be 95% confident that the observed $F$ in your study is larger than the $F$ you would have obtained if treatments accounted for 1% of the variance, you can also be at least 95% confident that the observed $F$ would be larger than that which would have been obtained for any $PV$ value between .00 and .01. If the observed $F$ is larger than the $F$ values expected 95% of the time when $PV = .01$, it must also be larger than 95% of the values expected for any smaller $PV$ value.

## An Example

Suppose 125 subjects are randomly assigned to one of five treatments. You find $F\,(4,120) = 2.50$, and in this sample, treatments account for 7.6% of the variance in the dependent variable. This $F$ is large enough to allow you to reject the traditional null hypothesis ($\alpha = .05$; the critical value of $F$ for testing this nil hypothesis is $F = 2.45$). This significant $F$ allows the researcher to reject the nil hypothesis, but since the nil hypothesis is almost always wrong, rejecting it in this study does not really tell the researcher much about the treatments.

Suppose also that treatments of this sort that account for less than 1% of the variance in the population have effects that can sensibly be labeled as "negligible". To test the hypothesis that the effects observed in this study came from a population in which the true effect of treatments is negligibly small, all the researcher needs to do is consult the noncentral F distribution with $df_{hyp} = 4$, $df_{err} = 120$, and $\lambda = 1.21$ (i.e., a good estimate of the noncentrality parameter $\lambda$ is given by $[120 * .01]/[1 - .01]$). In this noncentral $F$ distribution, 95% of the values in this distribution are 3.13 or lower. The obtained $F$ was 2.50, which is smaller than this critical value. This means that the researcher cannot reject this minimum-effect null hypothesis. That is, although the researcher can reject the hypothesis that treatments have no effect whatsoever (i.e., the traditional null), the researcher cannot reject the hypothesis that the effects of treatments are negligibly small (i.e., a minimum-effect hypothesis that treatments account for less than 1% of the variance in outcomes).

## Defining a Minimum Effect

The main advantage of the traditional nil hypothesis is that it is simple and objective. If a researcher rejects the hypothesis that treatments have no effect, he or she is left with the alternative that they have some effect. On the other hand, testing minimum-effect hypotheses requires value judgments, and requires that some consensus be reached in a particular field of inquiry. For example, the definition of a "negligible" effect might reasonably vary across areas, and there may be no set convention for defining which effects are so small that they can be effectively ignored, and which cannot. An effect that looks trivially small in one discipline might look reasonably large in another. However, it is possible to offer some broad principles for determining when effects are likely to be judged "negligible".

First, the importance of an effect should depend substantially on the dependent variables involved. For example, in medical research it is common for relatively small effects (in terms of the percentage of variance explained) to be viewed as meaningful and important (Rosenthal, 1993). One reason is that the dependent variables in these studies often include quality of life, and even survival. A small percentage of variance might translate into many lives saved.

Second, decisions about what effects should be labeled as "negligible" might depend on the relative likelihood and relative seriousness of Type I versus Type II errors in a particular area. As we will note in a section that follows, the power of statistical tests in the general linear model decreases as the definition of a "negligible effect" expands. In any particular study power is higher for testing the traditional nil hypothesis that treatments have no effect than for testing the hypothesis that they account for 1% or less of the variance in outcomes, and higher for tests of the hypothesis that treatments account for 1% or less of the variance than for tests of hypothesis that treatments account for 5% or less of the variance in outcomes. If Type II errors are seen as particularly serious in a particular area of research, it might make sense to choose a very small figure as the definition of a "negligible" effect.

On the other hand, there are many areas of inquiry in which numerous well-validated treatments are already available (See Lipsey & Wilson, 1993, for a review of numerous meta-analyses of treatment effects in the behavioral sciences. The Cochrane Collaboration has conducted hundreds of reviews in a range of health-related areas. See http://www.cochrane.org), and in these areas, it might make sense to "set a higher bar" by testing a more demanding hypothesis. For example, in the area of cognitive ability testing (where the criterion is some measure of performance on the job or in the classroom), it is common to find that tests account for 20–25% of the variance in the criterion (Hunter & Hunter, 1984; Hunter & Hirsch, 1987). Tests of the traditional null hypothesis (i.e., that tests have no relationship whatsoever to these criteria) are relatively easy to reject; if $r^2 = .25$, a study with $N = 25$ will have power of .80 for rejecting the traditional null hypothesis (Cohen, 1988). Similarly, the hypothesis that tests account for 5% or less of the variance in these criteria is easy to reject; if $r^2 = .25$, a study with $N = 55$ will have power of .80 for rejecting this minimum-effect hypothesis (Murphy & Myors, 1999). In this context, it might make sense to define a "negligible" relationship as one in which tests accounted for 10% or less of the variance in these criteria.

Utility analysis has been used to help determine whether specific treatments have effects that are large enough to warrant attention (Landy, Farr & Jacobs,

1982; Schmidt, Hunter, McKenzie & Muldrow, 1979; Schmidt, Mack & Hunter, 1984). Utility equations suggest another important parameter that is likely to affect the decision of what represents a negligible versus a meaningful effect – i.e., the standard deviation of the dependent variable, or $SD_y$. When there is substantial and meaningful variance in the outcome variable of interest, a treatment that accounts for a relatively small *percentage* of variance might nevertheless lead to practical benefits that far exceed the costs of the treatment.

For example, suppose a training program costs $1,000 per person to administer, and that it is proposed as a method of improving performance in a setting where the current $SD_y$ (i.e., the standard deviation of performance) is $10,000. Utility theory can be used to compare the costs with the expected benefits of this method of training. Depending on the effectiveness of the training program, researchers might conclude that benefits exceed costs (if the training is highly effective) or that costs exceed benefits (if the training is not very effective). Utility theory equations can also be used to determine the level of effectiveness the training program needs to meet so that benefits will at least equal costs, and it can be argued that this effectiveness level represents a very good definition of a minimum effect. That is, training programs that lead to more costs than benefits are not likely to be seen as effective, whereas training programs that lead to benefits that exceed their costs will be seen as effective. This break-even point represents a sensible definition of a minimally effective program.

Landy, Farr and Jacobs (1982) discuss the application of utility theory in evaluating performance improvement programs. They note that the overall benefit of such interventions can be estimated using the equation:

$$\Delta U = \left(r_{xy} * SD_y\right) - C \tag{3.1}$$

where:

$\Delta U$ = the projected gain in productivity associated with training
$r_{xy}$ = the correlation between training and performance
$C$ = the cost of the training program

In this example, $SD_y$ = $10,000 and $C$ = $1,000, so Equation 3.1 can be restated as:

$$\Delta U = \left(r_{xy} * \$10,000\right) - \$1,000 \tag{3.2}$$

A minimally effective training program will produce benefits that are equal to costs. The benefits of this training program depend on its effectiveness – i.e., the benefits are estimated by ($r_{xy}$ * $10,000). Therefore, in defining the minimum level of training effectiveness that will allow benefits to offset costs, all we need to do is to determine the value of $r_{xy}$ at which predicted benefits equal costs.

Benefits equal costs when $(r_{xy} * SD_y)$ equals C. Rearranging the terms in Equation 3.1, this break-even point is defined as:

$$(r_{xy} * SD_y) = C \text{ when } r_{xy} = C/SD_y \quad (3.3)$$

That is, benefits equal costs when the effectiveness of training $(r_{xy})$ is equal to $C/SD_y$. In our example, benefits equal costs when $r_{xy}$ is equal to .10 (i.e., $1,000/$10,000). If this value of $r_{xy}$ is squared, the point at which the benefits of training at least equal costs is when $PV = .01$. In other words, if training accounts for at least 1% of the variance in performance, benefits will at least offset costs, and this strikes us as a sensible definition of a minimum effect.

In many of the examples presented in this chapter and chapters that follow, we will use conventions similar to those described in Cohen (1988), describing treatments that have less than 1% of the variance as having small effects, and those that account for less than 5% of the variance in outcomes as having small to medium effects. Many of the tables presented in this book are arranged around these conventions. However, it is critical to note that the decision of what represents a "negligible" effect is one that is likely to vary across research areas, and that there will be many cases in which these conventions might not apply. The **R** code and Shiny Web app (https://murphy0921.shinyapps.io/ShinyPower/) described in Chapter 2 allow you to easily determine critical $F$ values and their $PV$ equivalents for minimum-effect tests that employ any operational definition of "negligible" that makes sense in the context in which research is being carried out, and we urge researchers to carefully consider their reasons for choosing any particular value as a definition of the minimum effect of interest.

### *Power of Minimum-Effect Tests*

As this example suggests, researchers should expect less power when testing the hypothesis that effects exceed some minimum value than when testing the hypothesis that the effect exactly zero. Switching from a central to a noncentral distribution as the basis of your null hypothesis necessarily increases the $F$ value needed to reach significance, thereby reducing power. In the sections that follow, we describe how to use the One-Stop F and $PV$ Tables (Appendices B and C), **R** code and our Shiny Web app to assess the power of minimum-effect tests and to use that information to aid in the design of more powerful studies.

## Using the One-Stop Tables to Assess Power to Test Minimum-Effect Hypotheses

In Chapter 2, we described the use of the One-Stop $F$ Table, presented in Appendix B and the One-Stop $PV$ Table, presented in Appendix C, in performing significance tests and power analyses for nil hypothesis tests. These same tables also include all the information you need to conduct significance tests and power analyses for tests of minimum-effect hypotheses, where negligible effects are defined as those that account for 1% or less of the variance, or as those that account for 5% or less of the variance in outcomes.

## Testing Minimum-Effect Hypotheses (PV = .01)

Rather than testing the hypothesis that treatments have no effect whatsoever, researchers might want to test the hypothesis that treatment effects are so small that they account for less than 1% of the variance in outcomes. If this $PV$ value represents a sensible definition of a "negligible" effect in a particular area of research, and researchers test and reject this hypothesis, they can be confident that effects are *not* negligibly small.

The fifth and sixth values in each cell of the One-Stop $F$ Table are the critical $F$ values needed to achieve significance (at the .05 and .01 levels, respectively) when testing the hypothesis that treatments account for 1% or less of the variance in outcomes. With $df_{hyp} = 3$ and $df_{err} = 50$, an $F$ of 3.24 or larger is needed to reject (with $\alpha = .05$) the hypothesis that the treatment effect is negligibly small (defined as accounting for 1% or less of the variance in outcomes).

The seventh and eighth values in each cell are $F$-equivalents of the effect size values needed to obtain particular levels of power (given an $\alpha$ level of .05 and the specified $df_{hyp}$ and $df_{err}$) for testing this minimum-effect hypothesis. The values in the table are 2.46 and 4.48, for power levels of .50 and .80, respectively.[1] This translates into $PV$ values of .13 and .21, respectively. That is, if treatments accounted for 13% of the variance in the population, a study with that uses minimum effect tests with $df_{hyp} = 3$, $df_{err} = 50$ and $\alpha = .05$ will have power of about .50 for detecting that effect. If treatments account for 21% of the variance in the population, the power of this study will be approximately .80.

## Testing Minimum-Effect Hypotheses (PV = .05)

There are many treatments that routinely demonstrate moderate to large effects. For example, well-developed cognitive ability tests allow researchers to predict performance in school and in many jobs with a relatively high degree of success (correlations in the .30–.50 range are common). Rather than testing the hypothesis that tests have no relationship whatsoever with these criteria (i.e., the traditional nil), or even that treatments account for 1% or less of the variance in outcomes, it might make sense to test a more challenging hypothesis — i.e., that the effect of this treatment is at least small to moderate in size. For reasons that are explained in the section that follows, there are many contexts in which it is useful to test the hypothesis that treatments explain 5% or less of the variance in the population. If a researcher can test and reject the hypothesis that effects explain 5% or less of the variance in outcomes, he or she is left with the alternative that treatments explain more than 5% of the variance. In most contexts, effects this large are likely to be treated as meaningful, even if smaller effects (e.g., those accounting for 1% of the variance) are not.

The ninth and tenth values in each cell of the One-Stop $F$ Table represent the critical $F$ values needed to achieve significance (at the .05 and .01 levels, respectively) when testing the hypothesis that treatments account for 5% or less of the variance in outcomes. With $df_{hyp} = 3$ and $df_{err} = 50$, an $F$ of 4.84 or larger is needed to reject (with $\alpha = .05$) the hypothesis that treatments account for 5% or less of the variance in the population.

The eleventh and twelfth values in each cell are $F$-equivalents of the effect size values needed to obtain specific levels of power (given an $\alpha$ level of .05 and the specified $df_{hyp}$ and $df_{err}$) for tests of this minimum-effect hypothesis. The values in the table are 4.08 and 6.55, for power levels of .50 and .80, respectively. This translates into $PV$ values of .19 of and .28, respectively. In other words, to have power of .50 for testing the hypothesis that treatments account for 5% or less of the variance in outcomes in a study with $df_{hyp} = 3$ and $df_{err} = 50$, the true population effect size will have to be fairly large ($PV = .19$). To achieve power of .80, the population effect will have to be quite large ($PV = .28$).

### *Using R Code and the Shiny Web App to Test Minimum-Effect Tests*

The **R** code and Shiny Web app (https://murphy0921.shinyapps.io/ShinyPower/) make it easy to conduct power analyses for minimum-effect tests and to determine the sample sizes needed to create a powerful study (i.e., one with power of .80 or more). They both ask you to enter a few pieces of information: (1) degrees of freedom for the hypothesis being tested and for error, (2) the alpha level, (3) the type of test you want to perform – for tests of the nil hypothesis, enter 0.00 as the minimum effect to test, for tests of the hypothesis that treatments account for 1% or less of the variance in Y, enter 0.01, etc., and (4) the assumed effect size in the population. You have the flexibility to test any kind of hypothesis that makes sense to you, so if your definition of a trivially small effect is one that accounts for anywhere between 0 and 2% of the variance, simply enter 0.02.

The program tells you the value of $F$ needed to reject the hypothesis you are testing, the equivalent $PV$ value, the power of the study and the $N$ needed to achieve power of .80.

---

## A Worked Example of Minimum-Effect Testing

A researcher randomly assigns 125 participants to one of five treatments, and reports $F(4,120) = 3.50$, and that treatments account for 10% of the variance (if you apply formula 6 in Chapter 2, you can confirm this transformation of $F$ to $PV$ – i..e., $PV = (df_{hyp} * F)/[(df_{hyp} * F) + df_{err}])$. What conclusions can you draw from this study?

First, it is possible to reject the nil hypothesis, at both the .05 and .01 levels; the critical values shown in Appendix B for $df_{hyp} = 4$ and $df_{err} = 120$ are 2.45 and 3.48, respectively. If the observed value of $PV$ (i.e., $PV = .10$), is taken as a reasonable estimate of the population $PV$, the study had plenty of power for nil hypothesis tests; the One-Stop $F$ Table (Appendix B), the One-Stop $PV$ Table (Appendix C), and the $df_{err}$ Needed Table (Appendix D) all show that the power of this study for testing the nil hypothesis is greater than .80. If you run the **R** code or the Shiny Web app, you will find that power equals .83.

Second, it is possible to reject the null hypothesis ($\alpha = .05$) that treatments account for 1% or less of the variance. Appendix B shows that an $F$ value of 3.13 or greater would be needed to reject this null hypothesis. If you set a more stringent alpha level (e.g., $\alpha = .01$), the observed $F$ will be smaller than the critical value of $F$ (i.e., $F = 4.40$). In other words, the researcher will not be able to reject, with a 99% level of confidence, the hypothesis that treatments account for 1% or less of the variance.

Appendix C shows that population effect sizes of $PV = .07$ and $PV = .11$ will be needed to achieve power levels of .50 and .80, respectively, in tests of this minimum-effects hypothesis. The observed $PV$ falls between these two values, and if this observed effect is used as an estimate of the population effect, this suggests that power is slightly lower than .80; the **R** code and the Shiny Web app confirms that power equals .70.

Finally, the results of this study would not allow researchers to reject ($\alpha = .05$) the hypothesis that treatments account for 5% or less of the variance. The critical value of $F$ for this null hypothesis test is 5.45. To have power of .80 for testing this hypothesis with $df_{hyp} = 4$ and $df_{err} = 120$, the population treatment effect would have to be quite large (i.e., $PV = .18$). In this study, the power for testing the hypothesis that treatments account for 5% or less of the variance is quite low (power = .265).

The most important critique of tests of the nil hypothesis revolves around the conclusion that the hypothesis that treatments have absolutely no effect, or that the correlation between two variables of interest is exactly zero is almost always false.[2] Personally, we think "almost always" is too conservative and the nil is never true, but it hardly matters whether $P(H_0)$ is exactly zero or some number very near zero. Either way, Type I errors are virtually impossible to make, and tests of the nil hypothesis do not tell you much that you did not really know. What they *do* tell you is whether your study has enough power to reject a hypothesis you already know is very likely to be false. A good power analysis will help you avoid doing studies that are likely to fail to reject the nil. More important, you can use the same methods to evaluate the power of any type of null hypothesis test you want to carry out.

Because it is very possible that the true effect of treatments is small, virtually everything you know (or thought you knew) about null hypothesis becomes relevant when conducting a minimum effect test. Type I errors are once again possible and Type II errors are less likely (because the null hypothesis will turn out to be true in many cases. Almost everything you are used to doing when testing null hypotheses stays the same – you compute the familiar $F$ statistic or transform whatever statistic you computed to $F$. The only thing that is different is that you use a noncentral $F$ table rather than the central $F$ table shown in most statistics books to determine whether you can reject the null hypothesis. The chapters that follow will demonstrate how to apply these methods to a range of statistical tests.

## Type I Errors in Minimum-Effect Tests

In tests of the traditional null (nil) hypothesis, the alpha level of the test (usually .05) indicates the likelihood of making a Type I error if the null hypothesis being tested is true. As we have explained in the previous chapter and in this one, the likelihood that the nil hypothesis is true is likely to be low, but in those rare cases where it *is* true, alpha gives a good indication of the probability of a Type I error. The situation is quite different when testing minimum-effect hypotheses, for two reasons.

First, there is often a realistic possibility that a minimum-effect null hypothesis will be true, meaning that Type I errors might also occur. Second, the idea of a null hypothesis being true takes on a different meaning when testing the nil hypothesis than when testing minimum-effect hypotheses. When testing the nil, *any* treatment effect, or *any* deviation from the hypothesis that the parameter being tested (e.g., a correlation coefficient) means the nil hypothesis is wrong. When testing a minimum-effect null hypothesis, "wrong" is no longer a binary question. That is, it is not sufficient to say that the hypothesis is either correct (e.g., that treatments account for 1% of the variance in outcomes or less) or wrong. Rather, two minimum-effect null hypotheses might both be correct, but one might be more nearly wrong than the other. Consider the hypothesis that treatments account for 1% of the variance or less. In one study, the population $PV$ is .0099. In another study, the population $PV$ equals .0001. The likelihood of a Type I error is always less than or equal to the alpha rate, and if the true effect is far from the upper bound of the null hypothesis being tested, the likelihood of a Type I error will be lower than if the true effect is close to the upper bound of the null hypothesis being tested. Furthermore, this trend is stronger when $N$ is large than when it is small.

72  Power Analyses for Minimum-Effect Tests

Table 3.1 Probability of Rejecting the Hypothesis That Treatments Account for 1% ($\alpha = .05$)

| Population PV | N | Probability of Type I Error |
|---|---|---|
| .0099 | 100 | .0495 |
| | 500 | .0488 |
| | 1,000 | .0484 |
| .0001 | 100 | .0087 |
| | 500 | .0015 |
| | 1,000 | .00003 |

Table 3.1 shows the probability of rejecting the hypothesis that treatments account for 1% or less of the variance in outcomes when the population $PV$ is either close to .01 or close to zero. As you can see, the probability of a Type I error is always smaller than the alpha level of the test, and if the population $PV$ is close to zero, the likelihood of incorrectly rejecting the hypothesis that treatments account for 1% or the variance in outcomes or less is very small. With a large sample size, the likelihood of a Type I error might become very close to zero.

Tests of the traditional nil hypothesis often involve a trade-off between Type I and Type II errors. For example, choosing a strict alpha level (e.g., .01 vs. .05) reduced the likelihood of Type I errors but *increases* the likelihood of a Type II error. When testing minimum-effect hypotheses, it is possible to take steps that simultaneously reduce the likelihood of Type I and Type II errors. In particular, large samples not only give you more power (reducing the likelihood of Type II errors), they also help to reduce the likelihood of Type II errors.

**Notes**

1 Remember, the equation for going from $F$ to $PV$, which is presented in Chapter 2, is $PV = (df_{hyp} * F)/[(df_{hyp} * F) + df_{err}]$, which in this case equals $PV = (3 * 2.46)/[(3 * 2.46) + 50] = .128$, which rounds to 13% of the variance explained.
2 It is possible for force two variables to be independent using techniques such as partial correlation, but it would be nonsensical to test the hypothesis that two variables that you have forced to be independent are independent.

# 4 Using Power Analyses

In Chapter 1, we noted that there are two general ways that power analysis might be used. First, power analysis is an extremely useful tool for planning research. Critical decisions, such as how many subjects are needed, whether multiple observations should be obtained from each subject, and even what criterion should be used to define "statistical significance" can be better made by paying attention to the results of a power analysis. Decisions about whether to pursue a specific research question might even depend on considerations of statistical power. For example, if your research idea involves a small (but theoretical meaningful) interaction effect in a complex experiment, power analysis might show that thousands of subjects would be needed to have any reasonable chance of detecting the effect. If the resources are not available to test for such an effect, it is certainly better to know this before the fact than to learn it after collecting your data.

Second, power analysis is a useful diagnostic tool. Tests of the traditional null hypothesis often turn out to be little more than roundabout power analyses. If you conduct a study, and all of the correlations among variables, as well as all of the planned and unplanned comparisons between treatments turn out to be statistically significant, this probably indicates that the sample was very large. If the sample is large enough, you will have tremendous power, and virtually any correlation or effect will be statistically different from zero. On the other hand, if none of a researcher's well-conceived hypotheses are supported with significant results, researchers might wish they had conducted a power analysis before launching the study and should carefully consider the power of their study before asking what is wrong with their ideas. If the power is too low, a researcher might never reject the null hypotheses, even in cases where it is clearly and obviously wrong.

In this chapter, we will discuss the major applications of power analysis. In Chapter 1, we noted that because statistical power is itself a function of three parameters, the number of observations ($N$), the criterion used to define statistical significance ($\alpha$), and the effect size (ES), it is possible to solve for any one of four values (i.e., power, $N$, ES, or $\alpha$), given the other three. The effect size parameter may be the most problematic because it represents a real but unknown quantity (i.e., the real effect of your treatments). Before discussing practical applications of power analysis, it is useful to examine more closely the methods that might be used in estimating effect sizes.

## Estimating the Effect Size

In Chapter 1, we noted that the exact effect size is usually unknown; if you knew precisely how treatment groups would differ, there would be little point in carrying out the research. Three general methods might be followed in estimating effect sizes in statistical power analysis. First, you might use inductive methods. If similar studies have been carried out before, you might use the results from these studies to estimate effect sizes in your own studies. At one time, this inductive method might have relied heavily on personal experience (i.e., whatever studies a particular researcher has read and remembers), but with the rapid growth of meta-analysis, it is often easy to find summaries of the results of large numbers of relevant studies (see, for example, Hunter & Hirsh, 1984; Lipsey & Wilson, 1993), already translated into a convenient effect size metric (e.g., $d$, $r^2$).

Second, you might use deductive methods, in which existing theory or findings in related areas are used to estimate the size of an effect. For example, suppose you want to estimate the effect of vitamin supplements on performance in long-distance races. Suppose further that you know that: (a) the vitamin supplement has a strong and immediate effect on the efficiency with which your body uses oxygen, and (b) efficiency in using oxygen is strongly correlated with success in such a race. It seems reasonable in this context to deduce that vitamin supplements should have a reasonably strong influence on outcomes of races.

Third, you might use widely accepted conventions that define what represents a "large", "medium", or "small" effect to structure your power analysis. As we will note below, analyses based on these conventions require very careful thought about what sort of effect you realistically expect or what sort of information about statistical power you really need. Nevertheless, the use of these conventions can help in carrying out useful and informative power analyses, even if there is no basis for predicting with any accuracy the size of the treatment effect.

### *Inductive Methods*

Inductive methods are best where there is a wealth of relevant data. For example, there have been hundreds and perhaps thousands of studies on the validity of cognitive ability tests in predicting performance in school and on the job (Hunter & Hirsch, 1987; Hunter & Hunter, 1984; Schmidt & Hunter, 1998). Similarly, Lipsey and Wilson (1993) reviewed numerous meta-analyses of psychological and educational interventions. This method is not restricted to the behavioral and social sciences; meta-analytic procedures are being applied in areas such as cancer research (e.g., Himel, Liberati, Laird & Chalmers, 1986); Lipsey and Wilson (1993) cited numerous meta-analyses of research on other medical topics; the Cochrane Collaboration (http://www.cochrane.org) has published hundreds of reviews in this area that include both systematic reviews of scientific literature in health-related areas and meta-analytic components. Some authors have conducted massive analyses of the results reported in entire areas of research (e.g., Bosco, Aguinis, Singh, Field & Pierce, 2015, examined over 140,000 correlations reported in studies in applied psychology over a 30 years), giving a solid foundation for estimating effect sizes in related work.

Suppose you were doing research on the effectiveness of programs designed to help individuals quit smoking and you have sufficient resources to collect data from 250 subjects. Lipsey and Wilson (1993) cite two separate meta-analyses that suggest relatively small effects on quit rates (for physician-delivered and worksite programs, $d = .34$ and $d = .20$, based on 8 and 20 studies, respectively). This body of research provides a reasonable starting point for estimating power; the average of the two $d$ values in these two meta-analyses is .274; the equivalent $PV$ value is .018. The $F$-equivalent for this effect size estimate, given a study comparing quit rates in a treatment group ($n = 125$) with those in a control group ($n = 125$) is $F(1,248) = 4.65$. Your power for testing the traditional null hypothesis ($\alpha = .05$) is therefore above .50 but below .80 (the $F$-equivalents for power of .50 and of .80 from the One-Stop $F$ Table are 3.81 and 7.86, respectively; using our Shiny Web app, described in Chapters 2 and 3 you will find power = .56). If you were testing the hypothesis that treatments accounted for 1% or less of the variance in outcomes in the population, your power would be well below .50 (the $F$-equivalent for power of .50 is 10.33; in this study, power equals .13).

## Using the One-Stop Tables and the R Code/Shiny Web App to Perform Power Analyses

Suppose you have 150 subjects available for a study. You can use the **R** code shown in Chapter 2 or our Shiny Web app (https://murphy0921.shinyapps.io/ShinyPower/) to carry out a range of analyses that are useful for understanding the statistical power of different studies that might be carried out and for understanding the limitations you might have to work with in planning and analyzing studies. For example, suppose you wanted to determine what sorts of effects could be detected in tests of the nil hypothesis with a power of .80.

If you want to compare two groups, $df_{hyp}$ and $df_{error}$ will be 1 and 148, respectively. If you use an alpha rate of .05, you could start a power analysis by using the One-Stop $PV$ Table (Appendix C) to assess the sensitivity of your study. If you look in the column for $df_{hyp} = 1$ and the row for $df_{error} = 150$ (the closest value in that table to 148), you will find that if group differences account for 5% of the variance in Y or more in the population, you will have power of .80 in testing the nil hypothesis. If you decide to test the minimum-effect hypothesis that groups account for 1% of the variance or less, you will have power of .80 only if group differences are considerably larger. Groups must account for 8.6% of the variance in Y in the population to achieve power of .80 for testing this minimum-effect hypothesis with 150 subjects and $\alpha = .05$.

Suppose you have good reason to believe that the effect size in the population is smaller than $PV = .05$. For example, you might consult a good meta-analysis of similar studies and find that the treatments you are studying are likely to account for 3% of the variance in Y. The analysis described in the preceding paragraph suggests that you will not have the desired level of power (the power for testing the nil hypothesis is .56 and for testing the 1% minimum effect hypothesis is .23). You could use the $df_{error}$ Needed Table (Appendix D) or the Shiny Web app to determine how many more subjects will be needed to achieve power of .80.

Enter the $df$ (1,148) and alpha (.05) levels. If you are testing the nil hypothesis, enter .00 for the minimum-effect size and .03 for the assumed effect size, and you will find that you need 257 subjects to reject the nil hypothesis. If you are testing the 1% minimum-effect hypothesis, you will need a substantially larger sample (1,060–1,110).

Finally, the **R** code and the Shiny Web app report the critical value of F (i.e., the F you will need to reject the null hypothesis you are testing) and the equivalent value of PV. If you proceed with 150 subjects, you will need to find that treatments account for 2.57% of the variance in your study. If your assumption that treatments account for 3% of the variance in the population is correct, you have about a 50–50 chance of finding differences large enough to reject the null with $\alpha = .05$). If you are testing the 1% minimum effect null hypothesis, you will need to find and observed PV value nearly twice as large (.053) to reject that null hypothesis with $\alpha = .05$.

Even though the study you have in mind has a reasonably large sample ($N = 150$), power is low, even for testing the traditional null hypothesis. Suppose that instead of testing a hypothesis where you expect treatments to have a small to moderate effect (i.e., PV = .03) you are studying the effects of smoking cessation programs on quit rates. As we noted earlier, these effects are quite small ($d = .24$ means that treatments account on average for about 1.4% of the variance in quit rates), and your power will be considerably lower than in the scenarios described in the boxed section above (e.g., your power for testing the nil hypothesis will be .32). The body of research in this area gives you good reason to expect small effects, and if you want to have adequate power for detecting these effects, you will need a much larger sample (e.g., you will need about 520–530 subjects to achieve power = .80 in tests of the traditional nil hypothesis).

### Deductive Methods

Deductive methods are the best when there is a wealth of relevant theory or models; Newtonian mechanics is probably the best example of an area in which numerous effects can be deduced based on a small set of theoretical statements and principles. However, there are areas in the social and behavioral sciences where the relevant theories or models are sufficiently well developed that sound inferences could be made about effect sizes. For example, the models most widely used to describe human cognitive abilities are hierarchical in nature, with specific abilities linked to broad ability factors, which in turn are linked to a single general cognitive ability (see Carroll, 1993, for a review of factor-analytic studies). If you wanted to estimate the validity of a new test measuring specific abilities as a predictor of performance in school, you could use what is known about the structure of abilities to make a reasonable estimate; tests that are strongly related to verbal or to general cognitive ability factors are likely to show moderate to strong relationships to school performance.

Suppose you used existing models of cognitive ability to estimate the validity of test scores for predicting performance in school and obtained a figure of .40. If you used a sample of 65 subjects, an expected correlation of .40 would yield an $F$-equivalent value of $F(1,63) = 12.00$, and your power for testing the traditional null would be .94; tests of the hypothesis that tests account for more than 1% of the variance in the population would also have power of .84. Because the expected effect is relatively large, it is easy to obtain adequate power, even with a small sample.

---

### Worked Example: Calculating $F$-equivalents and Power

If you expect the correlation between two variables to be .40, and you have 65 participants in your experiment, what level of power would you expect? There are several ways of attacking this question, but the most general of these involves transforming the correlation into its equivalent $F$ value.

1    The formula shown in Table 2.1 for transforming $r$ to $F$ is:

$$F(1, \text{df}_{\text{err}}) = \frac{r^2 \text{df}_{\text{err}}}{1 - r^2}$$

2   $df_{err} = N - 2 = 65 - 2 = 63$; $r^2 = .40^2 = .16$, so $F(1, 63) = (.16 * 63)/(1 - .16) = 12.0$. An $F$ value of 3.99 will allow you to reject the nil hypothesis, and an $F$ of 6.24 will allow you to reject the hypothesis that the population $PV$ is .01 or smaller. and in a context where you expect an $F$ closer to 12.0, you should not have much trouble rejecting either null hypothesis.

3   If you enter the df (1 and 63 for hypothesis and error, respectively) and an assumed effect size of $PV = .16$, you will find that you have power of .93 for testing the nil hypothesis and .89 for testing the hypothesis that the population $PV$ is .01 or less. You will also find that you can achieve power of .80 with sample sizes of 45 and 60 for testing the nil hypothesis and the 1% minimum effect hypothesis, respectively.

## *Effect Size Conventions*

As we noted in the preceding chapter, there are some widely accepted conventions for defining "small", "medium", and "large" effects. For example, a "small" treatment effect has been described as one in which treatments account for approximately 1% of the variance in outcomes, as one in which the difference between treatment and control group means is about one-fifth of a standard deviation, or as one in which a person randomly selected from the treatment group has a probability of .56 of having a higher score than a person randomly selected from the control group (see Table 2.2). None of these figures is sacred or exact (e.g., 2% of the variance might reasonably be described as a "small" effect"), but the conventions described in Table 2.1 do seem to be accepted as reasonable by many researchers, and they provide a basis for doing a power analysis even when the actual treatment effect cannot be estimated by inductive or deductive methods.

These conventions are broadly useful, but in specific research areas, the best definition of a small, medium, or large effect might change. For example, Bosco et al. (2015) examined 30 years of correlations in several major applied psychology journals, and they reported that the 20th, 50th and 80th percentiles of the distribution of these correlations were .05, .16, and .36. This suggests that in this domain, $PV$ values of .0025, .025, and .129 might be used to represent small, medium, and large effects in applied psychology.

When conventions are used to estimate effect sizes, it is usually best to base power analyses on small or small-to-medium effect sizes. As we noted in Chapter 1, a study with sufficient power to detect a small effect will have sufficient power for detecting medium and large effects as well. If you plan your study with the assumption that the effect is a large one, you run the considerable risk of missing meaningful effects that did not reach quite the magnitude you optimistically hoped to achieve. Thus, if the data and theory in a particular field do not provide a firm inductive or deductive basis for estimating effect sizes, you can always follow the convention and base your analysis on the assumption that the effects might very well be small. A study that has sufficient power to reliably detect small effects runs little risk of making a serious Type I or Type II error, regardless of the actual size of the treatment effect.[1]

## Four Applications of Statistical Power Analysis

The two most common applications of statistical power analysis are in: (1) determining the power of a study, given $N$, ES, and $\alpha$, and (2) determining how many observations will be needed (i.e., $N$), given a desired level of power, an ES estimate, and an $\alpha$ value. Both of these analyses are extremely useful in planning research and are usually so easy to do that they should be a routine part of designing a study. Power analysis may not be the *only* basis for determining whether to do a particular study or how many observations should be collected, but a few simple calculations are usually enough to help researchers make informed decisions in these areas. The lack of attention to power analysis and the deplorable habit of placing too much weight on the results of small-sample studies are long-standing problems that are well documented in the research literature (Cohen, 1962; Haase, Waechter & Solomon, 1982; Sedlmeier & Gigerenzer, 1989), and there is no good excuse to ignore power in designing your studies.

There are two other applications of power analysis that are less common, but no less informative. First, you can use power analysis to evaluate the sensitivity of your studies. That is, power analysis can tell you what sorts of effect sizes might be reliably detected in a study. If you expect the effect of a treatment to be small, it is important to know whether your study will detect that effect, or whether the study you have in mind only has sufficient sensitivity to detect larger effects. Second, you can use power analysis to make rational decisions about the criteria used to define "statistical significance".

## Calculating Power

Chapters 1 and 2 were largely devoted to explaining the theory and procedures used in calculating the statistical power of a study, and we will not repeat all the details laid out in those two chapters here. It is, however, useful to comment on problems or issues that arise in carrying out each of the steps of statistical power analysis.

Power analysis requires three decisions. First, you need decide about the alpha level you will use to decide whether to reject a null hypothesis. In many fields, particularly the social and behavioral sciences, $\alpha = .05$ is a widely accepted as a convention that the only decision many researchers need to make is whether to use a more stringent criterion, such as $\alpha = .01$. Our advice on this point is simple – *never choose $\alpha = .01$ over $\alpha = .05$ when testing the nil hypothesis*. If you accept the argument that the nil hypothesis is rarely true, there are no sensible scenarios in which the protection against Type I errors that $\alpha = .01$ give you will be more valuable than the increase in Type II errors that the choice of $\alpha = .01$ entails.

Although the choice of $\alpha = .05$ is simple and painless for many researchers, we will argue in a later section that different choices are often more sensible. However, if you use any other value to define "statistical significance" (e.g., .04 might be a perfectly reasonable choice in some settings; Ferguson, 2009), you will have to fight convention and defend your choice to a potentially hostile set of reviewers, readers, and editors (Labovitz, 1968).

Assume for now that you will use $\alpha = .05$ to define statistical significance. Your next set of choices have to do with study design. How many groups are being compared or how many variables are part of your research question? This will define $df_{hyp}$. How many subjects or observations do you plan to use? This will define $df_{err}$. In the section that follows, we will look at how you can use power analysis to make better choices about N, but sometimes you may have little choice. If there is a finite pool of resources to draw from, you might not have the luxury of planning the optimal sample, and you may need to determine how much power the study you are planning is likely to have.

Third, you need to make decisions about the effect size you expect in the population. As we laid out in the previous sections, you might make these decisions based on existing theory and research. You can probably make them with more confidence if there are good meta-analyses available in your research area. If you do not have a credible basis for estimating the population effect size, start your power analysis on the assumption it might be small (e.g., $PV = .01$).

Finally, decide about what type of null hypothesis to test. The default choice in most settings is to test the nil hypothesis, but it is so easy to test alternative hypotheses, such as the hypothesis that the effect in the population is trivially small (e.g., it accounts for 1% or less of the variance in Y) versus strength of the effect in the population is large enough that it is not trivial.

Once these decisions are made, there are three simple strategies for estimating power. First, you can transform your population effect size estimate into $F$ and use the One-Stop $F$ Table in Appendix B. Second, you can express your population effect size estimate in terms of $PV$ and use the One-Stop $PV$ Table in Appendix C. Both will give you information about whether your power will be .50 or above and/or be .80 or above. For example, suppose you want to test the nil hypothesis in a study with $df_{hyp} = 2$, $df_{err} = 80$ and $\alpha = .05$. You think that in the population, treatments account for 8% of the variance in Y. If you look at the One-Stop $PV$ Table, the power of this study is between .50 and .80. If the population effect size was .60, power would be .50. To reach power of .80, you need a population effect size of .11.

The simplest and most exact method is to use the **R** code or Shiny Web app discussed in Chapters 2 and 3. These would show that the power of this study, as currently planned, is .635. If you want to test a 1% minimum effect hypothesis, power drops to .467.

## Determining Sample Sizes

Rather than calculating the level of power a particular study had or will have, it is often useful to determine the sample or research design needed to achieve specific levels of power. To do this, you must first decide how much power you want. As we noted in Chapter 2, it is hard to justify a study design that yields power less than .50; when power is less than .50, the study is more likely to lead to an incorrect conclusion (i.e., it will not reject $H_0$, even though you are virtually certain this hypothesis is wrong) than to a correct one. Power substantially above .80 might be desirable, but it is often prohibitively difficult to obtain; in most analyses, the desirable level of power is likely to be .80.

Once the desired level of power is selected, determining sample sizes follows the same general pattern as the determination of power itself. That is, you need to think about the research design, the estimated effect size, the nature of the hypothesis being tested, and the significance criterion being used. Appendices B and C can be useful in determining the sample sizes needed to detect a wide range of effects, but a simpler and more direct option is to use Appendix D, which provides the $df_{err}$ needed to achieve power of .80 ($\alpha = .05$).

The rows of Appendix D correspond to effect sizes, described in terms of either the standardized mean difference ($d$) or the proportion of variance explained ($PV$). Table 4.1 provides formulas for translating several statistics, including $F$, into $d$ and/or $PV$ (these formulas are also shown in Appendix A).

Table 4.1 Translating Common Statistics into Standardized Mean Difference ($d$) or Percentage of Variance ($PV$) Values

Standardized Mean Difference ($d$)

$$d = \frac{2r}{\sqrt{1-r^2}}$$

$$d = \frac{2t}{\sqrt{df_{err}}}$$

Percentage of Variance ($PV$)

$$PV = \frac{t^2}{t^2 + df_{err}}$$

$$PV = \frac{df_{hyp}F}{df_{hyp}F + df_{err}}$$

$$PV = \frac{d^2}{d^2 + 4}$$

The values in Appendix D represent degrees of freedom ($df_{err}$) rather than exact sample sizes ($N$). Our reason for presenting a table of $df_{err}$ values rather than presenting a table of sample sizes is that $N$ is a function of both $df_{hyp}$ and $df_{err}$. In many applications, $N = df_{hyp} + df_{err} + 1$, but in complex multifactor designs (e.g., studies using factorial ANOVA), the total sample size depends on the number of levels of all design factors, and without knowing the research design in advance, it is impossible to put together an accurate $N$-needed table. In most cases, however, the sample size needed to achieve power of .80 will be very close to the $df_{err}$ value shown in Appendix D. Chapter 5 deals with applications of power analysis in multifactor studies.

Appendix D presents $df_{err}$ needed when testing the traditional null hypothesis and (in boldface) testing the hypothesis that treatments account for 1% or less of the population variance in outcomes. As you can see, larger samples are needed when testing this minimum-effect hypothesis than when testing the traditional (but sometimes trivial) hypothesis that treatments have no effect whatsoever.

To illustrate the use of Appendix D, consider a study comparing the effectiveness of four diets. Suppose you expect a small to moderate effect (e.g., the choice of diets is expected to account for about 5% of the variance in weight loss). To achieve power of .80 in testing the traditional null hypothesis (with $\alpha = .05$), you the value in the row of Appendix D that corresponds with $PV = .05$ and the column that corresponds to $df_{hyp} = 3$ (i.e., with four diets, there are three degrees of freedom) and find that the $df_{err}$ needed for power of .80 when testing the nil hypothesis would be 209. You would therefore need a sample of about 214 subjects (i.e., $N = 209 + 3 + 1$) to achieve this level of power, or about 54 per group.

Appendix D also shows the $df_{err}$ needed to achieve power of .80 in testing the hypothesis that treatments account for 1% or less of the variance in outcomes. To achieve a power of .80 in testing this minimum-effect hypothesis, you would need a sample of about 409 (i.e., $N = 405 + 3 + 1$), or about 100 per group. This sample is almost twice as large as the one needed to reject the traditional null hypothesis that treatments have no effect. However, if you put together a study that allows you to reject the hypothesis that treatments have a negligibly small effect (e.g., they account for less than 1% of the variance in outcomes), you will not only know that treatments have *some* effect, but you will also have a formal test of the hypothesis that the effect is large enough to warrant your attention.

You can use the **R** code or Shiny Web app to calculate the sample size needed to achieve power of .80. Simply enter $df_{hyp} = 3$, $\alpha = .05$, minimum effect size = .00 for the nil hypothesis or .01 for the 1% minimum effect null hypothesis. Enter .05 for the assumed effect size in the population. You can enter any number you want for $df_{error}$; the **R** code or Shiny Web app will tell you the $N$ needed to achieve power of .80. The Shiny Web app returns a value of 215, which is essentially identical to the value you get from Appendix D ($N$ needed = 214).

# A Few Simple Approximations for Determining Sample Size Needed

The $df_{err}$ Needed Table in Appendix D provides the information needed to determine the sample size needed to achieve specific levels of power in a specific study. However, there is an even simpler method to determining the approximate sample size needed, given the effect size (PV) that is assumed in a study.

In studies with $t$-tests, simple correlation, or comparisons between two groups, there is a single degree of freedom for the hypothesis being tested (i.e., $df_{hyp} = 1$). When $df_{hyp} = 1$, you can approximate the sample size needed (more specifically, $df_{error}$) to achieve a power of .80 with the simple formula:

$$df_{error} = \frac{7.75}{PV}$$

For example, if you assume that the difference between groups accounts for 5% of the variance in the population (i.e., $PV = .05$), this formula suggests that you will need a total sample of 156 (i.e., $df_{hyp} = 1$, $df_{error} = 155$) to achieve power of .80 in tests of the traditional nil hypothesis. Appendix D shows that you will need a sample of 152 (i.e., $df_{hyp} = 1$, $df_{error} = 151$) to achieve this level of power. Not bad for a quick approximation!

In more complex studies (e.g., multiple regression, ANOVA), where the degrees of freedom for the hypothesis to be tested (e.g., that three groups in a one-way ANOVA have the same mean) is greater than one, a more complex formula is needed, in particular:

$$df_{error} = \frac{5.26 + (3.24\sqrt{df_{hyp}})}{PV}$$

For example, if you assume once again that the difference between three groups in an ANOVA accounts for 5% of the variance in the population (i.e., $PV = .05$), this formula suggests that you will need a total sample of 200 (i.e., $df_{hyp} = 2$, $df_{error} = 197$) to achieve power of .80 in tests of the traditional nil hypothesis. Appendix D shows that you will need a sample of 189 (i.e., $df_{hyp} = 2$, $df_{error} = 186$) to achieve this level of power. Again, not bad for a quick approximation!

Suppose you want to test the hypothesis that differences between groups account for less than 1% of the variance in the population. In this case, the formula is:

$$df_{error} = \frac{df_{hyp} + (4.18/PV)}{\pi PV}$$

Assume that you believe that the difference between four groups accounts for 10% of the variance in the dependent variable and that you want to statistically test the minimum-effect hypothesis that differences account for at least 1% of the variance in the population. This formula suggests that you will need a total sample of 145 (i.e., $df_{hyp} = 2$, $df_{error} = 142$) to achieve power of .80 in tests of the traditional nil hypothesis. Appendix D shows that you will need a sample of 142 (i.e., $df_{hyp} = 2$, $df_{error} = 139$) to achieve this level of power. As the $df_{hyp}$ and/or the $PV$ you assume grow larger, the formula becomes more and more accurate, but for small $PV$ values, this formula will underestimate the sample size needed, and we do not recommend it for $PV$ values much lower than $PV = .05$.

## Determining the Sensitivity of Studies

It is often useful to know what sort of effect could be reasonably detected in a particular study. If a study can reliably detect only a large effect (especially in a context where you expect small effects), it might be better to postpone that study until the resources needed to obtain adequate power are available. The process of determining the effect size that can be detected, given specific values for $N$ and $\alpha$, together with a desired level of power again closely parallels the procedures described above. Appendices B and C, which specify the $df_{err}$ needed to achieve power of .80 in testing both traditional and minimum-effect hypotheses, are also quite useful for determining the type of effect that could be reliably detected in a study.

Suppose you are comparing two methods of mathematics instruction. There are 140 students available for testing (70 will be assigned to each method), and you decide to use an $\alpha$ level of .05 in testing hypotheses. Here, $df_{hyp} = 1$ (i.e., if two treatments are compared, there is one degree of freedom for this comparison) and $df_{err} = 138$ (i.e., $df_{err} = N - df_{hyp} - 1$). If you look down the $df_{hyp} = 1$ column of Appendix D, you find that 138 falls somewhere between the $df_{err}$ needed to detect the effect of treatments when $PV = .05$ and $PV = .06$ (or between $d = .46$ and $d = .51$). In other words, with 140 students, you would have power of .80 to detect a small to moderate effect, but you would not have this level of power for detecting a truly small (but perhaps important) effect.

Take another example. Suppose there are four cancer treatments and 44 patients. Here, $df_{hyp} = 3$ and $df_{err} = 40$. If you consult Appendix D, you will find that you have power of .80 to detect effects that are quite large (i.e., $PV = .24$); if the true effects of treatments are small or even moderate, your power to detect these effects will drop substantially.

Both examples above are based on tests of the traditional null hypothesis. If you want to test the hypothesis that treatments account for 1% or less of the variance in outcomes (which corresponds to $d$ values of .20 or lower), your sample of 140 students will give you power of .80 for detecting differences this large between the two treatments only if the true effect is relatively large (i.e., $PV$ = .09 or $d$ = .63; see Appendix D). You will have this level of power for detecting nontrivial differences among the four cancer treatments (with $N$ = 44) only if the true differences between treatments are truly staggering (i.e., $PV$ = .26; see Appendix C).

## Determining Appropriate Decision Criteria

As we noted earlier, the choice of criteria for defining "statistical significance" is often practically limited to the conventional values of .05 versus .01 (occasionally, social scientists use .001 or .10 as significance levels, but these are rare exceptions). The choice of any other value (e.g., .06) is likely to be met with some resistance, and the battle is probably not worth the effort. Because the choice among significance levels is constrained by convention, the steps involved in making this choice do not exactly parallel the processes laid out in the three preceding sections of this chapter. Rather than describing specific steps in choosing between .05 and .01 as alpha levels, we will discuss the range of issues that are likely to be involved in making this choice. This discussion reinforces the conclusion that *you should never use the .01 level (or any more stringent criterion, such as .001) when testing traditional null hypotheses, nor should you usually use other procedures designed to guard against Type I errors in testing this hypothesis*. As we will show below, the choice of the .01 significance criterion leads to a substantial reduction in statistical power, with virtually no meaningful gain in terms of protection against Type I errors. The same is true of most procedures designed to reduce Type I errors (see Zwick and Marascuilo, 1984, for a review of procedures used to control Type I error in testing multiple contrasts).

### Balancing Risks in Choosing Significance Levels

In Chapter 1, we noted that when testing the traditional null hypothesis, two types of errors are possible. A researcher who rejects the null hypothesis when it is true makes a Type I error ($\alpha$ is the probability of making this error if $H_0$ is true). The practical effect of a Type I error is that researchers could come to believe that treatments have some effect when in fact they have no effect whatsoever. Researchers who fail to reject the null hypothesis when it is false make a Type II error ($\beta$ is the probability of making this error when $H_0$ is in fact false, and power = $1 - \beta$). The practical effect of making a Type II error is that researchers might give up on treatments that have some effect.

The most common strategy for reducing Type I errors is to make it difficult to reject the null hypothesis (e.g., by using .01 rather than .05 as a criterion for significance). Unfortunately, this strategy also substantially reduces the power of your tests. For example, suppose you are comparing two treatments (with 200 people assigned to each treatment) and you expect a relatively small effect (i.e., $d = .30$, $PV = .022$, $F$-equivalent $= 8.95$). Using .05 as a significance criterion, your power would be .86; if $\alpha = .01$, power drops to .65 (Cohen, 1988). In other words, moving the cutoff for significance from .05 to .01 will cost you in terms of power. If you believe that the traditional null hypothesis is never or almost never correct, you will gain nothing in exchange, because the actual likelihood of a Type I error under alpha levels of .05 and 01, respectively, will be equal (i.e., it will both be zero, or very near zero regardless of your choice of $\alpha$).

This trade-off between Type I error protection and power suggests that in deciding which significance level to use, you must balance the risk and consequences of a Type I error with the risk and consequences of a Type II error. Nagel and Neff (1977) discuss a decision-theoretic strategy for choosing an alpha level that provides an optimum balance between the two errors. Cascio and Zedeck (1983) suggest that Equation 4.1 can be used to estimate the apparent relative seriousness (ARS) of Type I versus Type II errors in statistical significance tests:

$$\text{ARD} = \frac{p(H1)\beta}{[1 - p(H1)]\alpha} \quad (4.1)$$

where: $p(H_1)$ = probability that $H_0$ is false

According to Equation 4.1, if the probability that treatments have *some* effect is .7, alpha is .05, and the power is .80, your choice of the .05 significance criterion implies that a mistaken rejection of the null hypothesis (i.e., a Type I error) is 9.33 times as serious [i.e., $(.7 * .2)/(.3 * .05) = 9.33$] as a the failure to reject the null when it is wrong (i.e., a Type II error). In contrast, setting $\alpha$ at .10 leads to a ratio of 4.66 [i.e., $(.7 * 2)/(.3 * .10) = 4.66$], or to the conclusion that Type I errors are treated as if they are 4.66 times as serious as a Type II error (see also Lipsey, 1990).

The first advantage of Equation 4.1 is that it makes explicit values and preferences that are usually not well understood, either by researchers or by the consumers of social science research. In the scenario described above, an alpha level of .05 makes sense only if you think that Type I errors *are* over nine times as serious as Type II errors. If you believe that Type I errors are only four or five times as serious as Type II errors, you should set your significance level at .10, not at .05.

The second advantage of Equation 4.1 is that it explicitly involves the probability that the null hypothesis is true. As we have noted in several other contexts, if the nil hypothesis is almost always false, it is almost impossible to make a Type I error. Thus, the only circumstances in which you should use stringent significance criteria or adopt testing procedures that minimize Type I errors, etc. are those in which the null hypothesis has a reasonable probability of being true. This virtually never happens when testing the traditional null hypothesis, but it might occur when testing a minimum-effect hypothesis; few treatments have exactly zero effect, but there may be many that have very small effects.

### Should You Ever Worry about Type I Errors?

In testing the traditional null hypothesis, it is often virtually impossible to make a Type I error, no matter how hard you try, and your only concern should be maximizing power (and therefore minimizing Type II errors). In contrast, there might be good reasons for concern over Type I errors when testing a minimum-effect hypothesis. If the hypothesis to be tested is that the effect of treatments is trivially small, it is quite possible that the null *will* be true, and that you will learn something by testing it.

As Equation 4.1 suggests, it is impossible to sensibly evaluate the relative emphasis given to Type I versus Type II errors unless the probability that the null hypothesis is true can be estimated. Table 4.2 shows the relative seriousness with which Type I versus Type II errors are treated, as a function of both the alpha rate and the probability that the null hypothesis is true, in studies where power = .80. For example, if there is a 5% chance that treatments do have a negligible effect (i.e., the probability that the minimum-effect null hypothesis is true = .05), the decision to use an alpha level of .01 makes sense only if you believe that Type I errors are 380 times as serious as Type II errors. As you can see from Table 4.2, as alpha increases (i.e., as it becomes easier to reject $H_0$), the relative seriousness attached to Type I versus Type II errors goes down. If you believe that Type II errors are at all serious, relative to Type I errors, the message of Table 4.2 is clear – i.e., you should use a more lenient criterion for determining statistical significance.

It is easy to rearrange the terms of Equation 4.1 to compute the alpha level you *should* use to reach an appropriate balance between Type I and

Table 4.2 Relative Seriousness of Type I versus Type II Errors as Function of Alpha and Probability That the Null Hypothesis Is True (Power = .80)

| Probability that $H_0$ is true | Alpha .01 | .05 | .10 |
|---|---|---|---|
| .50 | 20.00 | 4.00 | 2.00 |
| .30 | 46.66 | 9.33 | 4.66 |
| .10 | 180.00 | 36.00 | 18.00 |
| .05 | 380.00 | 76.00 | 38.00 |

Type II errors, a balance that is described as the desired relative seriousness (DRS).[2] For example, if you decided that you wanted to treat Type I errors as twice as serious as Type II errors, the desired relative seriousness is 2.00; if the probability that the null hypothesis was true was .30, power was .80, and you wanted to treat Type I errors as if they were twice as serious as Type II errors, you should use an alpha level of .23. The desired alpha level (i.e., level which will yield the appropriate balance between Type I and Type II errors can be obtained from Equation 4.2.

$$\alpha_{desired} = \frac{\frac{p(H1)\beta}{[1-p(H1)]} * 1}{DRS} \tag{4.2}$$

where: $\alpha_{desired}$ = alpha level that will yield the desired relative seriousness of Type I and Type II errors

$DRS$ = desired relative seriousness of Type I and Type II errors

At the risk of being repetitive, we think it is important to take to heart what we regard as a critical point, that you should be concerned with Type I errors if and only if there is a realistic possibility that such an error can occur. One of the distinctions between the traditional null and minimum-effect tests is that there is a realistic possibility that the minimum-effect null hypothesis *will* be true, which suggests that serious decisions must be made about the appropriate balance between Type I and Type II errors. Equations 4.1 and 4.2 can help you make those decisions.

### Finding a Sensible Alpha

Suppose you believed that Type I errors were ten times as serious as Type II errors. Using Equation 4.2, it is possible to design a study in such a way that the likelihood of making Type I versus Type II errors reflects their relative seriousness, which in this case will translate into designing a study in which the probability of making a Type I error is one-tenth the probability of making a Type II error. In order to do this, you need to determine what level of power you want to achieve and make some estimate of the probability that the null hypothesis is *wrong* (i.e., $p(H_1)$). We have argued in several parts of this book that $p(H_1)$ will almost certainly be high (e.g., $p(H_1) = .90$) when testing the nil hypothesis. We have also argued that power should be high in most studies. If you assume that power $(1-\beta) = .80$ (and therefore, $\beta = .20$) and that Type I errors are ten times as serious as Type II errors, Equation 15 translates into:

$$\alpha_{desired} = [(.9 * .2)/(1 - .9)) * (1/10)] = .18$$

In other words, the alpha level that makes the most sense in this context is α = .18.

Suppose you think that the null hypothesis is almost always wrong and that p(H$_1$) should equal .99 rather than .90. Equation 15 translates into:

$$\alpha_{desired} = [(.99 * .2)/(1 - .99)) * (1/10)] = 1.98$$

An alpha level greater than 1.0 is impossible; so what does this calculation tell you? If you really believe that the null hypothesis is almost always wrong and that Type I errors are ten times worse than Type II errors, this calculation suggests that *you cannot set your alpha too high*. No matter what decision criterion you choose, performing significance tests under these circumstances does not make sense, given your beliefs about the relative seriousness of the two errors.

## Post-Hoc Power Analysis Should Be Avoided

As we noted in Chapter 1, some authors carry out a study and then wonder how much power that study had. To determine the power of the study they have already completed, they often use the observed *PV* as an estimate of the *PV* expected in the population. This method of post-hoc power analysis is *not* recommended (Levine & Epsom, 2001). The reason for recommending researchers avoid performing post-hoc power analyses is that they depend on the results of your study to estimate the expected *PV* in the population. These estimates are often imprecise, especially when samples are small, and it is better by far to rely on meta-analyses or theory to estimate population effect sizes. Don't wait until your study is done to start thinking about power. At that point, it is too much to do anything about low power or low levels of sensitivity. Doing a power analysis before you collect any data will save you lots of grief in the long run.

## Notes

1 As we noted in Chapters 1 and 2, Type I errors are a concern only when there is some possibility that the null hypothesis is true, which is virtually never the case in tests of the traditional null. Type II errors are still possible if the treatment effect is extremely small, but we would not regard this type of error (i.e., concluding that treatments have no effect when in fact they have a completely trivial effect) as very serious.
2 Keep in mind, however, that using non-conventional alpha levels will often require you to vigorously defend your choice, even in contexts where the "conventional" choice (e.g., .05, .01) makes no sense whatsoever.

# 5 Correlation and Regression

There is a clear link between correlation and regression analysis and the broader topic of power analysis because both are concerned fundamentally with effect sizes. Throughout this book, we have used the percentage of variance in the dependent variable that is explained by treatments, interventions, or other variables (i.e., $PV$) as our main effect size measure. If you square the correlation between two variables, X and Y, what you get is $PV$ – i.e., the proportion of variance in Y that is explained by X. Similarly, in multiple regression, where several X variables are used to predict scores on Y, the squared multiple correlation coefficient (i.e., $R^2$) is a measure of the proportion of the variance of Y that is explained. One of the substantial advantages of framing statistical analyses under the general linear model in terms of correlation (e.g., both $t$-tests and analyses of variance are easily performed using correlational methods) is that it is virtually impossible to make one of the most common mistakes that are made when testing the null hypothesis – i.e., forgetting to examine and report effect size measures. Correlations and squared correlations *are* effect size measures.

Power analysis is useful for both designing and understanding correlational studies. It provides a coherent framework for making decisions about sampling and the interpretation of both significant and nonsignificant findings. As with other analytic methods discussed in this book, the general message of power analysis when applied to correlation is that large samples are usually needed to provide adequate power. However, as the example discussed below will illustrate, large samples can lead to potentially confusing results, particularly when researchers depend on tests of the nil hypothesis to drive the interpretation of their results.

## The Perils of Working with Large Samples

McDaniel (1988) used very large samples to study the validity measures of pre-employment drug use as predictors of job suitability in the military. Validity coefficients for pre-employment use of drugs such as marijuana, cocaine, various stimulants, and depressants were calculated in samples ranging in size from 9,224 to 9,355 subjects. As you might expect, the

DOI: 10.4324/9781003296225-5

*Table 5.1* Predictive Validity of Pre-Employment Drug Use Reported by McDaniel (1988)

| Drug | N | Validity | PV | t | F |
|---|---|---|---|---|---|
| Marijuana | 9,355 | .07 | .0049 | 6.79 | 46.06 |
| Cocaine | 9,224 | .04 | .0016 | 3.85 | 14.81 |
| Stimulants | 9,286 | .07 | .0049 | 6.76 | 45.73 |
| Depressants | 9,267 | .07 | .0049 | 6.76 | 45.63 |

correlations were all "statistically significant", and an incautious reading of these "significance" tests might mislead readers when arriving at conclusions about the value of these tests.

Several of the validities reported by McDaniel (1988) are shown in Table 5.1. The One-Stop $F$ Table shown in Appendix B does not contain entries for $df_{err}$ of 9,000+, but if you look at the values of the $F$ needed to reject the traditional null (nil) hypothesis when $df_{err}$ are large, you will see that there is virtually no difference in the $F$ needed to reject the traditional null hypothesis at the .05 or .01 level, and an $F$ larger than 6.66 will allow you to reject the traditional null hypothesis with $\alpha = .01$. All of the $F$ values in Table 5.1 are considerably larger than 6.66, so all of the correlations in that table are significantly different from zero. However, as we will note below, the tests are significant because of the enormous samples involved, not because of the size of the effect.

Although significant in the traditional sense, McDaniel's (1988) validities are probably not meaningful in any substantive sense, and, as the author himself notes, employers would be better advised to base their employment decision on information other than the applicant's previous history of drug use. This study is an excellent example of the pitfalls of relying upon tests of the traditional null hypothesis. These correlations are all "significant", and it is all too easy to confuse "significant" with "meaningful". As we note below, these correlations are so small that they should probably be ignored altogether.

We titled this section "The Perils of Working with Large Samples" because significance tests can cause real problems for researchers who work with small samples (low power), but also can cause problems for researchers who work with large samples. In this study, all the correlations were significant, but none of them was very big. Large samples have many important advantages, notably their ability to provide more accurate estimates, but they can mislead researchers who rely on significance tests to tell whether their results are noteworthy and important.

## Traditional versus Minimum-Effect Tests

An examination of Table 5.1 shows just how small these validities are, accounting for less than half a percent of the variance in job suitability in each case. As the One-Stop $PV$ Table (Appendix C) shows, correlations this small will nevertheless be significant when samples are large. If there are 1 and 1,000 degrees of freedom, correlations as small as .007 will be significant at the .01 level. With more than 9,000 degrees of freedom, correlations would be significant even if $PV$ is smaller than .001. Regardless of whether they are statically significant, it would be hard to label validities in the .04–.07 range as meaningful or useful. However, if you use statistical significance as your guide, *any* correlation will be treated as meaningful if you just keep collecting data.

This example illustrates an important feature of minimum-effect null hypotheses. Unlike the traditional null, you cannot necessarily reject a minimum-effect null simply by collecting more data. If an effect is genuinely negligible, it will continue to be negligible no matter how many subjects you test. This phenomenon is reflected in the One-Stop $F$ Table (Appendix B) by the fact that critical $F$ values for the 1% and 5% minimum-effect nulls do not asymptote as $df_{err}$ increases. As you can see, as $df_{err}$ gets very large, the critical values for minimum-effect nulls keep pace with the increase in sample size and do not allow you reject the minimum-effect null hypothesis unless the effect you are studying is genuinely meaningful. Similarly, if you test the minimum-effect hypothesis that tests account for 1% of the variance or more in the criterion, you will need an observed $PV$ value of .015 to reject this null hypothesis if your sample is as large as 10,000, and no matter how large the value of $N$, the squared correlation required to reject this minimum-effect null will never fall below .01. Herein lies the key advantage of minimum-effect null hypothesis testing over traditional null hypothesis testing. When testing a minimum-effect null hypothesis, you cannot guarantee rejection simply by collecting a large sample.

## *Power Estimation*

As we noted above, the large samples employed in this study give you tremendous power in tests of the traditional null hypothesis. If we assume, for example, that pre-employment tests account for only one-half of one percent of the variance in job suitability (i.e., $PV = .005$), you would still have power of nearly 1.0 for all of the tests of the traditional null hypothesis reported in Table 5.1. It is little wonder that the author of this study rejected the traditional null hypothesis. Even if the true effect of pre-employment drug tests is absolutely trivial, this study would still provide plenty of power for rejecting the hypothesis that there is no effect whatsoever.

In contrast, if the hypothesis to be tested is that these tests account for 1% or less of the variance in suitability, McDaniel's study would not achieve power as large as .50 regardless of how large the sample becomes. If predictor tests account for 0.5% of the variance in the criterion and you test the minimum-effect hypothesis that they account for 1% of the variance, the likelihood of rejecting the null with samples of 9,000 or more is in fact *smaller* than the alpha level (usually .05 or .01). In Chapter 3, we noted that Type I error rates for minimum effect tests are generally smaller than the alpha rate of the test, especially when samples are large, and this is exactly what an analysis of the McDaniel (1988) study shows.

Suppose that the observed $PV$ value for marijuana tests reported by McDaniel is an accurate estimate of the population $PV$ and you use the same sample size as McDaniel ($N = 9,335$) to test the minimum-effect hypothesis that treatments account for 1% or less of the variance in the population. The probability of incorrectly rejecting the null hypothesis will be approximately .00002. When testing the traditional null hypothesis, the alpha level of the test indicates the probability of making a Type I error, but when testing a minimum-effect hypothesis, it is very difficult (and as $N$ increases, virtually impossible) to make a Type I error if the population effect is smaller than the minimum effect being tested.

## Multiple Regression

Students who apply for college admission usually submit scores on several tests (e.g., SAT general and subject tests), high school transcripts and other indicators of academic potential. One of the challenges faced by admissions officers is to put all of this information together to predict the applicants' future performance. One solution to this problem is to use multiple regression.

Multiple regression involves using several X variables to predict a criterion or outcome variable (Y). The X variables can be continuous (e.g., SAT scores) or discrete (e.g., membership in honor societies); multiple regression provides an analytic method of combining information from multiple predictors to do the best job possible predicting scores on Y.[1] In this particular context, multiple regression involved finding the weighted linear combination of test scores, high school grades, and other indicators that is most highly correlated with the criterion of success in college.

Power analysis can be easily applied to several types of regression. The examples below illustrate some of the possibilities.

### *Multiple Regression Models*

Beaty, Cleveland, and Murphy (2001) examined the relationship between four personality factors (neuroticism, agreeableness, extraversion, and conscientiousness) and evaluations of organizational citizenship behaviors. Their results suggest that these four personality variables account for about 3% of the variance in ratings of citizenship. Because their sample was large ($N = 488$), this $R^2$ value was judged to be significantly different from zero.

It is important to keep in mind that multiple regression involves finding the best possible combination of variables for predicting Y. That means that no other possible weighted combination of these four personality variables will account for more than 3% of the variance in citizenship. This $R^2$ is significantly different from zero, but it is worth asking whether it is large enough to allow you to conclude that the relationship between personality and organizational citizenship is anything but trivial. As we noted in our earlier discussions of minimum-effect hypotheses, it might be reasonable to define treatments, interventions, or predictors that account for 1% of the variance or less in the population as having trivially small effects. The the One-Stop $F$ Table, the One-Stop $F$ Calculator, or our Shiny Web app (https://murphy0921.shinyapps.io/ShinyPower/) can be used to test the hypothesis that the relationship between these four personality traits and evaluations of organizational citizenship is either trivially small ($H_o$) or large enough to be meaningful ($H_1$).

---

**Testing Minimum-Effect Hypotheses in Multiple Regression**

In this study, $N = 488$, $R^2 = .03$. This is a significant multiple $R^2$, but it is not clear whether it is large enough to be meaningful. One way to ask the question about whether we can be confident $R^2$ is large enough to be meaningful is to define the $PV$ value that represents a trivial effect (e.g., one that accounts for 1% of the variance or less) and test the hypothesis that in the population, the percentage of variance accounted for is larger than 1%. To carry out this minimum-effect test, all you need to do is to determine the degrees of freedom and use them to either guide your search through the One-Stop $F$ or $PV$ tables (Appendices B and C) or enter them into the Shiny Web app.

With a sample of 488 subjects and four predictor variables (i.e., p = 4), the degrees of freedom for the sample $R^2$ are 3 (p − 1) and 484 (N − p − 1). The Shiny Web app shows that to reach a .05 alpha level in tests of this minimum-effect hypothesis, $R^2$ must be .035, which is slightly higher than the observed $R^2$ of .03. If you use the One-Stop $PV$ Table (Appendix C), you will find the same result, that $R^2$ must be .035 to reject this minimum-effect null hypothesis.[2] In other words, you *cannot* reject the hypothesis ($\alpha = .05$) that the relationship between these four personality variables and evaluations of citizenship behavior is trivially small. However, you are clearly quite close to the type of effect needed to reject this hypothesis.

If the observed $R^2$ of .03 is a reasonable estimate of the population effect, the Shiny Web app also shows that with 3 and 484 degrees of freedom, your power for rejecting the minimum-effect hypothesis that $PV$ is .01 or smaller is quite low (.46). With a population $PV$ of .03, you would need a much larger sample ($df_{err} = 1,160$, $N = 1,161$) to achieve power of .80.

One of the properties of multiple regression is that adding new predictor variables always increases the value of $R^2$. For example, adding *any* variable to the four personality characteristics discussed in the example above will lead to an increase in the percentage of variance explained. As a result, it is always important in regression to know and consider how many predictor variables are in the regression equation. Explaining 20% of the variance in some important criterion variable with four predictors is considerably more impressive than explaining 20% of the variance with 15 predictors. As a result, virtually all assessments of power in multiple regression will require information about both the number of subjects (N) and the number of predictor variables (p).

## Hierarchical Regression Models

Bunce and West (1995) examined the role of personality factors (propensity to innovate, rule independence, intrinsic job motivation) and group climate factors (support for innovation, shared vision, task orientation, participation in decision making) in explaining innovation among health service workers. They carried out a 17-month, three-stage longitudinal study, and like most longitudinal studies, suffered significant loss of data (subject mortality) across time. Their N dropped from 435 at stage 1 to 281 and 148 at stages 2 and 3. Incomplete data reduced the effective sample size for several analyses further; several critical analyses appear to be based on a set of 76 respondents.

One analysis used stage 1 innovation, personality factors, and group factors as predictors in a hierarchical regression model, where the dependent variable was innovation at stage 2. The results of this analysis are summarized in Table 5.2.

The principal hypotheses tested in this analysis were that: (1) personality accounts for variance in stage 2 innovation not explained by innovation at stage 1, and (2) group climate accounts for additional variance not accounted for by personality and stage 1 innovation. Because stage 1 innovation is entered first in the equation, you might interpret tests of the changes in $R^2$ as personality and climate are added to the equation as reflecting the influence of these factors on changes in innovation over time.

*Table 5.2* Hierarchical Regression Results Reported in Bunce and West (1995)

| Predictor | $R^2$ | df | F | $\Delta R^2$ | df | F |
| --- | --- | --- | --- | --- | --- | --- |
| Innovation – Stage 1 | .18 | 1,75 | 16.73* | | | |
| Personality | .33 | 4,72 | 8.86* | .15 | 3,72 | 5.37* |
| Group Climate | .39 | 8,68 | 5.38* | .06 | 4,68 | 1.67 |

Note
* $p < .05$ in tests of the traditional null hypothesis.

The results suggest that personality has a significant effect, but that group climate does not account for variance above and beyond stage 1 innovation and personality.

### Power Estimation

The results in Table 5.2 show a pattern we have encountered in some of our previous examples: a relatively small sample combined with some reasonably large effects (e.g., stage 1 innovation accounts for 18% of the variance in stage 2 innovation), which makes it difficult to determine offhand whether the study will have enough power for its stated purpose. Some quick calculations suggest that this study does not possess sufficient power to test all the hypotheses of interest.

Rather than starting from the reported $R^2$ and $F$ values, suppose you used standard conventions for describing large, medium, and small effects to structure your power analysis. In hierarchical regression studies, the predictors are usually chosen to be relevant to the dependent variable (which means they should each be related to Y), and are therefore usually also intercorrelated (i.e., several variables that are all related to Y are likely to also be related to one another). As a result, you will generally find that the first variable entered will yield a relatively large $R^2$, and that $R^2$ will not increase as quickly as more variables are entered (Cohen & Cohen, 1983). This might lead you to expect a large effect for the first variable you enter, a smaller change in $R^2$ for the next variable, and a small change in $R^2$ for the last variable. In Chapter 2, we noted that $R^2$ values of .25, .10, and .01 corresponded to conventional definitions of "large", "medium", and "small" effects (see Table 2.2). These values turn out to be reasonably similar to the actual $R^2$ and $\Delta R^2$ values shown in Table 5.2 (i.e., .18, .15, and .06, respectively). Even more to the point, the overall $R^2$, which represents the sum of these conventional values (i.e., $R^2 = .36 = .25 + .10 + .01$) is very similar to the actual overall value (i.e., $R^2 = .39$) reported by Bunce and West (1995).

To estimate power for detecting $R^2$ values of .25, .10, and .01, given the degrees of freedom in this study, we first translate the $R^2$ values into F-equivalents, using equations shown in Table 2.1. The F-equivalents are 25.0, 3.69, and .27, respectively. There is plenty of power for testing the hypothesis that $R^2 = .25$ [$F(1,75) = 25.0$]; the critical table F for this level of power is 8.01. Even when testing the minimum-effect hypothesis that the first variable entered accounts for 5% or less of the variance, power far exceeds .80. If you enter $df$ values of 1 and 75 and an $R^2$ value of .25 into our Shiny Web app, you will find that power exceeds .99.

There is probably not quite enough power for testing the hypothesis that the second variable entered into the regression equation will have a medium effect [i.e., $R^2 = .10$, $F(3,72) = 3.69$]. Entering $df$ values of 3 and 73 and an $R^2$ value of .10 into our Shiny Web app, you will find that power equals .63.

If you assume that the third in a set of intercorrelated predictors will generally yield a small increment in $R^2$ (i.e., $\Delta R^2 = .01$), there is clearly not enough power to test that hypothesis. To achieve power of .50, you would need an $F$ of 1.68; assuming a small effect here, $F$ is only .27.

The conclusions one reaches by looking at these three conventional values closely mirror those that would be obtained if the actual $R^2$ values were used. The $F$ values for the first predictor (stage 1 innovation), second predictor (personality) and third predictor (group climate) are 16.73, 5.37, and 1.67, respectively. Power easily exceeds .80 in tests of the hypothesis that stage 1 innovation is related to stage 2 innovation (the critical $F$ for this level of power is 8.01). Power also exceeds .80 for testing the hypothesis that personality accounts for variance not explained by stage 1 innovation (the critical $F$ is 3.79; the observed $F$ is 5.37). Power is just below .50 for testing the hypothesis that group climate accounts for variance not accounted for by the other two predictors (the critical $F$ is 1.68; the observed $F$ is 1.67).

*Sample Size Estimation*

The sample is certainly large enough to provide a powerful test of the hypothesis that stage 1 innovation predicts stage 2 innovation. Power is also reasonably high for testing the hypothesis that adding personality to the equation yields a significant increase in $R^2$. However, a *much* larger sample would be needed to provide a powerful test of the last hypothesis – i.e., that group climate explains additional variance. We used the Shiny Web app (https://murphy0921.shinyapps.io/ShinyPower/) to determine the number of subjects needed to attain power of .80 for a test of an increase in $R^2$ of .01. A sample of over 1,190 would be needed, nearly 15 times as large as the sample collected by Bunce and West (1995).

## Power in Testing for Moderators

Many theories in the social behavior sciences take the form "X is related to Y, but the strength of this relationship depends on Z". This is a *moderated* relationship. For example, the relationship between cognitive ability and job performance is stronger for jobs that involve frequent and intense cognitive demands than for jobs that are relatively simple and not demanding (Gutenberg, Arvey, Osburn & Jenneret, 1983). That is, job demands moderate the relationship between cognitive ability and job performance.

One of the most widely replicated findings in research on moderator effects is that studies that search for moderators often lack sufficient power to detect them (Aguinis, Beaty, Boik & Pierce, 2005; Osburn, Callender, Greener & Ashworth, 1983; Sackett, Harris & Orr, 1986). Once you understand how moderators are defined and assessed, it is easy to see why power can be so low in so many studies.

Suppose you think cognitive ability (X) is related to job performance (Y), and that this relationship is moderated by job complexity (Z). The best method of testing the hypothesis that Z moderates the relationship between X and Y is to first determine whether Y is related to both X and Z (e.g., is there a significant and meaningful correlation between ability and performance and between job complexity and performance?). Next, a series of regression models is compared. In particular, the test for moderation involves computing the cross-product between X and Z (i.e., multiply each person's ability score by his or her job complexity score – symbolized by X * Y), then compare:

1. $R^2_{y.x,\ z}$ – i.e., the squared correlation between job performance and both ability and job complexity
2. $R^2_{y.x,\ z,\ x*z}$ – i.e., the squared correlation between job performance and ability, job complexity, and the cross-product between ability and job complexity

If there is a real moderator effect, adding the cross-product to a regression equation that contains both ability and job complexity considered alone will lead to an increase in $R^2$ (Murphy & Russell, 2017; Van Iddekinge, Aguinis, LeBreton, Mackey, & DeOrtentiis, 2021). That is, considering ability and job complexity jointly will provide new information over and above what can be gained by considering these two factors separately; the statistical test for a moderator involves testing the hypothesis that adding cross-product terms leads to a significant and meaningful increase in $R^2$. For example, if $R^2_{y.x,\ z} = .30$ and $R^2_{y.x,\ z,\ x*z} = .45$, this means that the joint consideration of ability and job complexity accounts for 15% of the variance in performance that cannot be accounted for by ability and job complexity considered alone.[3]

The most compelling explanation for the traditionally low levels of power for testing moderator hypotheses is that large increases in $R^2$ when cross-products are added to a regression equation are rare, meaning that most moderator effects are fairly small. Aguinis et al. (2005) reviewed 30 years of moderator research in the organizational sciences, and they reported that moderator effects typically accounted for less than 1% of the variance in outcomes. One of the recurring themes in power analysis, regardless of the statistical procedures being analyzed, is that small effects lead to low power unless samples are extremely large.

*Why Are Most Moderator Effects Small?*

There are statistical as well as substantive reasons to expect moderator effects to be small in most cases. In particular, the cross-product terms described above will almost always end up being highly correlated with the variables that were used to create this cross-product. For example, if the cross-product between ability and job complexity is used in testing for moderators, you should expect reasonably large correlations between ability and complexity considered alone and the cross-products between these two variables. If ability and complexity are positively correlated, the cross-product of these two variables *must* be highly correlated with the two variables considered alone.

As we noted in the preceding section, adding a new variable to a regression equation will always lead to some increase in $R^2$, but there is also a decreasing payoff in the sense each new variable that is added to a regression equation is likely to lead to increasingly small increases in $R^2$. In multiple regression, the increase in $R^2$ represents new information contributed by the variable you have added to the regression equation, and when the new variable to the regression (e.g., x * z) is correlated with variables already in the regression equation (e.g., both x and z considered alone), it is quite unlikely that it will add a great deal of new information. Even in situations where there are very strong moderator effects (i.e., where the relationship between X and Y changes substantially depending on the value of Z), it is uncommon to find large increases in $R^2$.

### Power Analysis for Moderators

Suppose you collected data from 500 employees in a range of jobs and correlated ability and performance scores. If you had highly reliable measures of ability and job performance, it would be reasonable to find ability-performance correlations in the range of .35–.60, with an average of about .50. This translates into an $R^2$ value of .25. A sample of $N = 500$ provides plenty of power for testing the hypothesis that ability is related to performance. For example, if you use the Shiny Web app and enter *df* of 1 and 498, alpha of .05, and 0 for the minimum effect size (i.e., you conduct a test of the traditional nil hypothesis), you will find that you achieve power of .80 with an $R^2$ value as low as .016. If you add job complexity to the regression equation, it is reasonable to expect an increase in $R^2$ of about .05–.10, and this will still give you plenty of power. If you enter .05 as the assumed population *PV*, with 1 and 497 degrees of freedom, the power of a test of the traditional nil hypothesis is greater than .99. If you add the cross-products of ability and job complexity into that regression equation, you might expect an increase in $R^2$ of about .005–.01. The Shiny Web app shows that you would need a sample of well over 1,500 employees to detect a change of .005 in $R^2$. Even if the change is a large as .01, you would still need a sample of over 750 employees.

A sample of $N = 500$ is relatively large, compared to the typical samples obtained in behavioral and social science research; depending on the topic, samples of 50–200 subjects are common. However, the sample in this study is not large enough to provide a powerful test of the moderator hypothesis unless the moderator effect is an unusually strong one.

Finally, tests of the traditional null hypothesis are challenging in moderator studies; very large samples are often needed to reliably detect moderators. If you test the hypothesis that moderator effects are large enough to be meaningful, these challenges multiply. Moderator effects are typically very small (Murphy & Russell, 2017), and no matter how large your sample, you may find it difficult to reject the hypothesis that these effects are large enough to be meaningful.

## Implications of Low Power in Tests for Moderators

Sometimes the relationships between variables are truly simple, but many relationships are complicated by the existence of moderator variables. For example, as noted above, the complexity of jobs moderates the relationship between cognitive ability and job performance. Ability is always positively related to performance, but that relationship is stronger in complex jobs than in simpler jobs. As a result, the best answer to the question "What is the correlation between ability and job performance?" is "it depends on the complexity of the job".

The low power of most tests for moderators has potentially important effects on research and applications of research in the behavioral and social sciences (Osburn et al., 1983; Sackett et al., 1986). In particular, lack of power in a search for moderators may lead researchers to underestimate the complexity of their data. For example, prior to Gutenberg et al.'s (1983) study showing that job complexity was a moderator of the ability-performance relationship, many researchers assumed that validity was invariant – i.e., that the validity of ability tests was identical across most if not all jobs. This was not necessarily because of the lack of variability in outcomes across jobs, but rather because of the lack of power in studies searching for moderators. Similarly, researchers sometimes assume that the tests they use are unbiased because tests of the hypothesis that sex, race, age, or other demographic characteristics of test takers do not moderate the relationships between test scores and important criteria. Unless very large samples are used, tests of the hypothesis that demographic variables moderate the effects of tests often have insufficient power, and the failure to identify moderators may say more about the quality of the study than about the lack of bias of the test. In general, the conclusion that any relationship between two variables in invariant may be difficult to test, because it can be very demanding to put together powerful tests of the hypothesis that moderator variables exist.

There are certainly statistical explanations for the low power of tests of moderation hypotheses, including weak measurement of the predictors and moderators and high correlations between them (Aguinis, 1995, 2004; Murphy & Russell, 2017). However, there is a substantive explanation that might be even more important. Careful reviews of research literature suggest that interaction effects are generally quite small, even when the moderator is a categorical variable, and that things get even worse when the moderator is a continuous one (Aguinis, Beaty, Boik & Pierce, 2005, Aguinis & Gottfredson, 2010). It may be that the reason we often fail to find moderators that our theories suggest should be present is that these moderation effects, if they exist at all, are so small as to be trivial in many cases. That is, one possible explanation for the generally weak track record of statistical tests of moderation hypotheses is that moderation is just not a very important concept, at least across much of the range of the variables that are measured in empirical studies. There may be some outliers (e.g., the performance of someone with great ability but no motivation may be near zero, as a moderation hypothesis would suggest, rather than near the mean, as an additive effects hypothesis might suggest), but these cases are much more common in statistics textbooks than in real life. It may be that our pursuit of moderators is much like the pursuit of unicorns – fun and interesting, but unlikely to be very fruitful in the end.

## If You Understand Regression, You Will Understand (Almost) Everything

One of the reasons we start our discussion of the application of power analysis with a chapter on correlation and regression is that multiple regression represents one of the most general models for analysis in the social and behavioral sciences (Cohen & Cohen, 1983; Cohen, Cohen, West & Aiken, 2002). *T*-tests, one-way analyses of variance, multifactor analyses of variance, discriminant analyses, and analyses of covariance can all be thought of as specialized applications of multiple regression. Therefore, if you understand multiple regression and the application of power analysis in multiple regression, you are well on your way to understanding much of what is presented in the chapters that follow.

There are four substantial advantages to understanding that much of what behavioral and social scientists know about data analysis represent variations on the general themes laid out in multiple regression analysis. First, a simple set of conceptual tools can be applied to an extraordinary range of analyses. Second, it really does not matter whether the variables are continuous or categorical, correlated, or orthogonal; the same set of concepts apply. Third, the effect size measure that is most central to multiple regression, $R^2$, or the percentage of variance explained represents an effect size that is almost universally relevant (and which, here, we refer to as *PV*). There is a large literature describing different effect size measures for different purposes, but none has the broad usefulness of *PV* (Kelly & Preacher, 2012). Percent of variance explained is not a perfect measure (there are no perfect measures), but it is to our way of thinking better than any of the plausible alternatives. Fourth, and finally, one of the most common weaknesses of studies in the behavioral and social sciences is the failure to report effect size estimates; too many studies report the results of significance tests and stop there. In multiple regression, it is hard to forget to report the strength of the effect; even the most careless of authors is unlikely to report that there is a significant $R^2$ without reporting the actual value of $R^2$, and if that author does forget this important statistic, only the most negligent reviewers and editors are likely to let this omission pass. In contrast, sophisticated researchers routinely fail to provide effect size estimates when doing something other than regression.

## Notes

1 The "best" set of predictions is defined here as predictions that minimize the sum of the squared differences between actual and predicted Y variables – i.e., the least-squares method.
2 If $df_{hyp}$ equals 3 and $df_{err}$ equals 500, the $R^2$ needed to reject the hypothesis that treatments account for 1% of the variance or less ($\alpha = .05$) is exactly .035. If you interpolate between the values shown in Appendix C for $df_{err}$ of 400 (.036) and the value shown for $df_{err}$ of 500 (.035), the best approximation for $df_{err}$ of 484 is that you will need an $R^2$ value of .035 to reject this minimum-effect null hypothesis.
3 Note that alternate measures of effect size have been proposed. For example, Liu and Yuan (2021) suggest that the best effect size estimate is one that evaluates the percentage of the explained variance in Y that is accounted for by the moderator as opposed to the percentage of the total variance uniquely explained the moderator.

# 6 $t$-Tests and the One-Way Analysis of Variance

One of the most common applications of statistics in the behavioral and social sciences is the comparison of means obtained from different treatments, from different groups, or from different measurements. For example, there are thousands of studies published each year that involve comparisons between a treatment group that has received some sort of intervention or special treatment and a control group that has not received this intervention. Other comparisons might involve assessing differences between scores obtained before some treatment or intervention (pretest) with scores obtained after that treatment (post-test). Other comparisons might contrast the average scores across several groups that receive different treatments or combinations of treatments.

## The $t$-Test

The $t$-test is commonly used to compare one group or one set of scores to another. For example, 200 subjects might be randomly assigned to treatment and control groups; the independent $t$-test can be used to statistically compare these average scores. Alternately, pretest and post-test scores might be collected from the same set of 200 subjects; repeated measures or dependent $t$-tests can be used to compare these scores. Finally, this same $t$ statistic can be used to compare the mean in a sample to some fixed or reference value. For example, you might have the hypothesis that the average SAT critical reasoning score in a sample is equal to the population mean of 500. This one-sample $t$-test is rarely encountered in the behavioral and social sciences; our discussion will focus for the most part on the independent and the dependent $t$-tests.

As with other test statistics, the evaluation of a $t$ statistic involves comparing the observed value of $t$ with the critical value of $t$ for the null hypothesis being tested. The critical value of $t$, in turn, depends on three things: the type of hypothesis being tested (e.g., traditional nil vs. minimum-effect test), the alpha level and the degrees of freedom. The degrees of freedom for the three different types of $t$-tests are shown in Table 6.1.

DOI: 10.4324/9781003296225-6

Table 6.1 Degrees of Freedom for the $t$ Statistic

| Type of test | Test Compares | df |
|---|---|---|
| One-Sample | Sample mean to some fixed value | $N - 1$ |
| Independent $t$ | Two sample means | $(n_1 - 1) + (n_2 - 1)$ or $N - 2$ |
| Dependent $t$ | Two scores from the same set of people | $N - 1$ |

It is not always clear why different tests have different degrees of freedom, but if you keep in mind what each test compares, it is easier to see why, for example, the one-sample test has different degrees of freedom than the two-sample test. In a one-sample test, the scores of $N$ people are used to estimate a population mean. Using our previous example, if you compare the mean in a sample of $N$ students who took the SAT critical reasoning test to the overall average score this test is designed to yield (i.e., a score of 500), the degrees of freedom are $N - 1$ because there is only one sample statistic being used to estimate its corresponding population parameter. That is, the hypothesis you are testing is that the mean score *in the population you sampled from* is 500. In the two-sample test, there are $n$ subjects in both the treatment and control group, and scores in these samples are used to estimate the means in the populations each group represents, yielding $(n - 1)$ degrees of freedom within each group, or $(n - 1) + (n - 1) = N - 2$ degrees of freedom for the study. In a dependent $t$-test, where the same $N$ people provide both sets of scores to be compared, the hypothesis being tested is that Post-test − Pre-test = 0. There are $N$ measures of this pre-post difference, yielding $N - 1$ degrees of freedom for estimating the population difference between pre-tests and post-tests.

where:

$N$ = number of individuals in sample
$n$ = number of individuals in each independent group

In very small samples, a $t$ table is often used in assessing statistical significance; for samples of 60 or more, the distribution of the $t$ statistic is very similar to a standard normal distribution. Thus, for example, the critical value for a two-tailed $t$-test with df = 100 is 1.96.

Rather than relying on the $t$ distribution, it is both easy and useful to convert $t$ to $F$, by simply squaring the value of $t$. In particular, $t^2(\text{df}_{\text{err}}) = F(1, \text{df}_{\text{err}})$. Throughout this text, we have noted that converting to $F$ simplifies power analysis, something that is trivially easy to do when working with the $t$ test.

Suppose 100 subjects are randomly assigned to a treatment designed to increase reading speed or to a control group. A researcher compares the means for these two groups and reports $t(98) = 2.22$. This translates into

$F$ (1, 98) = 4.94; applying the $F$ to $PV$ conversion formula shown in Chapter 2 (Equation 2.7 – $PV = (df_{hyp} * F)/[(df_{hyp} * F) + df_{err}])$ you find:

$$PV = (1 * 4.94)/[(1 * 4.94) + 98] = .047$$

This is both a relatively small sample and a moderate to small effect, suggesting that the level of power in this study will be low. The One-Stop $F$ Table confirms this. The $F$ needed to achieve power levels of .50 and .80 are 3.85 and 7.95, respectively. If you enter $df$ values of 1 and 98, a minimum effect size of 0.00 (i.e., you are testing the traditional null hypothesis) and an assumed population effect size of .047 in our Shiny Web app, you will find power equals .589 with α = .05.

---

### The $t$ Distribution versus the Normal Distribution

W.S. Gossett, writing under the pseudonym "Student", developed statistical methods for comparing the average values obtained in two small samples, in particular the $t$-test. The distribution of the $t$ statistic is similar in shape to the normal distribution, but it is a bit wider than the normal distribution when $N$ is smaller than 30 to 50. For example, in a normal distribution, a value of 1.96 is required to reject a two-tailed test of the null hypothesis ($p < .05$), whereas with a sample of 30 cases, a $t$ value of 2.04 is needed.

With samples of 100 or more, the differences between $t$ and the normal distribution are so small that they are not worth worrying about. Thus, for all practical purposes, we can think about $t$ statistics in terms of the normal distribution.

---

## Independent Groups $t$ Test

Suppose a researcher is interested in comparing the outcomes of two different programs designed to help people quit smoking. The researcher thinks that a good study can be done using 50 subjects (25 randomly assigned to each treatment). Power analysis might lead this researcher to rethink that assumption.

### Estimating Power for This Study

Suppose the researcher chooses to test a traditional null hypothesis, using an alpha level of .05. She is not sure whether there will be large or small differences between the programs and performs a power analysis that assumes small treatment effect (e.g., $PV = .01$). In the $t$-test, there is one degree of freedom for the hypothesis being tested (i.e., $df_{Hyp}$). The degrees of freedom for error (see Table 6.1) are:

$$df_{Err} = n1 + n2 - 2$$

$$df_{Err} = 25 + 25 - 2$$

$$df_{Err} = 48$$

To determine whether the proposed study has sufficient power, use the Shine Web app and enter 0 for ES, .05 for *alpha*, 1 for $df_{Hyp}$, and 48 for $df_{Err}$. This study has power of .10 to detect the small effect the investigator expects to find. That is, if there truly is a small difference between these two programs, the odds are about 1 in 10 that your study will find it, and about 9 in 10 that your study will miss it. To achieve a power of .80, you would need fairly strong effect (i.e., $PV = .148$), much larger than the small effect that the researcher assumed.

Suppose that the expected effect is a small one (e.g., $PV = .01$). As our Shiny Web app shows, the $N$ needed to achieve power of .80 is much larger than the sample the investigator has chosen. Alternately, you could use the $df_{error}$ Table presented in Appendix D. This table shows that when $PV = .01$ and $df_{hyp} = 1$, the $df_{error}$ required to achieve power of .80 is 775. Rather than running this study with 50 subjects, the investigator will need a sample in the 775–780 range to achieve power of .80.

### *Traditional versus Minimum-Effect Tests*

Clapp and Rizk (1992) measured placental volumes in 18 healthy women who maintained a regular routine of exercise during pregnancy. Nine of the women engaged in aerobics: four ran and five swam. The control group comprised 16 females who did not engage in such a regimen. Placental volumes were measured using modern ultrasound techniques at 16, 20, and 24 weeks' gestation. The results of the study are reproduced in Table 6.2.

Two aspects of this study are especially noteworthy. First, the sample is small (i.e., $N = 34$). In most cases, this would mean very low power levels. However, in this study, the effects are quite strong. In the three time

*Table 6.2* Placental Volumes Reported by Clapp and Rizk (1992)

| Week | Control (N = 16) | Treatment (N = 18) | t | F | PV | d |
|---|---|---|---|---|---|---|
| 16 | 106 (18) | 141 (34) | 3.68 | 13.55 | .29 | 1.94 |
| 20 | 186 (46) | 265 (67) | 3.96 | 15.66 | .33 | 1.71 |
| 24 | 270 (58) | 410 (87) | 5.45 | 29.66 | .47 | 2.41 |

Volumes are expressed in $cm^3$. Standard deviations are shown in parentheses. Note that in this table, *d* represents the mean difference divided by the control group SD. Use of pooled SD values yields somewhat smaller *d* values, but they will nevertheless exceed conventional benchmarks for "large" effects.

periods studied, exercise accounted for between 29% and 47% of the variance in placental volume ($d$ ranges from 1.71 to 2.41). Because the apparent effects of exercise were quite substantial, it should be easy to rule out the hypothesis that the treatment has no effect (traditional null), or even that the true effects of exercise are at best small (minimum-effect hypothesis).

This $t$-test in this study has 32 degrees of freedom; if the $t$ value is squared, the statistic is distributed as $F$ with 1 and 32 degrees of freedom. If you consult the One-Stop $F$ Table, you will find entries for 1 and 30 and for 1 and 40 degrees of freedom, but none for 1 and 32 degrees of freedom, meaning that you must interpolate to analyze these results. As can be seen in the One-Stop $F$ Table, critical $F$ values for the traditional null hypothesis at 1 and 30 degrees of freedom for $\alpha = .05$ and $.01$ are 4.16 and 7.56, respectively. The corresponding values at 1 and 40 degrees of freedom are as 4.08 and 7.31. If you interpolate between these two values, the $\alpha = .05$ and $.01$ critical values for 1 and 32 degrees of freedom when testing the traditional null hypothesis are 4.14 and 7.51. All the observed $F$ values in Table 6.2 are greater than 7.51, so we can reject the traditional null at the .01 level and conclude that regular exercise has *some* impact on placental volume.

You can conclusively rule out the possibility that exercise has *no* effect, and the data suggest that the actual effect is quite large. However, it is always possible that the *true* effect is small, and the large $PV$ and $d$ values observed here represent chance fluctuations in a process that usually produces only a small effect. You can use the One-Stop $F$ Table to test the hypothesis that the effects of treatments are negligible, or perhaps small to moderate. Again, if you interpolate between the values for 1 and 30 $df$ and 1 and 40 $df$, the critical $F$ ($\alpha = .05$) for the minimum-effect null hypothesis that the effect is small to medium in size (i.e., the null is that treatments account for no more than 5% of the variance in outcomes) is 9.49. All $F$ values in Table 4-1 exceed 9.49, so we can reject the minimum-effect null hypothesis that the effects of treatments are no greater than small to moderate, with a 95% level of confidence. We can also reject this hypothesis at the .01 level for mean placental volumes at 20 and 24 weeks (critical $F = 15.26$; observed $F = 15.66$ and 29.44 at 20 and 24 weeks, respectively).

## *Power Estimation*

This study has plenty of power for tests of the traditional null hypothesis or for tests that the effects are large enough to be meaningful (i.e., the population effect is .01 or larger). For example, if you use our Shiny Web app and enter $df$ of 1 and 32, a minimum effect size of .01 and an assumed population effects size of $PV = .29$ ($\alpha = .05$), you will find that power equals .895. With $PV$ values of .33 or .47, power approaches .95. That is, we can be at least 99% confident that the effects of treatments exceed our definition of "small to moderate".

In addition to placental volume, Clapp and Rizk (1992) examined a range of other dependent variables for the women in this study. If exercise could function as a viable treatment of fetal under- or over-growth, we might expect it to have some effect on the final birth weight of the babies as well. In this case, Clapp and Rizk (1992) reported the mean birth weight of babies for women in the control group as 3,553 gm (SD = 309) compared to 3,408 gm (SD = 445) for women in the treatment group. Reference to the One-Stop $F$ Table shows that when testing hypotheses about birth weight, you cannot even reject the traditional null hypothesis [$F(1, 32) = 1.19, p < .05$].

One possible explanation for this finding is low power. If the true effect of exercise on birth weight (as opposed to placental volume) is small, power will be well below .50 in this study. For example, if we assume that the true effect of exercise on birth weight meets the conventional definition for a small effect (i.e., $PV = .01$), the $F$-equivalent in this study is less than .32, whereas the critical $F$ for a power of .50 to reject the traditional null at the .05 level is 4.03. Thus the power of this study to detect traditionally significant differences in birth weight is much less than .50 (about .18, in fact). Of course, the power to detect substantively meaningful differences is even lower.

*Sample Size Estimation*

How many subjects would be required to reliably detect differences in birth weight because of exercise during pregnancy? To answer this question, we can refer to Appendix D, which allows you to determine sample sizes needed, given that $\alpha = .05$ and the desired level of power is .80. If you assume that the effect of exercise on birth weight is small ($PV = .01, d = .20$), you will need about 777 subjects in total (i.e., $N = df_{hyp} + df_{err} + 1$; for the $t$-test, $df_{hyp} = 1$) or about 389 subjects in each group (i.e., 777/2) to achieve this level of power in tests of the traditional null hypothesis.[1]

The power analyses conducted here suggest that the Clapp and Rizk (1992) study was well suited for answering questions about the effect of exercise on placental volume, which was its major focus. Power exceeded .80 for tests of both traditional and minimum-effect null hypotheses. However, the sample size is quite inadequate for answering questions about the effects of exercise on birth weight. Because a small effect might reasonably be expected here, huge samples are probably needed to provide adequate power.

## One-Tailed versus Two-Tailed Tests

In the two examples above, the null hypothesis being tested was that there were no differences between treatments or conditions, or that the differences were trivially small (e.g., accounted for less than 1% of the variance in outcomes). This represents a two-tailed statistical test, because differences in programs will lead you to reject the null regardless of the direction of the difference. For example, in Clapp and Rizk (1992), you would reject the null hypothesis if subjects in the treatment condition had higher scores than those in the control condition but would also reject the same null hypothesis if subjects in the control condition ended up with higher scores. This type of test is most common in the behavioral and social sciences, and it is called a two-tailed test because score differences at either end of the distribution (i.e., large positive differences or large negative differences) will cause you to reject the null hypothesis.

Sometimes, researchers have specific directional hypotheses – e.g., that scores in the treatment group are higher than scores in the control group. These one-tailed tests are more powerful than their two-tailed counterparts and are preferable in many circumstances. The key to working with one-tailed tests is that doubling the alpha level of a two-tailed test allows you to determine the critical values, power, sample size needed, etc. of a one-tailed test. That is, the results you obtained from analysis of the power of two-tailed tests at the $\alpha = .10$ level would be identical to those obtained for one-tailed tests at the $\alpha = .05$ level. Our Shiny Web app is particularly useful for working with one-tailed tests, because it allows you to enter virtually any alpha value you want.

---

### Re-analysis of Smoking Reduction Treatments: One-Tailed Tests

The analyses presented earlier in this chapter assumed that two-tailed tests were being performed – i.e., that two interventions designed to reduce smoking were being compared, without specifying *a priori* which one was likely to work better. Suppose the researcher had a good reason to believe that one treatment was indeed better. There would be substantial advantages to using one-tailed rather than two-tailed tests.

First, with degrees of freedom of 1 and 48 for the hypothesis being tested, power analysis for one-tailed tests, with $\alpha = .05$ Shiny Web app. If the true differences between programs is small (e.g., assume that the true $PV = .01$), power would still be quite low (.178) with only 50 subjects. To achieve power of .80 in this study, you would need a much larger sample ($N = 615$) or a much larger effect size (e.g., $PV = .12$ – you can arrive at this figure by trying our different "assumed effect size in the population" values).

---

### Repeated Measures or Dependent *t*-Test

Suppose 50 students take a pre-test designed to measure their knowledge of American history. They then go through a training program and complete a post-test measure of history knowledge three weeks later. One of the key

Table 6.3 Pre-Test and Post-Test Comparison

|  | Pre-Test | Post-Test |
|---|---|---|
| Mean | 110 | 116 |
| S.D. | 20 | 18 |

Note: The correlation between pre-test and post-test scores is .40, $t = 3.14$, $d = .315$.

assumptions of the two-sample $t$-test is that the scores being compared are obtained from independent samples, and that certainly is not the case in studies that rely on the dependent $t$. Rather, it is best to assume that multiple scores obtained from the same people (either on pre- and post-versions of the same measure or even on two distinct measures) are correlated, and this correlation affects both the structure of the significance test and the power of that test.

Table 6.3 presents the means and standard deviations for the pre- and post-test scores, as well as the correlation between pre and post. The formula for the dependent $t$ is:

$$t = \frac{\text{Mean 1} - \text{Mean 2}}{\sqrt{\frac{\text{var 1}}{N} + \frac{\text{var 2}}{N} - \frac{2r_{12}SD_1 SD_2}{N}}} \quad (6.1)$$

where:

Mean 1, Var 1, $SD_1$ – mean, variance, and standard deviation of first measure

Mean 2, Var 2, $SD_2$ – mean, variance, and standard deviation of second measure

Equation 6.1 presents a general formula for the $t$-test. When the means of independent groups are compared, the correlation between scores is necessarily equal to zero, and Equation 6.1 reduces to the more familiar formula for $t$ presented below:

$$t = \frac{\text{Mean 1} - \text{Mean 2}}{\sqrt{\frac{\text{var 1}}{N} + \frac{\text{var 2}}{N}}} \quad (6.2)$$

In our example of pre-test post-test comparisons, the difference between means is relatively small ($d = .315$), and the sample is also relatively small, but because of the increased power provided by a repeated-measured design, this difference is nevertheless statistically significant. In Chapter 7, we will discuss repeated-measures designs for comparing any number of scores (the dependent $t$ can only be used to compare two different scores) and will show a more general version of the dependent $t$ test that is based directly on the analysis of variance and the $F$ statistic. We will hold our discussions of power analysis for research designs that involve repeated measures until then. We will, however, note something that might be obvious from examining Equation 6.1. The larger the correlation between the two measures being compared, the smaller the standard error of the difference between means, and therefore the higher the power of statistical comparisons.

## The Analysis of Variance

The analysis of variance (ANOVA) was once the dominant method of analysis, and it is still an important analytic tool in the social and behavioral sciences. Students sometimes find the name, analysis of variance, confusing because this method is often used to test the null hypothesis that the means obtained from different groups or under different conditions are all identical, as opposed to the alternate hypothesis that some group means are different from others. This suggests that a more apt name might be "analysis of means". In fact, analysis of variance is a very good name, because the principal goal of this method of analysis is to help us understand *why* scores vary. In a traditional experiment, where subjects are randomly assigned to one of several treatments, ANOVA is used to determine how important differences between treatment means are for understanding the data obtained from an experiment. If treatments explain a substantial portion of the variance in scores, we might conclude that treatment effects are large and important. On the other hand, it is possible (particularly if the number of subjects is large) to conclude that there are significant differences between the means obtained in different treatments, even though the overall amount of variance explained by treatment effects is quite small.

The analysis of variance provides both significance tests and effect size measures. Significance testing is done using the $F$ statistic, making the application of the models developed in this book (all of which are based on the noncentral $F$ distribution) especially easy. Analysis of variance also provides an effect size measure, eta squared, which represents the proportion of variance (i.e., $PV$) explained by differences between groups of treatments. Again, this effect size measure fits neatly and easily into the frameworks presented here, most of which revolve around $F$ and $PV$.

The simplest application of ANOVA is called a one-way analysis of variance, in which there are two or more treatments or conditions that subjects might be assigned to and in which the main question is to determine how much of the variability in scores can be explained by the different treatments. This one-way analysis of variance is closely related to the independent-groups $t$-test, in that the $t$-test is used to compare scores in two groups or conditions, whereas the analysis of variance is used to compare average scores in any number of treatments. When there are only two groups, the one-way analysis of variance can be used in place of the $t$-test; as we noted earlier, the $F$ statistic that is used in significance testing in ANOVA is obtained simply by squaring the value of $t$. More complex analyses, which are discussed later in this chapter and in the chapters that follow, will allow us to ask a wider array of questions, but there is often considerable utility in asking the simple question "how important are the different treatments or conditions in explaining variability in subjects' scores". The one-way analysis of variance asks this question.

Suppose an investigator randomly assigns 60 rats to one of three groups. All rats run through a straight alley (runway) one time a day for a total of 30 days. Rats in the first group are given a small reward for each run through the runway. Rats in the second group are given a medium-sized reward for each run through the runway. Rats in the third group are given a large reward for each run through the runway. One hour after a given rat runs through the runway, each rat is given additional food. The total food per day for all rats is the same. That is, additional food for the small sized reward rats is more than additional food for the medium-sized reward rats. The additional food for medium-sized reward rats is more than additional food for large reward rats.

At the end of 30 days, training is completed, and each rat is given one last trial running through the alley; the time it takes each rat to make it to the end of the alley is recorded. The question here is whether the magnitude of reward during training will affect the dependent variable (time to run). Table 6.4 reports the results of an analysis of variance conducted on data from this experiment.

In this table, df represents the degrees of freedom for the hypothesis being tested and error (in one-way ANOVA, $df_{hyp} = k - 1$, where k represents the number of means being compared, and $df_{error} = N - k$), SS represents the sum of the squared deviations attributable to differences in treatment means and to variability within each of the treatment groups (which is used to estimate error), and MS represent mean squares (MS = SS/df) for treatment and error effects. These mean squares are sample estimates of the variance in the population that can be explained by group differences and by error (the variance is computed by taking the average of the squared deviations from the mean – hence, the term "mean square").

Eta squared (SS for the hypothesis divided by SS total) provides an estimate of the proportion of variance accounted for by differences between groups. The implication of finding that differences between groups account for 4.5% of the variance in scores is that the lion's share of variance (95.5%) is explained by something other than differences in the magnitude of rewards these rats

*Table 6.4* ANOVA Results – Rat Running Study

| Source | df | SS | MS | F | $eta^2$ |
|---|---|---|---|---|---|
| Magnitude of Reward | 2 | 32 | 16.0 | 1.34 | .045 |
| Error | 57 | 678 | 11.89 | | |
| Total | 59 | 710 | | | |

received during training. The $F$ ratio is calculated by dividing the mean square for differences between groups by the mean square for error (i.e., $F = MS_{hyp}/MS_{error} = 16.0/11.89 = 1.34$). In this study, an $F$ value of approximately 3.15 would be needed to reject the null hypothesis that differences in reward have no effect whatsoever, meaning that the effect here is nonsignificant.

## Power Analysis

The sample was not very large, and one likely explanation for the non-significant finding is that power is low. Using our Shiny Web app, enter 0 for minimum effect size (traditional null hypothesis), .05 for alpha, .045 for the assumed effect size in the population (i.e., the eta squared reported in the experiment is used here to estimate the population effect) and 2 and 57 for $df_{hyp}$ and $df_{error}$, respectively. You will find that this study has power of .27. If you want to achieve power of .80 in this study, you will need a much larger sample ($N = 210$). If you make the conservative assumption that in the population $PV = .01$ (i.e., assume that there is a small effect), you will need 959 rats to achieve this level of power.

---

### Retrieving Effect Size Information from $F$ Ratios

Unfortunately, many studies report the outcomes of significance tests without presenting the more important information about effect sizes. In the one-way ANOVA, it is quite easy to translate information about significance to information about effect size. In Appendix A and in Chapter 2, we presented a formula that bears repeating:

$$PV = (df_{hyp} * F)/[(df_{hyp} * F) + df_{err}]$$

As we noted in Chapter 2, this formula cannot be used for complex ANOVAs, but it is perfect for the one-way analysis of variance. Suppose two studies report non-significant results, the first reporting $F(1,500) = 3.52$, and the second reporting $F(1,32) = 3.70$. Would you regard the results in these two studies as similar? If you compute $PV$, you will find that differences between groups explained 6/10 of 1% of the variance in the first study, but 10.3% of the variance in the second. You might still pay attention to the fact that neither study produced significant results, but the most reasonable interpretation of these data is that the first study is dealing with a truly minuscule effect (with over 500 cases, there would be plenty of power to detect nontrivial effects), whereas the second study simply lacks the power needed to detect even moderately strong effects. These do not look like similar findings once effect size information is considered.

## Which Means Differ?

A significant F in the one-way analysis of variance indicates that the null hypothesis that all means are identical can be rejected, but it does not necessarily tell you which means are and which are not reliably different. A variety of procedures have been developed to provide a more detailed follow-up in studies that report significant differences between treatment means.

### *The Least Significant Difference (LSD) Procedure*

Suppose an investigator performs an ANOVA for an experiment in which 100 subjects are randomly assigned to one of four different treatments, and obtains a statistically significant F (e.g., $F_{(3,96)} = 5.0$). This F translates to a PV value of .20, which is a relatively strong effect. Knowing that ANOVA asks the global question of whether there are any differences between treatments, the investigator decides to use $t$-tests to compare all possible pairs of treatments, provided the F the ANOVA is statistically significant, and decides to use the same alpha level (e.g., .05) for all comparisons. This can easily be done using the least significant difference (LSD) procedure, which involves computing the smallest difference in treatment group means that will be judged to be statistically significant, using the formula:

$$LSD = t_{a/2}\sqrt{\frac{2MSE}{k}} \qquad (6.3)$$

where:

$t_{a/2}$ = square root of the value of F needed to achieve significance for 1 and df$_{error}$ degrees of freedom at the $\alpha/2$ level
MSE = mean square error
k = number of groups

The $t$ value used in this formula is simply the square root of the F needed to achieve statistical significance with the degrees of freedom of 1 and df$_{error}$ for an alpha level twice as high as the desired alpha for each comparison. The F tables presented in this book provide F values needed to reject null or minimum-effects hypothesis for $\alpha = .05$ or .01, but do not provide values for $\alpha = .10$ to find the value of F needed to reach statistical significance at the .10 level. These tables can be easily obtained by doing a web search for "F Table .10". You will find that for df 1 and 96, the F needed to reject the traditional null hypothesis with $\alpha = .10$ is 2.796, so the equivalent $t$ is 1.671 (the square root of 2.796).[2] Thus, the LSD value in this

## Power for the LSD Procedure

Our **R** code or our Shiny Web app makes it easy to estimate the power of the LSD procedure. Simple enter 0 for the minimum effect, .05 for alpha, .20 as the assumed effect size in the population (the observed eta squared is your best estimate of the population effect size), and the values of $df_{hyp}$ and $df_{error}$ (here, 1 and 96). In this study, power is quite high (.99) when comparing the means of each pair of treatments, rejecting the quite large effect size in this study.

## Ryan's Procedure

There are many alternatives to the LSD, many of which were designed to address the potential inflation of Type I error that results from conducting many different tests. As we have noted in several earlier chapters, the likelihood of Type I errors is much lower than most people think, especially when tests of the traditional null hypothesis are conducted. Nevertheless, it is useful to describe at least one of the methods that attempts to deal with uncertainty about error rates.

An investigator who wants to make the entire experiment the conceptual unit for all statistical tests can use Ryan's (1962) procedure. Ryan's procedure is a stairstep procedure in which potential $t$-tests are performed until the procedure indicates no further $t$-tests should be calculated. Each test is conducted using an alpha level that provides an overall alpha level of .05 for the experiment as a whole.

Consider the experiment with four groups. Assume that a traditional null hypothesis with an alpha level of .05 is used. Further assume that $df_{error}$ is 300 and means for the four groups are: M1 = 55, M2 = 23, M3 = 91, M4 = 12. The $F$ ratio ANOVA must be statistically significant before any means can be compared with Ryan's procedure. Suppose here that $F(3, 300) = 10.0$, which is easily significant. This procedure orders groups from the group with the lowest mean to the group with the highest mean: M4, M2, M1, M3. The first $t$-test compares Group 4 (lowest mean) and Group 3 (highest mean). Table 6.5 shows the four means ordered from lowest to the highest mean.

Table 6.5 Order Means Used in Ryans Procedure

| Steps | 1 | 2 | 3 | 4 |
|---|---|---|---|---|
| | M4 | M2 | M1 | M3 |
| | 12 | 23 | 55 | 91 |

Table 6.5 also shows the four steps between the lowest mean (M4) and the highest mean (M3).

The first *t*-test compares the means that are farthest away from one another, here M4 and M3, which are four steps apart. To preserve an overall alpha level of .05, the alpha levels for specific tests need to be adjusted. The "corrected" alpha level (alpha') for this test is:

$$\text{alpha}' = \frac{2\alpha}{J(R-1)} \tag{6.4}$$

where:

J = number of groups
R = number of steps between the two means.
alpha' in this example is $[(2 * .05)/(4 * (4-1))] = .0083$

Our Shiny Web app can be used to determine the critical value of F for this *t*-test. Enter 0 for the minimum effect size, 0.008 for alpha, and 1 and 300 for $df_{hyp}$ and $df_{err}$. In this study, the observed value of F (i.e., $F_{(3,300)} = 10.0$) is equivalent to $PV = .09$. If you enter this as the assumed effect size in the population, you will find that the critical value of F for significance testing is $F = 7.06$. The F from the ANOVA was larger than 7.06, which means you can reject the null hypothesis that M4 does not differ from M3. Note that you should report that M4 differs from M3 at the .05 level, not the .0083 level.

If the first *t*-test (M4 versus M3) was not statistically significant, no further *t*-tests would be performed. If the first *t*-test is statistically significant, two more *t*-tests are justified. In Table 6.5, there are two sets of means that are three steps removed from each other, M4 and M1 and M2 and M3. The corrected alpha' for both of these *t*-tests is once again given by equation 19 (i.e., alpha' = $[(2 * .05)/(4 * (3-1))] = 0.0125$).

If you enter 0 for the minimum effect, 0.012 for alpha, 1 and 300 for $df_{hyp}$ and $df_{err}$ and .090 for the assumed effect size, you will find that the critical value is $F = 6.31$, which allows you to reject the null hypotheses that M4 does not differ from M1 and that M2 does not differ from M3 (the observed F was 10.0). The power of this test is .99.

If means that are three steps apart are significantly different, *t*-tests are justified for means that are two steps apart (e.g., M4 versus M1, M4 versus M2, and M2 versus M1), and finally, significant results at two steps justify comparisons of adjacent means. Note that the critical value for F becomes less stringent as the investigator tests means that are closer together in terms of the number of steps. Also note that it is harder to find statistical significance for all *t*-tests with Ryan's procedure as opposed to the LSD procedure.

Ryan's procedure can be adapted for use with minimum effect hypotheses, by simply entering the appropriate minimum effect size when using the Shiny Web app. For example, in the analysis above, if you enter .01 for the minimum effect (i.e., you are testing the null hypothesis that mean differences are trivially small – that they account for less than 1% of the variance in the DV), 0.012 for alpha, 1 and 300 for $df_{hyp}$ and $df_{err}$ and .09 for the assumed effect size, you will find that the critical value of $F$ is 16.12, which is larger than the observed $F$, meaning that the differences in means described in the preceding paragraph is *not* significantly larger ($\alpha = .05$) than the effect size you would describe as trivial.

## Designing a One-Way ANOVA Study

Suppose you are planning a study that will use the analysis of variance to compare scores across 3–5 different treatments or groups. We hope that by now, you are convinced that you should use power analysis in planning the size of the sample you need. Table 6.6 shows the total sample sizes needed to achieve power of .80 if you assume the effect you are testing is small ($PV = .01$), small to moderate ($PV = .05$), or moderately large ($PV = .10$). As you can see, effect size is the critical variable here, not the number of groups or treatments being compared. You need a few more subjects to achieve adequate power with 4 groups than with 3 if the effect is assumed to be small (1,067 vs. 938), but even with only three groups, a large sample is needed. It is hard to overemphasize this point. If you are working in an area where you expect small effects, you will either need very large samples or some procedure that will allow you to substantially increase the power and precision of your study. In the chapters that follow, we discuss multifactor designs, repeated measures designs and multivariate procedures, all of which might lead to an increase in power. Nevertheless, if you work in an area where small effects are the norm, you will need to collect a lot of data to achieve a respectable level of power.

Even with small to moderate effects, relatively large samples (183–226) are needed. It is only when you are working in an area where you believe the effects are large that small samples will suffice, and if you have good reason to expect large effects, there may be little point in testing the null hypothesis.

*Table 6.6* Sample Sizes Needed for Power = .80, One-Way ANOVA

| | Groups | | |
|---|---|---|---|
| | 3 | 4 | 5 |
| **Effect is** | | | |
| Small ($PV = .01$) | 938 | 1,067 | 1,177 |
| Small to moderate ($PV = .05$) | 183 | 206 | 226 |
| Moderate ($PV = .10$) | 88 | 98 | 108 |

## Notes

1 As noted earlier, our Shiny Web app (https://murphy0921.shinyapps.io/ShinyPower/) yields a very similar estimate, that 780 subjects are required for power of .80.
2 For example, the first table I found using the Google search term F table .10 (https://www.statisticshowto.com/tables/f-table/) reported $F$ required values of 2.17 for $df = 3, 60$ and 2.12 for $df = 1, 1000$. If you interpolate between these values, using the formula Required $F = 2.12 + (4/40) * (2.17 - 2.12)$, you will find that the $F$ required to reject the traditional null hypothesis when $df = 3, 96 = 2.125$.

# 7 Multifactor ANOVA Designs

Experiments and quasi-experimental studies in the behavioral and social sciences often involve several independent variables that are manipulated or studied jointly. For example, an educational psychologist might be interested in the effects of both the amount of instruction (e.g., 2, 3, or 4 days a week) and the method of instruction (e.g., traditional lecture vs. hands-on learning) on the achievement of students. A factorial experiment is one in which the individual and joint effects of both the independent variables can be studied. So, for example, an investigator might randomly assign each of 240 students to one of the conditions illustrated in Figure 7.1.

If students are randomly assigned to these six conditions, you should end up with 40 students in each cell of Figure 7.1 (in this sort of study, it is common to use lower-case $n$ to designate the number of people in each condition and uppercase $N$ to designate the total number of participants). Sometimes, it is not possible to assign students at random. For example, it might be administratively impossible to move students around at random, but you might be able to assign classes or groups of students to each cell, creating a quasi-experimental study. The distinction between a true experiment and a quasi-experimental study is that experiments involve random assignment of each subject to conditions, whereas quasi-experimental studies do not allow every subject to be randomly assigned to conditions.

The methods used to analyze data from experiments and quasi-experiments are identical, but true experiments permit stronger inferences about causation.

## The Factorial Analysis of Variance

The analysis of variance (ANOVA) is a statistical technique that involves asking the question, "why do some people get high scores, and others get low scores on the dependent variable?" (or "why do scores vary?"). In a simple one-way ANOVA, in a study where subjects are assigned to different treatments and their scores on some common dependent variable are obtained, ANOVA answers this question by breaking down the variability of scores into two components, as illustrated in Equation 7.1.

DOI: 10.4324/9781003296225-7

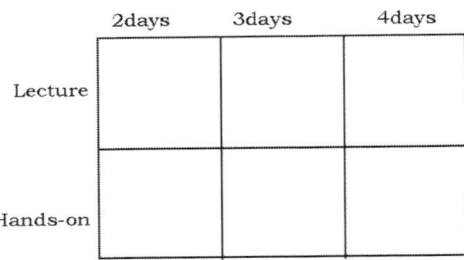

Figure 7.1 A Factorial Experiment.

$$\text{Variability in scores} = \text{Variability due to differences in treatment means} + \text{Variability within each of the treatments} \quad (7.1)$$

In a study where there is only one independent variable (e.g., where the independent variable is the number of days of instruction), Equation 7.1 corresponds to:

$$\text{Variability in scores} = \text{Variability due to treatments} + \text{Variability due to error} \quad (7.2)$$

In this sort of study, the mean square between ($MS_B$) provides an estimate of the variability associated with treatments, and the mean square within ($MS_W$) provides an estimate of the variability in scores among subjects who receive the same treatment (i.e., variability that is not due to the treatments). Thus, $MS_W$ represents variability due to error, i.e., the variability that you would expect if all the differences among scores were merely random, chance fluctuations. The statistic $F = MS_B/MS_W$ tells you whether the variability due to treatments is large relative to the variability in scores due to error, i.e., whether the variability due to treatments is larger than what you would expect if everything was due to chance. If the variability due to treatments represents more than chance fluctuations, you can conclude that your treatments had some effect. In a one-way ANOVA, all of the variability in scores is broken down into these two components (between vs. within), and as a result, it is easy to use the $F$ statistic to estimate the effect size, $PV$.

As we noted in Chapters 2 and 6, in a one-way ANOVA, $PV = df_{hyp} * F/[(df_{hyp} * F) + df_{err})]$. In factorial ANOVA, the questions asked and the breakdown of the variability in scores become more complex, and as a result, the estimation of $PV$ and of statistical power also becomes more complex. In this chapter, we will generalize the formula used for estimating $PV$ from the $F$ values obtained in a factorial ANOVA study.

Factorial experiments allow you to ask several questions simultaneously. For example, the study illustrated in Figure 7.1 allows you to ask three questions: (1) what is the effect of the number of days of instruction, (2) what is the effect of the style of teaching, and (3) do these two variables interact? (i.e., does the effect of the amount of instruction depend on the method of instruction used?). Different research designs might allow researchers to ask a wider range of questions (e.g., in a later section of this chapter, we discuss repeated-measures designs, which allow you to examine systematic subject effects and subject by treatment interactions), or might provide a more narrowly focused examination of the data (e.g., designs with nested factors can be more efficient, but may not provide an opportunity to ask all of the questions that are possible in a fully crossed factorial design). In multifactor ANOVA studies, it is common to have different levels of power for each of the hypotheses tested.

### *Different Questions Imply Different Levels of Power*

In the design illustrated in Figure 7.1, three distinct hypotheses can be tested via ANOVA, two main effect hypotheses (i.e., the hypothesis that the type of instruction and the amount of instruction influence learning), and one interaction hypothesis (i.e., the hypothesis that the effects of the type of instruction depend on the amount of instruction). All other things being equal, you will have more power for testing the hypothesis that the type of instruction matters than for testing the hypothesis that the amount of instruction matters. You will have more power for testing either of these main-effect hypotheses than for testing the hypothesis that the amount and type of instruction interact. To understand why this is true, it is necessary to think concretely about the type of question ANOVA is designed to answer.

As we noted in Chapter 6, in some ways, the name "analysis of variance" is an unfortunate one, because it can lead people to forget what ANOVA was designed to do. The analysis itself focuses on identifying sources of variability or variance (hence, the name ANOVA), but the reason for doing the analysis is often to compare means. For example, when I ask whether the main effect for type of instruction is large, what I am really asking is whether the mean score of people who received lecture training is substantially different from the mean score of people who received hands-on instruction. Similarly, when I ask whether the main effect for amount of instruction is large, what I really want to know is whether the scores of people who received more instruction are substantially different from the scores of people who received less instruction. Interaction hypotheses in the design illustrated in Figure 7.1 involve comparisons of individual cell means.

A general principle running through the power analyses discussed in Chapters 1–2 is that there is more power when hypotheses are tested in large samples than when similar hypotheses are tested in smaller samples. That is, power increases with $N$. In ANOVA, the level of power for comparing means (i.e., for testing hypotheses about main effects and interactions) depends largely on the number of observations that goes into calculating each mean.

Suppose, for example, that 240 subjects show up for the study illustrated in Figure 7.1. Tests of the main effect for type of instruction will compare the mean score of the 120 people who receive lectures with the mean score of the 120 people who receive hands-on instruction. Tests of the main effect of amount of instruction will compare three means, each based on 80 subjects (i.e., 80 people received 2 days of instruction, 80 received 3 days, and 80 received 4 days). Tests of the interaction effect will involve comparing cell means, each of which is based on 40 people (e.g., 40 subjects received 2 days of lecture, another 40 received 3 days of hands-on instruction, and so on). All other things being equal, power is higher for statistical tests that involve comparing means from samples of 120 people (here, tests of the main effect of instruction type) than for tests that involve comparisons samples of 80 people (here, tests of the main effect of amount of instruction). Statistical tests that involve comparisons among means obtained from samples of 40 people (here, tests of the interaction) will often have even lower levels of power.

In ANOVA, it is common to distinguish between the number of observations in each cell of a design (designated by the lowercase $n$) and the total number of observations in a study (designated by the uppercase $N$). In the study described above, the means that define the interaction are based on $n = 40$ subjects, and the means that define the main effect for the type of instruction and amount of instruction main effects are based on samples of $3n$ and $2n$ subjects (i.e., 120 and 80), respectively. If all other things were equal, you would expect more power the larger the sample. Unfortunately, all things are rarely if ever equal, and it is difficult to state as a general principle that tests of main effects will always be more powerful than tests of interactions, or that main effect tests that involve comparing fewer means (and therefore fewer $df_{hyp}$) will be more powerful than main effect tests that involve comparing more means (and therefore more $df_{hyp}$). The reason for this is that the effect sizes for each of the separate questions pursued in ANOVA can vary. For example, suppose that there is a very large interaction effect, a moderately large main effect for the amount of instruction, and a very small main effect for the type of instruction. This might lead to the highest level of power for tests of the interaction and to the lowest level of power for tests of the type of instruction main effect.

## Estimating Power in Multifactor ANOVA

The process of estimating the power of tests of main effects and interactions in ANOVA is a straightforward extension of the processes described in Chapters 1 and 2. In particular, power depends on sample size, effect size, and the decision criterion. For example, assume that you expected a small main effect ($PV = .01$) for the type of instruction, a moderately large main effect ($PV = .10$) for the amount of instruction, and an equally large effect ($PV = .10$) for the interaction. If $N = 240$ (i.e., there are 240 subjects randomly assigned to the six cells in this design), you would expect power to equal approximately .30 for the type of instruction main effect (assuming $\alpha = .05$). You would expect power levels of approximately .99 for the amount of instruction main effect and the type by amount interaction. In other words, power levels would be quite low for testing some effects and quite high for others in this study.

### Estimating PV from F in a Multifactor ANOVA

In a one-way ANOVA, all of the variability in scores is broken down into two components, variability due to treatments and variability due to error. Once you know the value of $F$ in a one-way ANOVA, it is easy to determine the value of $PV$, which reflects the proportion of variance due to differences between treatments. In a multifactor ANOVA, $F$ does not, by itself, give enough information to allow you to calculate $PV$. For example, in the study illustrated in Figure 7.1, the $F$ for the type of instruction main effect tells you how large this effect is relative to $MS_{error}$, but it does not tell you how much of the total variance is accounted for by differences in types of instruction. In a multifactor ANOVA, the $PV$ for one main effect depends in part on the size of the other main effects and interactions.

If you have the values of each of the $F$ statistics (and their degrees of freedom) computed in a multifactor ANOVA, is often possible to solve for the percentage of variance explained by each of the effects (we will describe exceptions below). Returning to our example from the study illustrated in Figure 7.1, suppose you found the following:

Type of Instruction — $F(1, 234) = 3.64$
Amount of Instruction — $F(2, 234) = 15.19$
Type X Amount Interaction — $F(2, 234) = 16.41$

These three sets of $F$ values and degrees of freedom provide all of the information needed to determine the $PV$ for each of these three effects, as well as the $PV$ for the error term.

To see how this information can be used to calculate PV, we start by noting that for any effect in this design:

$$F = MS_{hyp}/MS_{error} \tag{7.3}$$

and

$$F * df_{hyp} = SS_{hyp}/MS_{error} \tag{7.4}$$

And finally, the percentage of variance ($PV$) explained by any effect in an ANOVA model (e.g., Target) is simply the sum of squares for that effect divided by the sum of squares total, or:

124    *Multifactor ANOVA Designs*

$$PV_i = SS_i / SS_{Total} \tag{7.5}$$

In a design where there are two main effects (label these **A** and **B**) and one interaction (labeled **AB**), it is easy to show that:

$$PV_{error} = df_{error} / [df_{error} + (F_A * df_A) + (F_B * df_B) + (F_{AB} * df_{AB})] \tag{7.6}$$

If you know the percentage of variance explained by error, it follows that the rest of the variance is explained by the model as a whole (i.e., by the combined effects of the **A**, the **B** effect, and the **AB** interaction). That is

$$PV_{model} = 1 - PV_{error} \tag{7.7}$$

Finally, once you know the combined effects of **A, B,** and **AB**, all that remains is to determine the percentage of variance explained by each. These values are given by:

$$PV \text{ for } A = PV_{model} * [(F_A * df_A) / ((F_A * df_A) + (F_B * df_B) \\ + (F_{AB} * df_{AB}))] \tag{7.8}$$

$$PV \text{ for } B = PV_{model} * [(F_B * df_B) / ((F_A * df_A) + (F_B * df_B) \\ + (F_{AB} * df_{AB}))] \tag{7.9}$$

$$PV \text{ for } AB = PV_{model} * [(F_{AB} * df_{AB}) / ((F_A * df_A) + (F_B * df_B) \\ + (F_{AB} * df_{AB}))] \tag{7.10}$$

We illustrate the calculation of *PV* values in the boxed section below.

---

### Calculating *PV* from *F* and *df* in Multi-Factor ANOVA: Worked Example

Earlier in the chapter, we presented the following data from a study that presented *F* tests but did not present *PV* values. That is, we presented:

Type of Instruction (A) — $F(1, 234) = 3.64$
Amount of Instruction (B) — $F(2, 234) = 15.19$
Type X Amount Interaction (A × B) — $F(2, 234) = 16.41$

Applying Equation 7.6, you find:

$$PV_{error} = df_{error}/[df_{error} + (F_A * df_A) + (F_B * df_B) + (F_{AB} * df_{AB})]$$
$$= 234/[234 + (3.46 * 1) + (15.1 * 2) + (16.41 * 2)]$$
$$PV_{error} = 234/300.48 = .78$$

Applying Equation 7.7, you find:

$$PV_{model} = 1 - PV_{error}$$
$$PV_{model} = 1 - .78 = .22$$

Applying Equations 7.8–7.10, you find:

$$PV \text{ for } A = PV_{model} * [(F_A * df_A)/((F_A * df_A) + (F_B * df_B)$$
$$+ (F_{AB} * df_{AB}))]$$
$$= .22 * [(3.64 * 1)/((3.64 * 1) + (15.1 * 2)$$
$$+ (16.41 * 2)]$$
$$= .22 * .055 = .012$$
$$PV \text{ for } B = PV_{model} * [(F_B * df_B)/((F_A * df_A) + (F_B * df_B)$$
$$+ (F_{AB} * df_{AB}))]$$
$$= .22 * [(15.19 * 2)/((3.64 * 1) + (15.1 * 2)$$
$$+ (16.41 * 2)]$$
$$= .22 * .46 = .10$$
$$PV \text{ for } AB = PV_{model} * [(F_{AB} * df_{AB})/((F_A * df_A) + (F_B * df_B)$$
$$+ (F_{AB} * df_{AB}))]$$
$$= .22 * [(16.14 * 2)/((3.64 * 1) + (15.1 * 2)$$
$$+ (16.41 * 2)]$$
$$= .22 * .48 = .10$$

These effect sizes provide the basis for power analysis and for minimum-effect tests.

It is easy to extend Equations 7.7 through 7.9 to designs with three, four, or more factors. There are only two real limitations to these equations. First, they can only be used with fully crossed factorial designs, where an equal number of subjects is assigned to every possible combination of levels of A, B, etc. Thus, equations of this sort will not allow you to easily compute $PV$ from $F$ in nested designs or incomplete designs. Corresponding calculations are possible, but laborious. Second, and more important, these equations apply only to fixed-effect models, in which all significance tests involve comparing the MS for each effect in the model to $MS_{error}$. Models that include random effects require more complex significance tests, and it is not possible to develop a simple set of formulas that cover all possible combinations of fixed and random effects. Luckily, fixed-effect models are much more common in the behavioral and social sciences than random-effects or mixed models (e.g., the default method for virtually all statistical analysis packages is to use fixed-effects models for structuring significance tests), especially in designs that do not include repeated measures.

## Factorial ANOVA from Means and Standard Deviations

Sometimes, researchers conduct studies in which an analysis of variance might be a very appropriate and helpful tool for analyzing the data, but they decide not to conduct (or to report) an ANOVA. If they present some basic descriptive statistics, it is often possible to reconstruct the main results that would have been obtained if ANOVA had been performed. Suppose, for example, that advertising researchers randomly assigns 120 subjects to one of three types of persuasive messages, built around the classic taxnomy first proposed by Aristotle that classifies persuasive messages in terms of ethos, pathos, or logos. Persuasive messages built around ethos focus on the credibility of the speaker. Persuasive messages built around pathos focus on the emotions evoked by the message. Persuasive messages built around logos focus on the logical appeal of the message. Each subject is also randomly assigned listen to messages designed to persuade them to either buy a product or sign up for a new course. Each subject rates the degree to which they were persuaded by the message. This study creates the factorial ANOVA design shown in Figure 7.2.

The researchers do not perform an ANOVA, but they do provide the means and standard deviations needed to reconstruct the results that would have been obtained if they had performed an ANOVA, and more important, to conduct power analyses for both traditional mull hypothesis tests and minimum effect tests. Suppose first that the researcher presents the overall mean and standard deviation of the persuasiveness ratings ($M = 5.0$, $SD = 3.2$). The researchers also present the graph shown in Figure 7.3, which shows the means for each condition.

| Type | Target | |
|---|---|---|
| | Product | Course |
| Ethos | | |
| Logos | | |
| Pathos | | |

n = 20 in each cell

*Figure 7.2* Study Design.

Multifactor ANOVA Designs 127

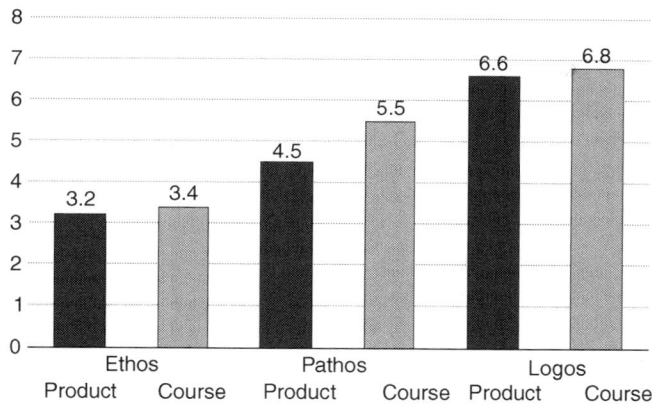

*Figure 7.3* Study Results.

*Table 7.1* The Shell of an ANOVA Table Showing Sources of Variance in This Design

| Source | df | Sum of Squares | Mean Squares | F | PV |
|---|---|---|---|---|---|
| Type | | | | | |
| Target | | | | | |
| Type X Target | | | | | |
| Error | | | | | |
| Total | | | | | |

From the means shown in Figure 7.3, we can calculate the means for ethos, pathos, and logos (3.3, 5.0, and 6.7, respectively) as well as for product and course (4.76 and 5.24, respectively). These cell means, row means, column means, and the standard deviation of the DV give you all of the information needed to determine the outcomes of an analysis of variance, assuming that the number of subjects in each cell is equal.

It helps to lay out a traditional ANOVA table, which gives considerable guidance regarding the information needed to calculate $F$ values and $PV$ values. In this study, you can start constructing an ANOVA table that lays out the sources of variance in this design, as shown in Table 7.1.

### Reconstructing ANOVA Results from Descriptive Statistics: A Worked Example

Our first step is to work out the degrees of freedom shown in Table 7.1. There are 120 subjects, so the total number of degrees of freedom is 119. There are three types of messages (pathos, ethos, logos), so the $df$ for message type is 2. There are two targets (product, course), so the *df* for the message target is 1. You obtain the *df* for the Type X Target interaction by multiplying the *df* for Type and Target and you obtain the *df* for error by subtraction (i.e., 119 − 2 − 1 − 2 = 114).

Next, note that the standard deviation of the persuasiveness rating (3.2) is the square root of the variance (10.24), and the variance multiplied by $N$ represents the Total Sum of Squares (i.e., 10.24 * 120 = 1,228.8).[1] The analysis of variance breaks this Total Sum of Squares into the parts that: (a) can be explained by message Type, message Target, and the Type X Target interaction; (b) cannot be explained by these effects, and is therefore error.

The Sum of Squares for Type can be obtained using the formula:

$$SS_{type} = n_i \left[ \begin{array}{l} (Ethos\ Mean - Overall\ Mean)^2 \\ + (Logos\ Mean - Overall\ Mean)^2 \\ + (Pathos\ Mean - Overall\ Mean)^2 \end{array} \right]$$

where $n_i$ equals the number of subjects who receive each type of message (here, 40). That is,

$$SS_{type} = 40 * [(3.3 - 5)^2 + (5 - 5)^2 + (6.7 - 5)^2]$$
$$= 40 * 5.78 = 231.20$$

Similarly, the Sum of Squares for Target can be obtained using the formula:

$$SS_{Target} = n_j * [(Product\ Mean - Overall\ Mean)^2$$
$$+ (Course\ Mean - Overall\ Mean)^2\}$$

where $n_j$ equals the number of subjects whose messages targeted a product or a course. That is,

$$SS_{Target} = 60 * [(4.76 - 5)^2 + (5.23 - 5)^2] = 60 * .052 = 3.17$$

Finally, the Sum of Squares for the Type X Target Interaction is given by:

$$SS_{Type\ X\ Target}$$
$$= n_{ij} * \sum (cell\ mean_{ij} - row\ mean_i - column\ mean_j + overall\ mean)^2$$

where $n_{ij}$ represents the number of subjects in each cell of this design (20). That is,

Table 7.2 Complete ANOVA Table for This Study

| Source | df | Sum of Squares | Mean Squares | F | PV |
|---|---|---|---|---|---|
| Type | 2 | 231.2 | 115.6 | 13.36 | .18 |
| Target | 1 | 3.17 | 3.17 | .36 | .002 |
| Type X Target | 2 | 8.53 | 4.26 | .49 | .006 |
| Error | 114 | 985.9 | 8.68 | | |
| Total | 119 | 1228.8 | | | |

$SS_{Type \ X \ Target} = 20 * (.4269) = 8.53$

With this information, we can fill in the ANOVA table, as is shown in Table 7.2. The Sum of Squares error is obtained by subtraction (i.e., 1228.8 − 231.2 − 3.17 − 8.53 = 985.9). Mean Squares are obtained by dividing Sums of Squares by their degrees if freedom. F statistics are obtained by dividing each of the Mean Squares by the Mean Square for Error, and PV values (eta squares) are obtained by dividing the Sum of Squares for each effect by the Sum of Squares total.

Table 7.2 shows that the Type effect is statistically significant. If you consult the One-Stop F Table in Appendix B, the observed F is larger than the F needed with $df = 2, 120$ to reject the traditional nil hypothesis with $\alpha = .05$ (3.07), the hypothesis that treatments account for less than 1% of the variance in the population (4.74), and the hypothesis that treatments account for less than 5% of the variance in the population (9.64). This makes sense because the effect of message type is reasonably large (i.e., $PV = .18$). In contrast, the effect sizes for Target and the Type X Target interaction are very small, and the F values are too small to allow you to reject any type of null hypothesis.

## Power Estimation

We can use the *df* and *PV* values to evaluate the power of this study and to make decisions about how to better design this study to achieve higher levels of power. First, with $df = 2, 114$ and $PV = .18$, here is plenty of power (.99) for detecting the message type effect. With a sample as small as 49, you would still have power of .80 for detecting this effect. Similarly, power equals .98 for testing the hypothesis that the effects of type of message is trivially small (i.e., message type accounts for 1% of the variance in ratings of persuasiveness or less) and equals .77 for a much more stringent minimum effect hypothesis. − i.e., that message type accounts for 5% or less of the variance in ratings of persuasiveness.

What about the target main effect? This effect is very small ($PV = .002$) and very large samples ($N = 3{,}920$) would be needed to have power of .80 to reject the traditional null hypothesis. If you wished to test a minimum-effect hypothesis, the likelihood of rejecting the null hypothesis that treatments account for 1% or less of the variance in ratings of persuasiveness would be very low. If $\alpha = .05$, there is a less that 2% probability of rejecting this null hypothesis *because the null hypothesis is, in this case, true*. That is, it is possible to make a Type I error, but the likelihood of doing so in this case is considerably smaller than the nominal alpha level. The same is true for the Type X Target interaction. The effect is very small ($PV = .006$), and the hypothesis that treatments have a trivial effect is true, meaning that the likelihood of rejecting this hypothesis is at most .05, and in this case, is once again considerably smaller (.029).

In tests of the hypothesis that the target or Type X Target effects are trivially small (i.e., that they account for 1% of the variance or less), it does not matter how large the sample gets, you will never have a probability greater than the alpha level (.05) of rejecting these hypotheses.

---

**Eta Squared vs. Partial Eta Squared**

In the analysis of variance, two effect size indicators are widely used. The first is eta squared ($\eta^2$). The value of eta squared for the **A** main effect in a **A** × **B** ANOVA is simply Sum of Squares for A/Sum of Squares Total (i.e., $SS_A/SS_{Tot}$), which is simply the percentage of the total variance explained, or $PV$. We find this to be a very useful effect size measure, but it is not the only game in town.

In a complex experiment, with many factors and many interactions, dividing up the total variance into all of its component parts might not always be informative, especially if the goal is to compare the importance of the **A** effect across experiments that vary in their complexity and in the nature and number of factors included. Some researchers prefer the partial eta squared ($\eta_p^2$), which is given by $SS_A/(SS_A + SS_{error})$. For example, the widely used statistical program SPSS reports the values of the partial eta squared, although it does not always make it explicit that it is a partial eta rather than a regular eta squared. The partial eta squared cannot be directly interpreted as a percentage of variance figure, and it is possible to have partial eat squared values in the same experiment that sum to more than 1.0, and for this reason, we are not wild about this specific effect size measure.

One other effect size measure is sometimes reported, omega squared ($\omega^2$). The omega squared is an estimate of the population variance accounted for. Unfortunately, it is difficult to calculate omega squared correctly in complex designs or in designs that include both fixed and random effects, and the most widely used formula

$[\omega^2 = (SS_A - (df_A * MS_{error}))/(SS_{Tot} + MS_{error})]$ only applies in the simplest fixed-effects designs.

If you consider the ease of computing and interpreting eta squared vs. partial eta squared or omega squared, there is no contest. Eta squared is a simple description of the percentage of variance accounted for by a particular variable or set of variables in a study, and we continue to find it the most useful effect size estimator.

## General Design Principles for Multifactor ANOVA

There are many variations on the multifactor ANOVA designs described here. In this chapter, we have focused on fully crossed designs, which have the following characteristics: (1) all combinations of treatments are included (for example, in Figure 7.1, both types of instruction are present in the 2-day, 3-day, and 4-day conditions) and (2) an equal number of subjects per cell. These two characteristics of a fully crossed design mean that it will be possible to estimate the main effects of type of instruction and length of instruction as well as the interaction between these two factors and that the main effects and interaction terms will all be independent. Here, independent means that each main effect and interaction explains unique variance, and that the combined variance explained by the two main effects and the interaction is exactly equal to $1 - PV_{error}$.

A fully crossed design is the simplest and most informative experimental design, in the sense that it allows you to cleanly separate the main effects, interaction, and error term, but it is also the most demanding, in terms of the total sample required. Depending on the research question you hope to answer, the fully crossed design might not be the most efficient one. The simplest alternative to a fully crossed design is to employ a nested design, in which you do not examine all possible combinations of factors. To illustrate a nested design, consider a research team that is interested in the effects of hospital type (urban vs. rural) and disease type (e.g., stomach cancer, prostate cancer, pancreatic cancer, and lung cancer) on the cost of post-surgery recovery. Two designs that might answer this question. A fully crossed design would examine all combinations of disease and hospital types and would look like the one shown in the first panel of Figure 7.4. The second panel of Figure 7.4 shows a potentially more efficient design that nests diseases within hospitals. This nested design has only half the cells as is needed for the fully crossed design, which means that with the same total N, the nested design will have twice as many subjects per cell.

The choice between fully crossed and nested designs depends on whether the researcher has a reason to believe that disease type interacts with hospital type. For example, suppose that you have reason to believe that recovery times for more invasive surgeries (e.g., lung) are not only longer

**Factorial Design**

|  | Stomach | Prostate | Pancreas | Lung |  |
|---|---|---|---|---|---|
|  |  |  |  |  | Urban |
|  |  |  |  |  | Rural |

**Nested Design**

| Urban | | | Rural | |
|---|---|---|---|---|
| Stomach | Prostate | Pancreas | Lung |
|  |  |  |  |

*Figure 7.4* Factorial vs. Nested Designs.

than those for less invasive types (e.g., prostate), but that the difference is larger at urban than at rural hospitals. This would imply that there are both main effects for disease and interactions between disease type and hospital type, and a nested design would not allow you to estimate these interactions or to separate them from main effects.

The choice of a fully crossed design makes sense when you have a good reason to believe that interactions are present, and even when you are not sure whether they are present but want to make sure that an undetected interaction does not cause you to misinterpret your findings. The choice of a fully crossed design also implies a commitment to large samples. In general, interaction effects tend to be small, both in experiments and in field studies (Aguinis, 1995, 2004). This is especially true for higher-order ($\mathbf{A} \times \mathbf{B} \times \mathbf{C}$) interactions. If you are seriously interested in interactions, in a context where you expect them to be small in the first place, and where they are based on comparisons between specific individual cells of a design, the numbers needed to conduct a mutifactor ANOVA with adequate power to detect interactions can be quite large. As we note in Chapter 8, one way to address the problem of the need for huge sample sizes is to employ research designs in which multiple observations are obtained from each subject.

To illustrate the sort of data required to conduct a powerful study that focuses on interactions, think about a study in which type of instruction (lecture vs. interactive), type of material (verbal vs. numerical), and frequency of class meetings (daily vs. one long meeting each week) are varied, and in which the core hypothesis is that if lectures are used for numerical content, daily meetings work best, but for numerical content of got interactive instruction, it does not matter how frequently you meet. This is a 3-way interaction hypothesis in a $2 \times 2 \times 2$ study. Assume also that you expect a moderately small effect (e.g., $PV = .03$). You will need over 225 subjects in this experiment. If you expect a small 3-way interaction (e.g., $PV = .01$), you will need over 725 subjects.

## Fixed, Mixed, and Random Models

Earlier in this chapter, we mentioned the concept of fixed versus random factors. The distinction between the two is not always clear, but it can have important implications for the analysis of variance and for power analysis. To illustrate the difference between fixed and random factors, consider an experiment in which subjects are randomly assigned to perform a simple motor task (e.g., keeping a cursor on a moving target) under various types of distraction. Two factors are varied: noise level (three different levels – 70, 90, 110 db) and room temperature (four different levels 60, 67, 80, 90 degrees). The data suggest that both noise and temperature (as well as the combination of the two) are likely to influence performance. The question is whether noise and/or temperature are fixed or random factors.

The best way to determine whether a factor is fixed or random is to read the discussion section of an article. A factor is fixed if the inferences that are made are restricted to the levels included in the study. So, if the inference is about the difference between 70, 90, and 110 db of noise, noise is a fixed factor. On the other hand, if the discussion says, "the more noise, the worse the performance", the inference goes beyond the levels of noise included in the experiment and uses these as a sample from a broader population of values that might have been included. If the researcher draws inferences about levels of noise not included in the experiment, noise is a random factor.

As this example implies, it is not necessarily the variable itself or even the design of an experiment that causes factors to be classified as fixed versus random. Sometimes, a variable will necessarily be fixed because all possible levels have already been included in the experiment (e.g., if gender is a factor in a design), but in most cases, it is the inferences drawn by the researcher and not the variables or the design that matters most.

The analysis of variance is simplest when all factors are fixed, and virtually all analyses in the social and behavioral sciences are conducted on this basis (fixed effects are the default assumption of virtually every piece of ANOVA software). Calculations for SSs, $dfs$, and MSs are the same for fixed, mixed, and random models (mixed models include both fixed and random factors), but there are important differences in the way significance tests are conducted. In particular, the use of random effects leads to questions about the meaning of error in the $F$ ratio. In a fixed-effect study, $F$ ratios are created by dividing the mean square of interest by the mean square for error. In studies that involve random factors, the choice of error terms for constructing $F$ rations depends on which factors are fixed and which are random.

If both factors in the study described above are treated as random factors, the $F$ ratio for the noise factor is given by $F = MS_{noise}/MS_{noise \times temperature}$ and the $F$ ratio for the temperature factor is given by $F = MS_{temperature}/MS_{noise \times temperature}$. That is, the denominator for the $F$ ratio for each of

*Table 7.3* Denominators for Fixed, Mixed, and Random Models

|  | Fixed | Mixed | | Random |
|---|---|---|---|---|
|  |  | Factor A Fixed<br>Factor B Random | Factor A Random<br>Factor B Fixed |  |
| Factor A | $MS_{err}$ | $MS_{AB}$ | $MS_{err}$ | $MS_{AB}$ |
| Factor B | $MS_{err}$ | $MS_{err}$ | $MS_{AB}$ | $MS_{AB}$ |
| AB Interaction | $MS_{err}$ | $MS_{err}$ | $MS_{err}$ | $MS_{err}$ |

Note: that all three models use $MS_{err}$ for the denominator in testing the **AB** interaction.

these random effects is the mean square for the noise × temperature interaction (the error term for testing this interaction is $MS_{err}$).

Table 7.3 shows what term is used as a denominator for fixed, mixed, and random models in experiments that involve two factors (**A** and **B**).

Because it is difficult to determine in advance the exact structure of the *F* test in research designs that might involve various mixes of fixed and random factors (the status of some factors as fixed or random might not even be determined until the data have been collected), it is difficult to develop power analyses for all variations of ANOVA designs. The power analyses developed here, and in other texts that discuss power analysis for ANOVA, typically assume that all factors are fixed, a common assumption, but one that not always made explicit in the behavior and social sciences.

## Note

1 The variance is the average of the squared deviations from the mean, so the variance multiplied by $N$ is the sum of these squared deviations.

# 8 Studies with Multiple Observations for Each Subject: Repeated-Measures and Multivariate Analyses

There are a several research designs that include repeated measures, sometimes in combination with between-subject factors. This chapter discusses variations on the multi-factor designs discussed in Chapter 7 that include repeated measures and ends with a discussion of the multivariate analysis of variance.

## Randomized Block ANOVA: An Introduction to Repeated Measures Designs

Several years ago, a colleague interviewed for a job at a university where animal learning researchers decided to break the mold. Instead of the usual laboratory with rats, pigeons, and other small animals, researchers had a laboratory where alligators tried to solve a flooded maze. It is a good bet that the researchers did not get more alligators every time a power analysis told them their $N$ was too small.

There are many research areas where it is virtually impossible to use large numbers of subjects, but where it is practical to obtain multiple observations from each subject. Sleep research often involves multiple nights in a large sleep laboratory, with a large number of observations each night. Vision research often involves hundreds of trials per subject. Although the advice that researchers should aim for as many subjects as possible applies even in these areas, so that researchers can generalize their conclusions to the wider population, a research design that involves a large number of participants is often beyond the resources of an investigator. Repeated measures designs allow researchers to obtain a large amount of information from a relatively small number of subjects, by obtaining several scores from each subject.

In Chapter 5 we discussed a one-way ANOVA study in which rats received a small reward, medium reward, or large reward for every trial in a straight alley runway. Each rat contributed one score for only one level of the independent variable. Suppose the design for the study is changed so that each of 50 rats experiences all three conditions. That is, a rat might receive small rewards for 30 days; on day 30 this rat is tested. Next, the rat might receive medium rewards for 30 days; on day 60, the rat is tested

DOI: 10.4324/9781003296225-8

Table 8.1 Sources of Variance in a Randomized Block Study

| Source | df | Comment |
|---|---|---|
| Rats | 49 | differences in the average performance across rats |
| Reward | 2 | differences in the average performance across conditions |
| Error | 98 | variability in scores not explained by rat or reward effects |

Note: With 50 rats and 3 data points per rat, the total number of observations is 150.

again. Finally, that rat might receive large rewards for 30 days and finally take one last test run. Each rat provides three scores for analysis, time to run through the straight alley after 30 days of each level of reward. In a good design, the order in which each rat experiences the three reward conditions would be randomized or counterbalanced from rat to rat, to control for practice effects.

Many books refer to subjects in a repeated measures study as blocks; the design of this experiment is often referred to as a randomized block design. This design allows you to measure individual differences and to remove variability from block to block (i.e., from rat to rat) from the error term, increasing statistical power. In particular, a randomized block design breaks down the variability of scores into variability that can be explained by differences between treatments, variability that can be explained by differences between blocks and error variance. In this study, the three sources of variance and their degrees of freedom are shown in Table 8.1.

The statistic $F = MS_{rat}/MS_{error}$ is used to determine if variability among rats is large relative to variability caused by error. The statistic $F = MS_{reward}/MS_{error}$ is used to determine if variability caused by treatments is large relative to variability for error. From the perspective of power analysis, there are at least two advantages to this randomized block design. First, it increases the total number of observations (i.e., there are only 50 rats, but 150 data points in the study). Second, it removes variability associated with systematic differences among the rats from the error term. If some rats just run faster than others, irrespective of reward levels, this variability in speed would represent another source of error in a simple one-way ANOVA. In a randomized block design, this variability is removed from the error term, which might dramatically increase the power of statistical tests that compare $MS_{reward}$ to $MS_{error}$.

This simple study illustrates some of the advantages of a repeated-measures design, as well as some of the complexities. First, power analysis suggests that the larger the $N$, the more power there is for a statistical test. In a between-subjects study, there is no ambiguity because $N$ refers to the number of subjects. In a study involving repeated measures, $N$ usually refers to the number of observations with several observations coming from each subject, and different ways of producing the same number of observations might have different implications for statistical power. For example, in this study, 50 rats each provided 3 observations. An alternative design might employ 3 rats, each of which provides 50 observations. These two designs provide the same number of observations (i.e., 150), but would not necessarily yield equivalent levels of power.

When considering a repeated measures design, the key concern is often whether a measurement taken at time 1 will directly influence a second measurement at time 2. For example, in a study with 3 rats, each of whom runs the maze 50 times, you would have to be concerned about the possibility that the very act of running the maze over and over would change the rat's response, because of exercise and experience, independent of any effects of reward. This might not be a big deal with 3 trials, but with 50 trials, it could have a substantial effect.

Sometimes, carry-over effects may rule out the use of repeated measures. For example, many studies involve some level of deception, or at least leave subjects in the dark about what is being studied, and once they are debriefed, their responses to the same stimuli or problems in the future might be substantially different. Other studies might involve an intervention that is designed to create a long-lasting change in subjects (e.g., teaching them a new skill) that cannot simply be undone if the intervention is repeated.

Sometimes there may be little direct carry-over, but there can be incidental effects that limit the possibility of collecting a many repeated measures. In our earlier example involving rats, there is likely to be some number of trials between 3 and 50 that induce real changes in the ability of rats to negotiate a maze, and there may be an upper limit on the number of observations that can be obtained without changing the phenomenon being studied.

## Independent Groups versus Repeated Measures

In situations where it is feasible to collect repeated measures, these designs can achieve relatively high levels of power with a smaller number of subjects and even a smaller number of observations than a comparable study that collects only one observation from each subject. Table 8.2 illustrates a

*Table 8.2* Number of Observations Required to Achieve Power = .80 (Maximum $d$ = .50, $r$ = .20)

| Between-Subject | | Repeated Measures | | |
| --- | --- | --- | --- | --- |
| Number of Treatments | Number of Subjects | Number of Treatments | Number of Subjects | Number of Observations |
| 2 | 128 | 2 | 53 | 106 |
| 3 | 237 | 3 | 65 | 195 |
| 4 | 356 | 4 | 74 | 296 |
| 5 | 485 | 5 | 82 | 410 |
| 6 | 624 | 6 | 88 | 528 |

key difference between a repeated measures study and studies that use between-subject designs in which each subject provides one observation – i.e., that you can achieve high levels of power with a relatively small number of subjects if it is possible to obtain multiple observations from each subject. Suppose studies to be compared use a .05 alpha level and a small to moderate treatment effect is expected (e.g., the maximum difference between treatment means is $d = .50$). Further assume there are relatively small correlations between repeated measures (i.e., $r$ values of .20 or lower). Table 8.2 shows the number of subjects that would be needed for a between-subjects study versus a repeated measures study to achieve power of 0.80 (Maxwell & Delaney, 1990).

A comparison of these two designs provides interesting information. First, if total number of observations is held constant, researchers obtain more power from the repeated measures than from a between-groups design. With two treatments, only 53 subjects with two observations per subject are required to provide power of .80. A between-groups design requires 128 subjects to achieve the same level of power. If there are four treatments, power of .80 is obtained with 74 subjects in repeated-measures and 356 subjects in a between-subjects design.

Suppose the correlation between observations is higher (e.g., the average correlation between observations is .40 rather than .20). As Table 8.3 shows, the higher the correlation between observations, the larger the statistical advantage repeated-measures designs provide. In Chapter 6, we noted that the correlation between observations was part of the formula for the standard error term in the dependent $t$-test, and that large correlations meant smaller standard errors. This principle carries over to repeated measures designs. For example, with three treatments, 50 subjects are required to achieve power in a repeated-measures design when the average correlation among measures is $r = .40$, whereas 65 subjects would be required when $r = .20$.

There are several ways to explain why repeated measures studies are more powerful than similar between-subjects studies. First, as a comparison

Table 8.3 Number of Subjects Required to Achieve Power = .80 (Maximum $d = .50$, $r = .40$)

| Between-Subject | | Repeated Measures | | |
|---|---|---|---|---|
| Number of Treatments | Number of Subjects | Number of Treatments | Number of Subjects | Number of Observations |
| 2 | 128 | 2 | 40 | 80 |
| 3 | 237 | 3 | 50 | 150 |
| 4 | 356 | 4 | 57 | 228 |
| 5 | 485 | 5 | 63 | 315 |
| 6 | 624 | 6 | 68 | 408 |

of Tables 8.2 and 8.3 suggests, correlation among pairs of repeated measures makes a big difference. The analysis of variance involves breaking down variability in scores into variance caused by treatments and variance caused by error. If observations for a given subject are consistently high or consistently low, there is a high correlation for each pair of treatments and less error variance. Usually, there is less random variability in data from a repeated measures design than in comparable independent groups design.

The decreased variability in repeated measures or correlated measures designs is analogous to the central idea in traditional psychometric theory. By grouping together several intercorrelated observations, it is possible to get a highly reliable measure of where a subject stands on the dependent variable. Repeated measures designs allow researchers to take advantage of the fact that several correlated observations from a subject are more reliable as compared to a single observation from a subject. In addition, repeated measures designs allow for the identification and removal of sources of variance in scores that are treated as error in a between subject design. In particular, these designs allow for the estimation and removal of systematic subject effects that cannot be estimated or controlled in between-subject designs.

Going back to an earlier example of a study in which subjects are asked to complete a complex psychomotor task under different sorts of distraction, noise (70, 90, 110 db) and room temperature (60, 67, 80, 90 degrees), suppose you have enough time and money to collect 120 observations. Table 8.4 shows the effects that can be estimated (along with their degrees of freedom) in a between-subjects study (using 120 subjects) and in a within-subjects study in which 10 subjects participate in all 12 conditions.

The repeated-measures design allows you to find out whether there are systematic subject effects (some subjects might be better at this sort of task than others), subject by noise interactions (some subjects may be more distracted by noise than others) and subject by temperature interactions (some subjects may be more distracted by hot or cold than others). These subject effects can all be estimated and removed from the residual error term. In contrast, between-subject designs lump all of these effects into

*Table 8.4* Sources of Variance in Between versus Within-Subjects Designs

| Between | df | Within | df |
|---|---|---|---|
| Noise (N) | 2 | Noise | 2 |
| Temperature (T) | 3 | Temperature (T) | 3 |
| N × T | 6 | N × T | 6 |
| Error | 108 | Subjects (S) | 9 |
| | | N × S | 18 |
| | | T × S | 27 |
| | | Error | 54 |

"Error". If there are meaningful subject effects and subject by treatment interactions, repeated measures designs will allow you to remove them from error, while between-subjects designs will treat these systematic effects as part of the overall error term. Anything you can do to reduce error is likely to increase statistical power, and one reason for the power of repeated measure designs is that they allow you to isolate these subject effects.

Finally, there is a more general statistical explanation for the power of repeated-measures designs. In Chapter 2, we discussed the noncentral $F$ distribution and noted that as the noncentrality parameter gets larger, the mean of the $F$ distribution increases and the variance grows larger as well. One thing that affects the noncentrality parameter is the effect size; it is this relationship that allowed us to create $F$ tables for minimum-effect $F$ tests. Thus, the larger the effect of your interventions and treatments, the stronger the upward shift in the distribution of $F$ values you expect to find in your study.

Vonseh and Schork (1986) note that in repeated measure studies the value of the noncentrality parameter depends on both the effect size and the correlations among repeated measures. The precise effects of the correlations among repeated measures on the distribution of $F$ are complex and nonlinear, but in general, the higher the correlation among your measures, the more noncentral the distribution of $F$ becomes. Repeated-measures studies tend to yield larger values of $F$, and therefore more power.

### *Why Doesn't Everyone Use Repeated-Measures Designs?*

Repeated-measures designs are more powerful, more efficient, and easier to implement than between-subjects designs. For example, suppose a power analysis tells you that 400 subjects are needed to obtain a reasonable level of power. You might go out and recruit 400 individuals, getting a single score from each one (a between-subjects design). Alternately, you might need to recruit only 10 subjects and get 40 observations from each (in fact, this will yield more power than a between-subjects design with $N = 400$). Given the practical and statistical advantages of repeated-measured designs, you might wonder why anyone uses between-subjects designs.

Although repeated-measures designs are attractive in many ways, they often turn out to be inappropriate for many research topics. Consider, for example, a study examining the effectiveness of five different methods of instruction for teaching second-grade students basic mathematics. A repeated-measures design that exposes all students to all five methods will run into a serious problem – i.e., carry-over effect. That is, whatever the students learn with the first method they are exposed to will affect their learning and performance on subsequent methods. By about the fourth time students have gone over the material (using each of four methods), they probably will have learned it, and will certainly be getting sick of it. The first four trials will almost certainly affect outcomes on trial number five.

In general, many research questions make a repeated-measures design difficult to implement. Studies of learning, or interventions designed to change attitudes, beliefs, or the physical or mental state of subjects might all be difficult to carry out in a repeated-measures framework. If the interventions you are studying change subjects, it might be difficult to interpret data when several different interventions are implemented, one after the other. The fact that human beings have good memories probably compromises most repeated-measures designs using human participants to some extent. Similarly, some research paradigms involve tasks that are complex or time consuming, and even if it is theoretically feasible to repeat the task under a variety of conditions, there may be serious practical barriers to obtaining multiple observations from each subject. Some studies allow for mixed designs, in which some of the factors studied represent repeated measures and others represent between-subject factors, and these designs often represent the best compromise between the power and efficiency of repeated-measures designs and the independence of observations that characterizes between-subject designs.

One of the apparent strengths of repeated-measures designs is also a potential weakness – i.e., the ability to obtain manyobservations from a small number of subjects. For example, in some areas of physiological psychology, it might be common to obtain several hundred observations from each subject, and it is not unusual to see studies where the number of subjects is 10–20, and sometimes samples are even smaller. These are not small-sample studies in the traditional sense, because the number of observations is potentially huge (e.g., in some areas of vision research, studies in which 10 subjects each provide 400 trials are common). However, the small number of subjects does raise important concerns about the generalizability of your results. Even if true random sampling procedures are used, you have to be concerned about the possibility that the results obtained from one group of 10 subjects might be quite different from those obtained from another group of 10. In between-subject studies, you might not have the same set of worries. A random sample of 400 subjects is almost certain to provide results that generalize to the parent population, and a second random sample of 400 is almost certain to provide converging findings. In some cases where it is feasible to obtain hundreds of observations from each of a few subjects, you might conclude that it is better to obtain a handful of observations from a larger set of subjects.

## Complexities in Estimating Power in Repeated-Measures Designs

The application of ANOVA (and of power analysis) in repeated-measures designs can be complicated because these designs lead to violations of important statistical assumptions that underlie the analysis of variance.

In particular, research designs that involve obtaining multiple measures from each respondent can lead to violations of assumptions of independence or sphericity (i.e., the assumption that the variances of the differences between all possible pairs of repeated measures are equal). To obtain accurate results in such designs, both the degrees of freedom and your estimate of the noncentrality parameter must be adjusted by a factor "epsilon" ($\varepsilon$), which reflects the severity of violations of the assumption of sphericity [e.g., the best estimate of the noncentrality parameter ($\lambda$) in repeated-measures designs is given by ($\varepsilon * df_{err} * PV/(1 - PV)$); see Algina and Keselman, 1997, for discussions of sphericity and power]. However, this correction factor is not always reported, and cannot often be calculated based on the results that are likely to be reported in a journal article.

A conservative approach to this problem is to make a worst-case correction (Greenhouse & Geisser, 1959). The factor epsilon ranges in value from a maximum of 1.0 (indicating no violation) to a minimum of $1/(k-1)$, where $k$ represents the number of levels of the repeated-measures factor. When the degrees of freedom for factors involving repeated measures are multiplied by epsilon, it is possible to obtain a conservative test of significance by comparing the obtained $F$ to the critical value of $F$ using the epsilon-adjusted degrees of freedom.

For example, in this study, data are collected from each subject on three occasions, meaning that Occasion is a repeated-measures factor, and the Group X Occasion factor has a repeated-measures component. The worst-case estimate of epsilon is that epsilon = .5 [i.e., epsilon = $1/(3-1)$]. To use this worst-case estimate of epsilon here, you would multiply the degrees of freedom for the Occasion effect and the Occasion X Group effects by .5 (i.e., you would use degrees of freedom of $df_{hyp}$ = 1, $df_{err}$ = 57 and $df_{hyp}$ = 2, $df_{err}$ = 114, respectively, to test the Occasion and Occasion X Group effects rather than the actual degrees of freedom of $df_{hyp}$ = 2, $df_{err}$ = 57 and $df_{hyp}$ = 4, $df_{err}$ = 114). Similarly, you would need to adjust the estimate of $\lambda$, also multiplying your estimate of $\lambda$ by .5.

In practice, a good estimate of power can be obtained by simply multiplying both the degrees of freedom and $PV$ by this worst-case estimate of epsilon (rather than directly adjusting your estimate of $\lambda$). If you compare the power estimate you obtain without any epsilon correction (the power estimates in this example did not include any correction for violations of the sphericity assumption) to the power estimate obtained making a worst-case assumption about violations, you will have a pretty good idea of the range of power values you could reasonably expect.

## Mixed Designs: Split-Plot Factorial ANOVA

A good deal of the early development of the analysis of variance happened in areas related to agricultural research, and this is sometimes the best explanation for the terminology still used in ANOVA. The split-plot factorial

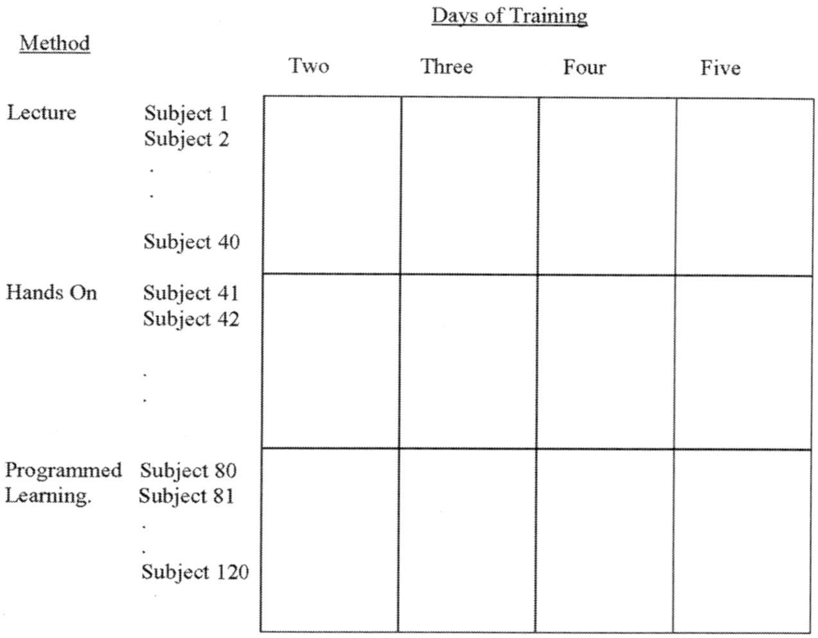

| Method | | Days of Training | | | |
|---|---|---|---|---|---|
| | | Two | Three | Four | Five |
| Lecture | Subject 1<br>Subject 2<br>.<br>.<br>Subject 40 | | | | |
| Hands On | Subject 41<br>Subject 42<br>.<br>.<br>. | | | | |
| Programmed Learning. | Subject 80<br>Subject 81<br>.<br>.<br>Subject 120 | | | | |

*Figure 8.1* A Split-Plot Factorial Design.

design was originally developed to help analyze data obtained when plots of agricultural land were divided for various uses.

Suppose a researcher randomly assigns each of 120 students to one of three methods of instruction (lecture, hands-on, or programmed instruction) and assesses their performance after two, three, four, and five days of training, as shown in Figure 8.1. Methods of instruction represents a between-subjects factor; 40 different students are assigned to each level of this factor. Training time, on the other hand, is a repeated-measures factor. That is, each student is tested four times. In a split-plot design, each participant is exposed to one level of the between-subjects factor and to all levels of the within-subjects factor.

One of the distinguishing features of the split-plot factorial design is that it allows you to estimate variability that is due to systematic subject differences. However, because each subject receives only one method of instruction, subject differences are partially confounded with group differences. In analyzing data from this design, we use the concept of nesting and analyze variability attributable to systematic subject differences *within groups*. The analytic layout and degrees of freedom for this design are shown in Table 8.5. Note that there are 480 total observations (i.e., 120 subjects, 4 observations per subject), and that there is no specific term

Table 8.5 Sources of Variance, df, and Error Terms for Split-Plot Design

| Source | df | Error Term for F Statistic |
|---|---|---|
| Methods of Instruction (M) | 2 | $MS_{S/M}$ |
| Subjects within Methods (S/M) | 117 | |
| Training Time (T) | 3 | $MS_{S/M \times T}$ |
| M × T | 6 | $MS_{S/M \times T}$ |
| S/M × T | 351 | |
| TOTAL | 479 | |

that is labeled "error". Rather, different terms in this model are used in constructing specific $F$ statistics.

In the split-plot factorial design, tests of the significance of the between-subjects factor use variability attributable to subjects, nested within levels of that factor, as a basis for constructing the $F$ statistic. In particular, the significance test for the methods of instruction factor compares the mean square for methods with the mean square for subjects nested within methods (i.e., $F = MS_M/MS_{S/M}$). The rationale here is that all the subjects within each training method received the same treatment. Therefore, variability within these treatment groups cannot be due to treatments, but rather must be due to non-treatment factors, which are treated as sources of error.

Significance tests for repeated-subject factors (or for interactions that include repeated subject factors) use the interaction between subjects and treatments as a basis for forming the $F$ test. For example, the significance test for the training time factor compared variability attributable to training time with variability that is explained by the interaction between subject differences and training time (i.e., $F = MS_T/MS_{S/M \times T}$). Similarly, the $F$ test for the methods of instruction by training time interaction involves computing $F = MS_{M \times T}/MS_{S/M \times T}$.

You might notice in Table 8.5 that the number of degrees of freedom for subjects nested within methods of instruction has a value that does not seem intuitively obvious (i.e., df = 117). Because subjects are nested within three different methods of instruction, the best way to determine degrees of freedom is to calculate the df within each of the three training methods (40 subjects receive each training method, so there are 39 degrees of freedom within each method), then multiplying by the number of methods. This yields df = 3 * 39 = 117. The degrees of freedom for the interaction between training time and subjects nested within methods of instruction is given by multiplying 117 by the number of methods (i.e., df = 3 * 117 = 351).

In other ANOVA designs, power has been affected by the degrees of freedom for both the hypothesis and the error term, and the same is true for split-plot factorial designs. That is, in evaluating power, the first step is to compute $df_{hyp}$ and $df_{error}$.

## Estimating Power for a Split Plot Factorial ANOVA

As in other ANOVA designs, power depends on ES, alpha, $df_{hyp}$, and $df_{error}$, and $PV$ for each main effect and interaction. Suppose, for example, that you decide to test minimum-effect hypotheses, testing the null hypothesis that the main effects and interactions account for 1% or less of the variance in responses. On the basis of prior experience and existing research, you believe that methods of instruction will have a relatively small effect (e.g., $PV = .05$) and that training time and the M × T interaction will have a moderately large effect (e.g., $PV = .10$) on responses. To assess the power of the test of the methods of instruction main effect, use our Shiny Web app and enter .05 for the assumed effect size, .01 for the minimum effect (i.e., this is a test of the the hypothesis that methods of instruction account for 1% of the variance or less), .05 for alpha, and $df_{hyp} = 2$, $df_{err} = 117$. In this study, power for the methods of instruction main effect is equal to .35. To obtain power of .80, you will need approximately 400 subjects.

To assess power for the time of training effect, enter .10 for the assumed effect size, .01 for the minimum effect (i.e., this is a test of the hypothesis than methods of instruction account for 1% of the variance or less), .05 for alpha, 01. In this study, power for the time of training main effect is a bit larger than .99. The power for studying the M × T interaction is also .99, because the assumed effect size (i.e., $PV = .10$) and the degrees of freedom are the same for each test.

We should note that the results for time and M × T effects are likely to be underestimates of the true power because each of these effects involves repeated measures. As we explain below, the degree to which power is underestimated depends substantially on the correlations among repeated measures.

## Power for Within-Subject versus Between-Subject Factors

Studies that use repeated measures usually have more power than comparable between-subjects studies, but there are several factors that influence this generalization, and an exact description of the effects of including repeated measures on power can be complex. First, as Bradley and Russell's (1998) analysis of split-plot designs shows, the use of repeated measures tends to increase power for some statistical tests and decrease power for others. In general, tests of repeated-measures factors have higher power and tests of between-subjects factors have lower power than they would in comparable studies that relied exclusively on between-subjects designs. The increase in power for repeated-measures factors is proportional to the square root of $(1 + \rho)$, where $\rho$ represents the average correlation between repeated measures, whereas the decrease in power for between-subject factors in split-plot designs is proportional to the square root of $(1 - \rho)$. If the correlation between measures is small the effects of using repeated measures on the power of either type of test will be small.

Second, the effects of adding additional repetitions of measures to a study can often be smaller than you might think (Overall, 1996). In particular, adding a new set of measures does not always add much new information, and as the redundancy of new information increases, the contribution of additional measurements to the power of a study decreases. This effect is comparable to one seen in psychometrics, where adding a few new items to a short scale can dramatically increase reliability, whereas adding the same number of new items to a longer scale may have no discernable effect.

One result of the effects of correlations among observations on power analysis is that the tables and software provided with this book will tend to produce conservative estimates of power and of sample-size requirements for repeated-measures factors. On the whole, we think it is best to err on the side of designing studies with more rather than with less power, so the relatively small bias that can be produced as a result of failing to take the correlations among observations into account when calculating power does not strike us as a bad thing.

## Split-Plot Designs with Multiple Repeated-Measures Factors

Suppose 120 subjects are randomly assigned to one of three exercise programs (e.g., strength-oriented, cardiovascular, combined). Forty subjects in each program follow two different diet regimens (high protein, high fiber), and three different cycles of exercise and rest (massed exercise, frequent short breaks, fewer but longer breaks). As with the simpler slit-plot design illustrated in Figure 8.1, this design includes both a between-subjects factor (exercise) and within-subjects factors (diet, break cycle).

Table 8.6 illustrates the sources of variance, the error terms for testing each source, and the degrees of freedom. In general, power analysis for this design proceeds very much in the same way as power analyses for the slit-plot design illustrated in Figure 8.1 and in Table 8.5. Estimates of effect sizes are still needed, and once the effect size and degrees of freedom are established, power estimation proceeds in the same way as in the other designs described in Chapters 7 and 8. That is, power always depends on the effect size, the decision criterion and the degrees of freedom. A final factor that needs to be considered is the correlation among the repeated measures. Our **R** code or Shiny Web app will underestimate the power of statistical tests that involve repeated measures (i.e., an effect that includes diet or break cycles, such as the D, D × B, E × B, E × D effects, etc.).

Table 8.6 Split-Plot Design with Two Repeated-Measures Factors

| Source | df | F Test |
|---|---|---|
| Exercise | 2 | $MS_E/MS_{S/E}$ |
| Subjects/Exercise | 117 | |
| Diet | 1 | $MS_D/MS_{S/E \times D}$ |
| Break Cycle | 2 | $MS_B/MS_{S/E \times B}$ |
| D × B | 2 | $MS_{D \times B}/MS_{S/E \times D \times B}$ |
| E × D | 2 | $MS_{E \times D}/MS_{S/E \times D \times B}$ |
| E × B | 4 | $MS_{E \times B}/MS_{S/E \times D \times B}$ |
| D × E × B | 4 | $MS_{D \times E \times B}/MS_{S/E \times D \times B}$ |
| S/E × D | 117 | |
| S/E × B | 234 | |
| S/E × D × B | 234 | |

The degree to which power estimates are conservative will depend on the level of correlation among repeated measures.

## The Multivariate Analysis of Variance

Throughout this book, we have concentrated on a family of statistical methods that all have one feature in common — i.e., that there is some specific variable that represents the focus of the analysis, the dependent variable. In many studies, researchers are likely to collect data on several dependent variables and are likely to conduct multivariate rather than univariate analyses. For example, suppose you are interested in determining whether there are differences in the outcomes of four specific training programs that have been proposed for teaching pilots to use a new global positioning technology. You might evaluate their performance in terms of the number of errors made in applying this technology, in terms of the amount of time needed to learn the technology or in terms of increased efficiency in planning and sticking to flight routes. Rather than conducting three separate analyses of variance (i.e., one for errors, one for time spent and one for flight planning), you are likely to use multivariate methods that combine these three dependent variables into a single analysis. Specifically, you are likely to carry out a multivariate analysis of variance (MANOVA).

There is a substantial literature dealing with power analysis in MANOVA (e.g., Maxwell & Delaney, 1990; Stevens, 1980, 1988, 2002; Vonesh & Schork, 1986). In a very general sense, the issues in determining the power of MANOVA are quite similar to those that affect the power of univariate ANOVAs. First, the power of MANOVA depends on $N$, the effect size and the alpha level, just as in univariate ANOVA. Large samples, strong effects, and lenient alpha levels lead to high levels of power, whereas small samples, small effects, and stringent alpha levels lead to lower levels of power. Second, a principle already noted for repeated-measures ANOVA also applies to MANOVA. In general, power is higher when the variables being examined (here, the multiple dependent variables that are combined to carry out the MANOVA) are highly correlated than when they are uncorrelated. Finally, the power of MANOVA depends on the number of dependent variables. A study that uses four or five dependent measures will tend to have more power than a study that uses one (leading to a univariate ANOVA) or two. Stevens (1988, 2002) presents useful tables for estimating power in MANOVA.

A simple, if somewhat conservative, approach to power analysis in MANOVA is to conduct a power analysis for the dependent variable that is expected to produce the smallest effect. If you have power of .80, for example, in a univariate ANOVA for this DV, you can be sure that you will have power greater than .80 for a MANOVA that combines this variable with other dependent variables for which stronger effects are expected.

Finally, it is useful to note that MANOVA can be used as a substitute for many repeated-measures analyses, ranging from dependent $t$-tests (when there are only two measures) to factorial repeated-measures designs. If there are several dependent variables, MANOVA presents a more powerful alternative to ANOVA or the $t$-test. In general, MANOVA will have slightly more power than would be obtained when analyzing the same data using a repeated-measures design (i.e., one that treats the dependent variables as a set of repeated measures).

# 9 Power Analysis for Multilevel Studies

It is common to collect data from individuals who are part of some larger group or unit. For example, in educational research, you might collect data from 150 children grouped into five classrooms. In organizational research, you might collect data from 500 employees in eight separate organizations. This sort of hierarchical data structure creates many potential problems and ambiguities in data analysis, but it also presents opportunities for answering important questions about why scores vary (e.g., how important differences are between schools vs. differences between individuals).

Consider our education example. Suppose you find that there is substantial variability in educational attainment. This might reflect differences between children, but it might also reflect differences between classrooms. Perhaps some classrooms have better teachers, better resources, or even students who, on average, are better prepared to do well. Perhaps some classes are taught in such a way that students get more benefit from specific activities or teaching strategies. Suppose further that what works well in one classroom might not work as well in another. A data collection design in which there are many individuals in each of several organizations or classrooms allows you to ask questions that are sometimes ignored (e.g., do classroom differences matter), but it also creates a series of important challenges.

The biggest challenge in the analysis of data of this sort is that the data might violate one of the common assumptions of regression or other similar methods of data analysis – i.e., the assumption that observations are independent. We usually assume that one child's score is independent of the scores achieved by other children, but it is possible that students within a classroom are more similar in performance than children in different classrooms. This is especially likely if the level of ability or performance varies substantially from one classroom to the next. This potential lack of independence leads to inaccurate estimates of error variance and, therefore, to inaccurate statistical tests.

This chapter deals with power analysis for multilevel analyses – i.e., analyses where data can be classified in terms of multiple, hierarchically-organized levels of analysis, such as the analysis of students who are drawn from (and therefore nested within) different classrooms or employees from

DOI: 10.4324/9781003296225-9

many different organizations. These analyses are referred to in different ways in particular research literatures, including hierarchical linear modeling and random coefficient modeling, but we find the term "multilevel" the most useful and the most descriptive. A large literature has grown describing the use of multilevel models (e.g., Aguinis, Gottfredson & Culpepper, 2013; Bikel, 2007; Bliese & Hanges, 2004; Hox, Moerbeek & van de Schoot, 2018; Raudenbush & Bryk, 2002), and numerous papers and chapters describing the application of power analysis to multilevel modeling have been published (Mass & Hox, 2005; Meink & Vanderplass, 2012; Scherbaum & Ferreter, 2009; Snijdgers, 2005).

## What Do Multilevel Analyses Tell You?

Suppose 150 children from five classrooms participate in a study that looks at the links between time spent reading (X) and academic achievement (Y). The data are organized into two levels: (1) Level 1 – students within classrooms, and (2) Level 2 – classrooms.[1] Multilevel analyses ask two big questions: (1) what is the relationship between time spent reading (X) and academic achievement (Y)? and (2) is this relationship similar or different in different classrooms? Figure 9.1 illustrates several different ways this analysis might turn out. The simplest possibility (panel 1) is that the relationship between X and Y is essentially identical regardless of classroom, and in this case, it would probably be safe to ignore the fact that students are nested within classrooms in most analyses. Another possibility (panel 2) is that the slope of the relationship between X and Y is the same across classrooms, but because of differences in the means and/or X and Y, the intercepts differ.[2] Still another possibility (panel 3) is that the intercepts are similar but that the relationship between X and Y varies across classes. Finally (panel 4), it is possible that the slopes and intercepts differ from one classroom to the next.

As Figure 9.1 suggests, questions about slopes have to do with the possibility that the very nature of the relationship between X and Y changes as you move from one classroom to the next. Changes in the slope from one classroom to the next might suggest that in some classrooms there is real payoff for spending more time reading but in others time spent reading has little to do with achievement. This type of change in the relationship between X and Y is often referred to as a "moderated" relationship. As we noted in Chapter 5, there is a substantial number of studies dealing with moderator variables and methods of detecting them, most of which deal with simple-level data rather than multilevel analyses, but some firm conclusions can be drawn from this literature. Most important, moderator effects, to the extent that they exist at all, tend to be very small (Aguinis, Beaty, Boik & Pierce, 2005; Murphy & Russell, 2017; O'Boyle, Banks, Carter, Walter & Yuan, 2019). As a result, the statistical power of moderator studies tends to be very low, and relatively large samples (often 500–3,000) may be needed to attain power of .80 or more for detecting

Power Analysis for Multilevel Studies   151

1. The relationship is the same in all classrooms

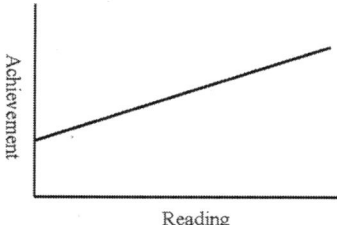

2. Similar slopes, but different intercepts

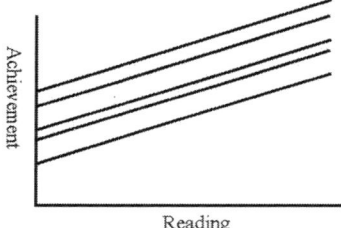

3. Similar intercepts but different slopes

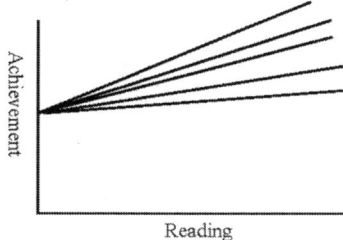

4. Different intercepts and slopes

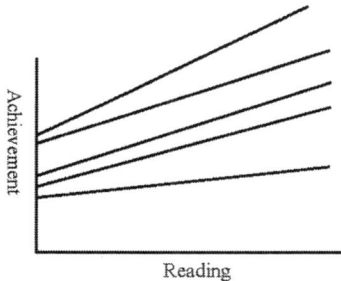

*Figure 9.1* Possible Patterns of Relationships Between Time Spent Reading and Achievement.

moderator effects. There have not yet been enough large-scale meta-analytic reviews of multilevel studies to determine the likelihood that slopes will vary across organizations, classrooms, or other higher-level groupings, but the large literature showing that moderator effects are typically small leads us to believe that meaningful variation in slopes is not likely in many multilevel studies.

Questions about differences in intercepts, on the other hand, are mainly concerned with the possibility that the students in some classrooms spend more time reading than others (i.e., that $\bar{X}$ differs) and/or that they end up performing better than others (i.e., that $\bar{Y}$ differs). That is, differences in intercepts do not imply that the fundamental relationship between X and Y differs across classrooms, but rather that classrooms differ in their averages on X and/or Y. On the whole, intercept differences are probably more likely to be found in multilevel studies than slope differences.

In this chapter, we emphasize a step-by-step procedure to evaluate the possibility that allowing intercepts, slopes, or both to vary across groups might increase the predictive power of our models. In a later section of this chapter, we use this model-comparison process to structure power analysis for tests of both the nil hypothesis and of the hypothesis that allowing slopes, intercepts, or both to vary might lead to trivial increases in the fit of the prediction model. If you read the current research literature, in which multilevel models are used to analyze data, you will find that this step-by-step approach is not necessarily the most common approach to analyzing multilevel models. Instead, many researchers appear to fit the most complex model and rely on the interpretation of significance tests for individual parameters in the model to guide their interpretation of the data. We believe this is a mistake. First, the most complex model is often not appreciably better than simpler alternatives, and the fact that you *can* fit a complex model to data is rarely a good argument that you *should* fit one. Second, as models become more complex, researchers can easily lose sight of what the data mean (Murphy, 2021). Our bias is to keep models as simple as possible and to introduce complexities such as slopes or intercepts that vary across organizations only when necessary.

## The Multilevel Equation

To illustrate how multilevel analyses are structured, let's expand the scope of the study described in the previous section, looking at two predictors of academic success, time spent reading (X1) and time spent on math problems (X2).

In our multilevel example, we can start by creating separate model models for studying students within classrooms (Level 1) and for studying the effects of classrooms (Level 2). The Level 1 model is built to examine the link between reading time (X1), math problems (X2), and performance in each classroom. This Level 1 model takes the form of a regression model:

$$Y_{ij} = \beta_{0j} + \beta_{1j}X_{ij} + \beta_{2j}X_{2j} + \varepsilon_{ij} \qquad (9.1)$$

Where:

$Y_{ij}$ is the achievement level of student $i$ in classroom $j$
$\beta_{0j}$ is the intercept in classroom $j$
$\beta_{1j}$ is the regression weight for variable X1 (reading time) in classroom $j$
$\varepsilon_{ij}$ is a residual (part of Y not explained by X) or error term that is normally distributed with a mean of zero and the variance of $\sigma^2$

The Level 1 equation is designed to predict or account for the achievement level of individuals in each group. The Level 2 equation attempts to account for possible differences in the intercept and slopes of the Level 1 equation across classrooms. For example, suppose that some classrooms have better resources (Z) than other classrooms, such as better computers, or more teacher aids. These might help you predict and make sense of classroom differences in both intercepts and slopes. Thus, the intercept in each group can be modeled, using Equation 9.2:

$$\beta_{0j} = \gamma_{00} + \gamma_{01}Z_j + U_{0j} \qquad (9.2)$$

Similarly, the slopes for reading and math can be estimated using Equations 9.3 and 9.4:

$$\beta_{1j} = \gamma_{10} + \gamma_{11}Z_J + U_{1j} \qquad (9.3)$$

$$\beta_{2j} = \gamma_{20} + \gamma_{21}Z_J + U_{2j} \qquad (9.4)$$

Equations 9.2, 9.3, and 9.4 use characteristics of each classroom (here, resources) to predict the slopes and intercepts when reading and math time are used to predict achievement in that classroom. So, the intercept in each classroom ($\beta_{0j}$) is a combination of the overall intercept for the sample of 150 children ($\gamma_{00}$, or gamma$_{00}$), a group effect ($\gamma_{01}$) which serves as a coefficient or weight for the classroom resources variable $Z_j$ (in multilevel models, regression coefficients at Level 1 are symbolized by $\beta$ while coefficients at Level 2 are symbolized by $\gamma$) and an error term ($U_{0j}$), which represents variation in Level 1 intercepts not captured by the grouping variable (i.e., resources). $U_{0j}$ is normally distributed with a mean of zero and a variance of $\tau_{00}$ (tau$_{00}$).

The X1-Y slope in each classroom ($\beta_{1j}$) is a function of the overall slope for the 150 students ($\gamma_{10}$) and the way resources in that classroom affect the X1-Y slope ($\gamma_{11}$), plus the error term ($U_{1j}$), which is normally distributed with a mean of zero and a variance of $\tau_{11}$. The equation for the X2-Y slope has the same basic structure.

## 154 Power Analysis for Multilevel Studies

*Table 9.1* Common Terms and Symbols in Multilevel Analysis

---

Level 1 – refers to individuals within each of the units included in your study
Level 2 – refers to higher-level units, such as classrooms or organizations
Level 1 terms
  $\beta$ – a coefficient. $\beta_{0j}$ is the intercept in group j, $\beta_{1j}$ is the regression coefficient linking X and Y in group j
  $\varepsilon$ – error term for each classroom
  $\sigma^2$ – error variance within each classroom
Level 2 terms
  $\gamma$ – a coefficient. $\gamma_{00}$ is the intercept for the total population; $\gamma_{10}$ the overall slope for the population; $\gamma_{0j}$ is the coefficient for the group effect for intercepts, which indexes how much intercepts differ; and $\gamma_{11}$ is the group effect for slopes
  U – error terms for Level 2 equations
  $\tau$ – variance of errors for Level 2 equations
General Terms
  Fixed Effect – effects that are the same for all individuals, such as the population intercept and slope
  Random Effect – effects that vary across groups of units, such as different intercepts or slopes for different groups
  Variance Components – Level 1 variance ($\sigma^2$) and Level 2 variances ($\tau_{00}$, $\tau_{11}$ – variance in slopes and intercepts from group to group, respectively) and covariances ($\tau_{10}$ – covariance between group intercepts and group slopes)

---

For students familiar with regression, multilevel analysis might require a substantial learning curve, in part because the two methods sometimes use different symbols for conceptually similar things. Table 9.1 summarizes some of the more common symbols encountered in multilevel analysis and their meaning. For new users, the terminology can be confusing, but with some practice you will see that there is some value to, for example, clearly identifying Level 1 vs. Level 2 coefficients or error terms.

Unlike multiple regression, which uses Ordinary Least Squares (OLS) regression to solve for slopes and intercepts, and which uses $F$ tests to evaluate the significance of $R^2$ or changes in $R^2$, multilevel analyses normally use iterative procedures for solve for the coefficients in Equation 9.4, and they rarely use $R^2$ or $F$ to evaluate the performance or the fit of various models.[3] This creates some challenges to applying the models developed and illustrated in Chapters 1–8, which rely on the $F$ distribution and the non-central $F$ distribution, but as we show in this chapter, the concepts laid out in Chapters 1–8 generalize well to multi-level modeling, which usually relies on the chi square statistic rather than $F$ to evaluate significance.

Although it makes sense to discuss the Level 1 and Level 2 equations separately, multilevel modeling combines them into a single equation and solves everything at once. Equations 9.1, 9.2, 9.3, and 9.4 can be combined into a single equation that represents the full multilevel model:

$$Y_{ij} = \gamma_{00} + \gamma_{10} reading_{ij} + \gamma_{20} math_{in} + \gamma_{01} resources + \gamma_{11} resources \times reading$$
$$+ \gamma_{21} resources \times math + U_{1j} reading + U_{2j} math + U_{0j} + \varepsilon_{ij}$$
(9.5)

If you are new to multilevel models, the material laid out in this section can seem overwhelming. It helps substantially to keep Bikel's (2007) advice in mind – *it's just regression*. That is, multilevel analysis may look very different from what you have done in studies using ordinary least squares (OLS) regression, but the core ideas are very much the same. Multilevel analysis focuses on two main questions: (1) are your predictors related to Y and (2) are those relationships similar or different from classroom to classroom or organization to organization?

## Are Multilevel Models Necessary? – The Intraclass Correlation

The fact that data are collected from individuals in different classrooms, organizations, or other types of clusters does not mean that these higher-level groups make a difference, and in many cases, it is probably safe to ignore the multilevel structure of the data. The *intraclass correlation* ($\rho$) is used to assess the relevance of group-level factors, and if this intraclass correlation is small (e.g., less than .20), it might be reasonable to ignore groups and simply use OLS regression [However, Bikel (2007) notes that multilevel models can provide more accurate estimates of some parameters than OLS even when group effects are small.]

The intraclass correlation is given by Equation 9.6. It is a comparison of the group-level error variance to the total error variance (group plus individual level), where $\sigma^2_{u0}$ is the Level-2 error variance component and $\sigma^2_e$ is Level-1 error variance component.

$$\rho = \frac{\sigma^2_{u0}}{\sigma^2_{u0} + \sigma^2_e}$$
(9.6)

If you are using **R** to analyze multilevel data (e.g., the R package lme4 is widely used for this purpose), the variance components you need to calculate the intraclass correlation are given in the section of the results labelled "Random intercepts". This will give a variance for the group level variable ($\sigma^2_{u0}$) and for the individual level ($\sigma^2_e$). For example, in an analysis of data obtained from several schools, the output included the results shown in Table 9.2. We have highlighted the two variance components needed to calculate $\rho$, $\sigma^2_{u0}$, which equals .1988 and $\sigma^2_e$, which equals .9592, giving a $\rho$ value of .178. This is a relatively small intraclass correlation coefficient, suggesting that group effects might not be very important in this study and that it might be possible to ignore them altogether.

Table 9.2 Information Needed to Compute the Intraclass Correlation – R lme4 package

| Random Intercepts | | |
|---|---|---|
| Random effects: | | |
| Groups Name | Variance Std. | Dev. |
| school (Intercept) | **0.1988** | 0.4459 |
| Residual | **0.9592** | 0.9794 |
| Here, $\rho = \frac{.1988}{.1988 + .9592} = .178$ | | |

It is always useful to calculate ρ. Even in cases where you decide to continue with a multilevel model when ρ is small, it will give you some clear and useful information about the relative importance of group vs. individual-level effects.

## An Illustration of Multilevel Analysis

A data set that includes 465 students four schools, where scores on quantitative ability are used to predict academic performance,[4] can be used to illustrate multilevel analysis. We should start by telling you that there is tremendous variability in the ways multilevel analyses are conducted and reported in the research literature. We believe that multilevel analyses are most interpretable if they proceed in a stepwise fashion, starting with the simplest model possible and assessing how well it fits the data and them adding additional terms. The simplest possible model is the "intercept-only" model that proposes that there is no real variation in achievement, and that in the population, everyone has the same score. Like the null hypothesis, the intercept-only model is not something people genuinely believe in. Rather, this provides a baseline. As more information is added to the model, the fit of the model to the data (i.e., the ability of the model to predict achievement) should improve. If the information is useful for predicting scores on Y, the fit of the model should improve substantially. If the information does not really help you predict Y, there should be only small changes in fit.

Several statistics can be used to express the fit of the model, but the most general and in many ways the most useful is "deviance". The smaller the deviance, the better the job the model does in explaining stress. The baseline model predicts that the score on Y in the population is the same for everyone. The next step is to add the Level-1 predictors of gender and experience. This model will almost certainly do a better job predicting stress than the intercept-only model and, as long as models are nested (i.e., the more complex model includes everything in the simpler one, plus more predictors or more terms), we can directly compare the deviance statistics. The change in deviance as you add more terms to a model is distributed as

*Table 9.3* Preliminaries to a Multilevel Analysis

1. **Are predictors related to Y?**
   Fit an OLS regression model – $R^2 = .41$, $F_{(1,463)} = 327.00$, $p < .01$, a = 1.16, b = .80
2. **Do schools differ?**
   ICC = 35.40/(35.40 + 31.63) = .528
3. **Preliminary conclusions**
   Quantitative ability is related to performance, but this relationship might not vary much from school to school

Chi squared, with degrees of freedom equal to the number of new terms in the model.

A good starting point for a multilevel analysis is to first ask: (1) does quantitative ability have anything to do with performance, and (2) does performance vary across schools. Table 9.3 displays the results of these two first analyses. First, ability is linked to performance; using OLS regression, we find $R^2 = .41$, $F_{(1,463)} = 327.00$, $p < .01$, a = 1.16, b = .80. However, there appear to be differences from school to school in performance; the intraclass correlation coefficient shown in Table 9.3 is .528. Values of ICC. of 20 or larger are often interpreted as large enough to indicate meaningful variation, and the ICC value shown in Table 9.3 suggests considerable variation across schools in performance. Table 9.4 shows the results of the next stage of multilevel analysis. The first model (baseline – intercept only)

*Table 9.4* Comparing and Making Sense of Alternative Multilevel Models

|  |  | df | *Deviance* | *Change* |
|---|---|---|---|---|
| Baseline – intercept only |  | 462 | 2945.1 |  |
| **Does Ability Make a Difference?** |  |  |  |  |
| Intercept | Slope |  |  |  |
| Random | Fixed | 461 | 2889.1 | 56.0 [a] (1)** |
| Fixed | Random | 461 | 2927.6 | 17.5 [a] (1)** |
| **What Sort of Difference?** |  |  |  |  |
| Random | Random | 458 | 2887.6 | 1.50 [b] (3) |
|  |  |  |  | 40.0 [c] (3)** |
| **Slopes, Intercepts, and Descriptive Data by Group** |  |  |  |  |
| School | Intercept | Slope | $r_{ability, performance}$ | Mean $_{Ability}$ Mean $_{Performance}$ |
| 1 | −2.32 | .029 | .20 | 6.11  5.10 |
| 2 | −3.52 | .049 | .44 | 6.64  4.09 |
| 3 | −2.02 | .025 | .35 | 6.28  5.88 |
| 4 | 7.87 | −.110 | .35 | 16.95  18.76 |

Note
a – compared to baseline
b – compared to random intercept, fixed slope
c – compared to fixed intercept, random slope
Degrees of freedom are shown in parentheses

Note
** $p < .01$

serves as a baseline for this phase of the analysis. Subsequent models help in determining whether allowing the intercept, slope, or both to vary might improve the fit of the prediction model.

Table 9.4 includes three models that use quantitative ability to predict performance, one in which the slope between ability and performance is fixed (i.e., the same in all schools) but the intercepts are allowed to vary, one in which the slope between ability and performance is allowed to vary but the intercepts are fixed (i.e., the same in all schools) and finally, one in which both slopes and intercepts vary. Both the fixed-intercept, random-slope model and the random-intercept, fixed-slope model fit significantly better than the intercept-only baseline. This is to be expected; you know from Table 9.5 that there is a relationship between quantitative ability and performance, so models that include this relationship should generally fit better than models that do not.

The most important and most informative comparisons are the final ones shown in Table 9.4. A model that allows intercepts and slopes to vary does *not* fit better than one in which slopes are the same for all schools, but in which intercepts are allowed to vary. In other words, allowing slopes to vary across schools does not lead to a better fit than forcing all schools to use the same slope coefficient, suggesting that the relationship between quantitative ability and academic performance is similar in form across schools. On the other hand, a model that allows intercepts and slopes to vary *does* fit better than one in which slopes are the same for all schools, but in which intercepts are allowed to vary. Taken together, these results suggest that intercepts vary from school to school but that slopes do not, and this impression is confirmed by examining the intercepts and slopes for each school shown in Table 9.4. All the slopes are relatively flat, but the intercept in school 4 is much larger than the other intercepts because of differences in the mean levels of quantitative ability and performance. In other words, the relationship between the X variables and Y looks more like panel 2 of Figure 9.1 than panels 1, 3, or 4. This is not a surprise, given the large ICC value shown in Table 9.3.

You can get a clear sense of why intercepts varied but slopes did not by looking at a few simple descriptive statistics in each school. First, the correlation between quantitative ability and performance does not vary greatly from school to school. The correlations shown in Table 9.4 are not all the same, but they are all in the moderate range. In contrast, the means for ability and performance vary greatly, with much higher means on both ability and performance in school 4. In OLS regression, the intercept is simply the mean on Y minus b (the regression coefficient) times the mean on X, and these large differences in the mean on both X and Y, together with the relatively small values for the regression coefficient shown in Table 9.4, will lead to differences in intercepts across schools.

## Remember, *It's All Regression*

The models described in Table 9.3 may look very different from the sort of regression models we are used to when applying OLS regression, but as Bikel (2007) illustrates, this kind of data can be analyzed using familiar regression methods, and the questions that are answered using OLS regression are very similar to those that might be answered using more standard regression methods. To illustrate how this is done, let's simplify our analysis a bit, just to illustrate how this might be done using OLS regression. We want to emphasize that the study analyzed in Tables 9.3 and 9.4 could be analyzed in similar ways using OLS regression; simplifying the study just makes it easier to describe the approach.

Suppose there are two schools, Jefferson High School and St. Joseph's, and you are looking at the relationship between quantitative ability (Ability) and performance (Y). First, create a variable (School) to represent the school each student attends (0 = Jefferson, 1 = St. Joseph's)[5]. Second, create a cross-product term by multiplying scores on Ability with scores on School (A × S). Then, use the method of moderated multiple regression (Aguinis, 2004) to study questions very similar to those studies using multilevel modeling. These questions are laid out in Table 9.5.

The questions asked in moderated multiple regression are similar in many ways to those asked in multilevel modeling. We would argue that the moderated multiple regression approach might be superior for several reasons. First, it asks some questions that are not given much attention in multilevel modeling, such as how much better the prediction is when various terms are added to the prediction model. Second, it allows you to use the familiar statistical methods and familiar statistics, such as $R^2$, changes in $R^2$, and $F$ in evaluating your data. Third, using OLS regression methods allows you to easily apply the minimum-effect methods described throughout this book. For example, the usual test for changes in $R^2$ as you add new material to a prediction model test the nil hypothesis, that the change in $R^2$ is zero. Suppose you were interested in the question of

*Table 9.5* Using Moderated Multiple Regression to Study Multilevel Data

| Regression Models | Questions Asked |
|---|---|
| $R^2_{Y.School}$ | Does stress vary from school to school? |
| $R^2_{Y.Ability}$ | Is performance related to quant. ability? |
| $R^2_{Y.Ability, School} - R^2_{Y.School}$ | Does ability explain variance in performance not explained by schools? |
| $R^2_{Y. Ability, School} - R^2_{Y. Ability}$ | Do schools explain variance in performance not explained by ability? |
| $R^2_{Y. Ability, School, A \times C} - R^2_{Y. Ability, School}$ | Is there a moderator effect? Does the relationship between ability and performance vary from school to school? |

Table 9.6 Results of Study Described in Table 9.5

| | Value |
|---|---|
| $R^2_{Y. \text{Ability}}$ | .30 |
| $R^2_{Y. \text{School}}$ | .15 |
| $R^2_{Y. \text{Ability, School}}$ | .40 |
| $R^2_{Y. \text{Ability, School}} - R^2_{Y.\text{School}}$ | .25 |
| $R^2_{Y. \text{Ability, School}} - R^2_{Y. \text{Ability}}$ | .10 |
| $R^2_{Y. \text{Ability, School, A} \times \text{S}} - R^2_{Y. \text{Ability, School}}$ | .41 |

whether the change in $R^2$ is large enough to care about (e.g., the new information accounts for at least 1% of the variance in Y not explained by the previous material).

Framing multilevel modeling in terms of OLS regression not only makes the whole topic less threatening, because it can be seen as a simple extension of familiar methods, but it also gives you considerable flexibility in terms of what hypothesis you decide to test (e.g., is there *any* change in $R^2$ vs. is the change in $R^2$ large enough to be meaningful?).

Table 9.6 displays the findings of the study described above. First, there is a meaningful relationship between ability and performance ($R^2_{Y.\text{Ability}} = 30$). Second, performance varies meaningfully across schools ($R^2_{Y.\text{School}} = .15$). Ability and schools, considered together account for 40% of the variance in performance, but they overlap somewhat. Ability accounts for 25% of the variance not explained by schools (i.e., $R^2_{Y.\text{Ability, School}} - R^2_{Y.\text{School}} = .25$), while schools account for 10% of the variance not explained by ability (i.e., $R^2_{Y. \text{Ability, School}} - R^2_{Y. \text{Ability}} = .10$). Finally, the moderation hypothesis, that the relationship between ability and performance (i.e., the slope) varies across schools receives little support; the increase in $R^2$ when the ability by schools cross-product (i.e., A × S) is added to a regression equation that includes ability and skills is very small ($R^2$ increases from .40 to .41).

The method of analysis illustrated in Table 9.6 is less elegant and less efficient than typical multilevel methods, but we would argue that it is considerably easier to interpret and considerably more information. Nevertheless, it is fair to note that OLS regression methods are not often used in the analysis of multilevel models. In the sections that follow, we will discuss power analysis as it applies to the methods typically used in these studies, applying the methods developed in this book to allow researchers to ask similar questions, such as whether the increase in fit of a model as you add more predictors is large enough to be worth the added complexity or is so small that it should be considered as trivial.

## Effect Sizes in Multilevel Analysis

It is relatively rare for authors of multilevel studies to provide effect size measures, and the evaluation of multilevel models often depends on significance tests. There is an important exception to this rule, which we will describe below.

One reason for the reluctance to report effect size estimates is the belief that there is no exact equivalent to $R^2$ or some similar effect size measure. Several pseudo-$R^2$ have been proposed (e.g., Snijders and Bosker, 1999, describe indices based on the one minus the variance not explained in full models compared to the variance not explained in a null baseline model) and they are not difficult to obtain when analyzing multilevel data in R (see the boxed section below), but their uptake has been uneven. There are indications, however, that things are starting to change. Rights and Serba (2019, 2020) have shown how to obtain meaningful $R^2$ measures in multilevel modeling and have illustrated the use of nested regression models. In many ways, their methods are conceptually similar to the use of the nested models shown in our discussion of moderated regression earlier in this chapter.

It is important to note that the $R^2$ measures developed by Rights and Serba are conceptually related to the $R^2$ obtained in the more traditional applications of regression (i.e., they are measures of explained outcome variance divided by total outcome variance), but these indices are not mathematical identical to $R^2$ in ordinary least squares regression. They are, however, similar in many ways, particularly in their application to nested models. First, they allow researchers to examine four general ways of explaining the variance in Y, by Level-1 predictors by fixed slopes, by Level-2 predictors by fixed slopes, by variation in slopes across clusters and by variation in intercepts across clusters. Furthermore, these sources of variance are often additive. For example the proportion of total outcome variance explained by all predictors via fixed slopes is equal to the proportion explained by Level-1 predictors via fixed slopes plus the proportion explained by Level-2 predictors via fixed slopes (i.e., $R^{2(f)}_t = R^{2(f1)}_t + R^{2(f2)}_t$), which means that subtracting Level-1 variance explained from total variance explained tells you how much variance is explained by Level-2 variances (i.e., $R^{2(f)}_t - R^{2(f1)}_t = R^{2(f2)}_t$).

It is too soon to tell whether the methods developed by Right and Serba (2019, 2020) will change the lamentable practice of reporting the results of MLM analyses without saying anything meaningful about how well the various models explain the data. We are cautiously optimistic that reporting standards for multilevel analyses will gradually improve to the point where effect size estimates are commonly reported.

---

### R code for obtaining $R^2$ and pseudo-$R^2$ estimates

To illustrate how pseudo-$R^2$ can be obtained in **R**, we will use an example of a study that looks at math achievement (mathach) in schools. We start with a baseline null model that has no predictors but does allow for different intercepts for each school, and then create a full model that adds three Level-1 predictors to the model (ses, sector, meanses), each of which is related to socioeconomic status. The **R** code is shown below.

install.packeges("sjstats")
library(sjstats)
install.packages("lme4")
library(lem4)
nullmodel<- lmer (mathach ~1 + (1|school), data=mydata,REML= FALSE)
fullmodel<- lmer (mathach ~ ses + sector +meanses + (1|school), data=mydata,REML=FALSE)
r2(fullmodel,nullmodel)

This code will produce the two pseudo-$R^2$ estimated shown in Equations 9.2 and 9.3. The Level-1 estimate is labeled "R-squared (tau-00)"; the Level-2 estimate is labeled "Omega-squared" in the section of output labeled "R-squared (tau-11)".

Rights and Serba developed an **R** function, **r2mlm** that reads raw data and MLM parameter inputs from an MLM analysis (using **R, SAS** or some other platform to perform the basis MLM analysis). The **R** code that defines this function and an example of applying this function are provided in the online Appendix A of Rights and Serba (2019; see https://supp.apa.org/psycarticles/supplemental/met0000184/Supplemental_Materials.pdf). **r2mlm** is easiest to use when it is integrated with **R**-based multilevel modeling (e.g. using a package such as **lme4**). For example, in a multilevel study of the relationship between salary and feelings of control of teachers and satisfaction in several schools, where Level-1 measures are labeled "_c" and Level-2 measures "-m", the code below will produce $R^2$ estimates.

model_nlme <- lme(satisfaction ~ 1 + salary_c + control_c + salary_m + control_m +
    s_t_ratio, random = ~ 1 + salary_c + control_c | schoolID, data = teachsat,
    method = "REML", control = lmeControl(opt = "optim"))
r2mlm(model_nlme)

## Power for What?

Power analyses of multilevel models often focus on the power to detect individual and group-level effects. Several **R** packages have been developed that allow you to perform power analyses for multilevel models, including **powerlmm**, **longpower** (developed to perform power analyses in multilevel longitudinal models) and **simr**. In most multilevel studies, you are likely to find different levels of power for Level-1 and Level-2 effects, and one of the major concerns in much of the literature on the power of multilevel tests concerns balancing out power for these different levels.

For example, Maas and Hox (2005) suggest that the number of groups if often more important than group size in determining the accuracy of estimating Level-1 vs. Level-2 effects. Thus, it is often better, from the perspective of multilevel modeling, to sample 150 children from 30 classrooms than 150 children from 5 classrooms. The number of higher-level units does not substantially affect power of first-level tests. Meinck and Vandenplas (2012) note that the optimal balance between number of groups and number of subjects per group depends on the size of the intraclass correlation coefficient, and that when $\rho$ is large (i.e., groups are very different), the payoff for increasing the number of subjects within groups is smaller. Kreft (1996, cited in Scherbaum & Ferreter, 2009) offered a 30/30 rule of thumb, leading to a minimum total sample size of 900, no matter what type of effect is studied. That is, they suggest that overall power is maximized when there are about 30 groups, each of which includes about 30 subjects, giving a total sample size of approximately 900. Hox (1995) provided an alternative rule of thumb, suggesting sample sizes of 50 clusters and 20 individuals per cluster as appropriate for multilevel modeling (giving a total sample size of approximately 1,000).

Scherbaum and Ferreter (2009) suggest that like most rules of thumb, these rules much be applied with caution, and that, as noted in Chapters 1–8, power depends on the likely effect sizes, meaning that larger samples might be needed when the likely effects are small and smaller ones might be fine if the effect sizes are likely to be large. They review several statistical packages for estimating power in multilevel models.

Studies of the power of multilevel studies to detect Level-1 vs. Level-2 effects virtually always rely on traditional null hypothesis tests. That is, the test the hypothesis that the effects of Level-1 or Level-2 variables is exactly zero, and if they reject the null hypothesis that the variables of interest have no effect, they go on to interpret and draw conclusions from those tests. In this book, we have advocated testing a different null hypothesis, that the effects of treatments, variables, groups, etc. is either so small that it can be safely ignored (which in many cases if defined in terms of effects that account for 1% of the variance or less), in large part because tests of the traditional null hypothesis that the effects of treatments, groups, variables, etc. are precisely zero are largely pointless. If you apply this minimum-effect logic to multilevel analyses, it is likely that the rules of thumb outlined here are probably too conservative. That is, if you wish to test the hypothesis, for example, that Level-2 effects are not trivially small rather than testing the hypothesis that they are precisely zero, somewhat larger samples will typically be needed.

At the beginning of this chapter, we described how multilevel models are used to test various types of hypotheses about whether and how prediction models vary across groups (e.g., can the same intercept or slope be used for everyone or do these vary across slopes). In our view, the most important and the most informative tests in multilevel modeling are those that involve comparing nested models, starting with some sort of baseline, and then asking how the fit of the model improves as various types of terms (Level-1 predictors, group intercepts, etc.) are added to the model. The approaches developed here to conducting tests of minimum-effect null hypotheses generalize quite easily to tests of changes in model fit and provide a straightforward way of asking whether the improvements in fit as we add information to multilevel prediction models are large enough that we can confidently reject the hypothesis that, in the population, these changes are trivially small.

## Using Changes in Model Fit as a Basis for Power Analysis in Multilevel Modeling

One of the principal concerns of multilevel modeling is the comparison on models. For example, does a prediction model that allows intercepts or slopes to vary from across clusters (e.g., classrooms, organizations) fit better than a model that uses the same intercept or slope for everyone. These comparisons are made using Chi-square tests; when you add new terms to a model, deviance will always go down, and the change in deviance is distributes as $\chi^2$ with degrees of freedom equal to the number of new terms you have added to the model.

The traditional Chi-square test asks whether the change in deviance is different from zero.

We can test minimum-effect hypotheses to test the hypothesis that the change in deviance is not trivially small using a noncentral Chi-square distribution, in the same way we have used the noncentral $F$ distribution to perform minimum effect tests in previous chapters.[6] For example, suppose you had a sample of 300 people and were comparing two nested models where the full model added three parameters (e.g., group variables) to a simpler model. If you define a small effect using Cohen's (1988) benchmarks (the effect size measure $w = .10$ – the Chi-square effect size measure $w$ is interpreted in the same way as the correlation coefficient $r$ in terms of what defines a small, medium, or large effect) and you also assume that the true effect is moderately large (Cohen, 1988, suggests $w = .30$ to define a moderately large effect), you can use the **R** code below to test the hypothesis that the change in deviance is equal to or larger than what would traditionally be defined as a small effect, and to evaluate the power of the study for testing this hypothesis.

# R code for calculating critical Chi-squared values and power for minimum-effect comparisons of models

As in all minimum-effect power analyses, you start by specifying the type of test you want to carry out, by specifying your definition of a trivially small effect. Luckily, the effect size most widely used for the Chi-square test, $w$, can be interpreted using similar guidelines to the correlation coefficient. A small effect is most conventionally defined as one that corresponds to $r = .10$ or to $PV = .01$, and we can use the same $w$ value of .10 to define a small effect here. We must also specify the degrees of freedom for the test ($df_{hyp}$), $N$, and the desired alpha level. Finally, specify the assumed effect size in the population. Again, using the conventional effect size definition for the correlation coefficient $r$, if we think there is a moderate effect in the population, we specify $w = .30$. The **R** code below gives you the critical value for this minimum-effect Chi-squared test and the power of that test.

```
## minw is max value of w that defines a small effect
## assumedeff is assumed value of w in the population
N=300
dfhyp=3
minw=.10
alpha=.05
Mineff = minw*minw
assumedeff=.30
assumedeffsq= assumedeff* assumedeff
lambda1<-N*Mineff
cumper=1-alpha
noncenchisq<-qchisq(cumper, dfhyp, ncp=lambda1)
lambda2 = N * assumedeffsq
imp<-pchisq(noncenchisq, dfhyp,ncp=lambda2)
power<-1-imp
## critical value of chi square (alpha = .05) for this min effect test
noncenchisq
## power of this minimum effect chi square test
power
```

In this study ($df = 3$, $N = 300$, alpha = .05, minw = .10, assumed $w$ in the population = .30), the critical Chi-square value for testing the change in deviance is 14.18 (the comparable critical value for testing the traditional null hypothesis would be 7.81). If the change in deviance when these three parameters are added to the model is greater than 14.18, you can be at least 95% certain that the change in deviance is as large or larger than what you would conventionally describe as a small effect (i.e., $w = .10$ or $PV = .01$). This study would give you power of .951 for testing this hypothesis. If someone else was to run a similar study using a smaller sample, power would be lower. For example, in a similar study with $N = 100$, power would drop to .555.

Suppose the population effect size is a lot smaller, say $w = .15$. It would be difficult to achieve a high level of power in testing the hypothesis that the effect of treatments was small ($w = .10$) because the assumed effect size is also small. In this study, the power for this minimum-effect hypothesis test would be .19, and you would need a much larger sample (over 2,500, assuming that there are the same number of groups as in the earlier calculations) to achieve power of .80 for testing this hypothesis.

### *The Ambiguity of* N

In a multilevel study, the same number of cases might lead to different levels of power depending on whether there are many people in each of a small number of groups or many groups, each with a small number of members. To understand why, it is useful to think concretely about the types of hypotheses that are tested in multilevel models.

Consider how random intercept models are tested. First, you assess the fit of a fixed-intercept, fixed-slope model. You then allow the intercepts to vary and see how much the fit improves. The difference in the deviance of these two models is distributed as Chi squared, with degrees of freedom equal to the number of groups. As a result, if you have many groups, you need to evaluate this change in deviance in relation to the Chi-squared distribution with the same large number of degrees of freedom. The Chi-squared distribution has a mean equal to the number of degrees of freedom and a variance equal to twice the degrees of freedom. As a result, when the number of groups is large, the Chi-squared value needed to reject a nil hypothesis or a minimum-effect hypothesis will be correspondingly large. The Chi square needed to reject a nil hypothesis or a minimum-effect hypothesis will be considerably smaller if the number of groups is small. As a result, the same $N$ might lead to different levels of power depending on how the study is designed, and it does not make sense to compute the $N$ needed to achieve power of .80 when comparing nested multilevel models.

### *Power Analysis for our Quantitative Ability Study*

In Tables 9.3 and 9.4, we illustrated the analysis of a study examining the relationship between quantitative ability and academic performance in four schools ($N = 465$). In that study, we concluded that: (a) the predictor made a difference – all models that included quantitative ability as a predictor fit better than the baseline, and (b) allowing the intercepts to vary across schools increased the fit of the model but allowing slopes to vary across schools did not. These conclusions make sense given the descriptive statistics shown in Table 9.4, but it is possible that we did not detect variation in slopes because our study had insufficient power.

As you can see from Table 9.4, two models that include a predictor are compared the baseline model, one with fixed intercepts and random slopes and the other with random intercepts and fixed slopes. Both are tested by examining the change of deviance has as the predictor is added to the baseline, and both have one degree of freedom for the hypothesis that adding the predictor increases the fit of the model. Similarly, the comparison of the random intercept, random slope model to models that fix either intercepts or slopes have one degree of freedom for the hypothesis being tested. Using the **R** code shown previously, we can evaluate the power of these comparisons, both when testing the nil hypothesis (that there is no change in fit) or a minimum-effect hypothesis (e.g., that the change in fit is so small as to be trivial).

Remember that the effect size used with the Chi-squared statistic ($w$) can be interpreted similarly to a correlation coefficient. This suggests that a $w$ of .10 corresponds to the traditional definition of a small effect (i.e., $PV$ = .01). We can assess the power of this study for comparing nested models and determine whether the failure to detect variation in slopes in this study might be due to low power.

Power depends on $df_{hyp}$, $df_{error}$, alpha, and the assumed effect size in the population. Table 9.7 shows the power of this study for comparing models ($\alpha$ = .05) depending on whether we are testing the nil hypothesis or the hypothesis that the increase in fit is so small that it might be treated as trivial (i.e., the equivalent of $PV$ = .01 or less) for population $w$ values ranging from .10 (small effect) to .30 (a moderately large effect).

Suppose you are testing a nil hypothesis, that there is no difference in fit when nested models are compared. Table 9.6 suggests that power will be low (.57) if the change in fit in the population is small, but that this study, with $N$ = 465, will have plenty of power for detecting changes in model fit if these changes are not too small in the population.

Suppose instead that you want to test the hypothesis that increases in model fit are not trivially small. Table 9.6 suggests that this study might not have sufficient power unless the increase in fit in the population is moderately large ($w$ = .25, $PV$ = .06). Suppose that the population effect is

Table 9.7 Power for Detecting Increases in Model Fit – Quantitative Ability Study

| Hypothesis | Population Effect Size (w) | | | | |
|---|---|---|---|---|---|
| | .10 | .15 | .20 | .25 | .30 |
| nil | .57 | .89 | .99 | .99 | .99 |
| 1% | .05 | .29 | .69 | .94 | .99 |

small, corresponding to $PV = .01$. If you are testing the hypothesis that treatments account for 1% of the variance or less, the probability of rejecting the null will be equal to .05, because in this case, rejecting this minimum-effect null will amount to a Type I error.

If you are testing the nil hypothesis, you would probably conclude that your inability to detect variation in slopes across the four schools studied is not due to low power. Unless the population variation in slopes is quite small, you should have plenty of power in this study. On the other hand, this study does not have much power for testing the hypothesis that changes in model fit are large enough that they cannot reasonably be labelled as trivially small. You would probably need a sample of about $N = 2,500$ to detect changes in fit corresponding to $w = .15$ (the equivalent $PV$ would be .02) with a power of .80 or above. We should caution, however, that power might be difficult if the number of schools stays the same but the number of students per school increases than if the number of schools increases but the number of students per school stays the same. Thus, $N = 2,500$ should be considered only an approximation.

## Sample Size – Some General Guidance

How large should samples be to achieve adequate power in multilevel studies? There is no simple answer to this question, in part because a sample with 900 subjects drawn from 5 organizations might not yield the same power for all tests as a sample of 900 subjects drawn from 90 organizations. We can, however, give some general advice.

First, while increasing the number of units (e.g., drawing a few subjects from each of many organizations versus drawing many subjects from a few organizations) can increase the power of some tests of Level-2 effects, we advise against these designs. Small gains in power are, in our opinion, offset by large losses in interpretability. It is hard enough to make concrete sense of results when there are separate slopes and/or intercepts for five different organizations. When each of 90 organizations has its own slope and/or intercept, the interpretability of results goes out the window. If interpretability is an important consideration, we advise keeping the number of units as small as reasonably possible and focusing on the overall sample size rather than on the number of separate organizations, departments, or other units sampled.

In this chapter, we focus on model comparisons. Regardless of whether one is testing the nil hypothesis or a specific minimum-effect hypothesis, the most important question in determining the sample size needed to achieve adequate power (e.g., power of .80) is the difference between the hypothesis being tested and the assumed effect size in the population. For example, the traditional definition of a small effect in tests of the nil hypothesis is one where treatments account for 1% of the variance in outcomes. If the number of degrees of freedom in the hypothesis are small (e.g., you are comparing models that differ by only a few parameters), you might need a sample of $N = 800$ to achieve a power of .80.

If you are testing the hypothesis that treatments account for 1% of the variance or less, the corresponding definition of a small effect might be the case where treatments account for 2% of the variance in outcomes. You will need a very large sample (up to $N = 4,000$) to achieve a comparable level of power. If the assumed effect is moderately large (e.g., treatments account for 10–11% of the variance, samples of 90–120 will give adequate power for testing nil of 1% minimum-effect hypotheses. If the assumed effect is large (25% of the variance explained or more), samples as small as 30 will yield plenty of power.

## Notes

1 Multilevel studies might involve more than two levels. For example, employees might be sampled from several different organizations, and in each organization, from several departments. However, the norm in multilevel studies is to focus on data that are organized into two levels – i.e., groups (classrooms, organizations) and individuals within groups.
2 In simple linear regression, the slope is given by $b = r_{xy}(SD_y/SD_x)$ and the intercept is given by $a = \bar{Y} - b\bar{X}$.
3 However, Bikel (2007) illustrates how many multilevel analysis can be tackled using OLS, and notes the strong conceptual similarities between OLS regression and multilevel analysis.
4 This data set can be accessed at https://www.dropbox.com/s/xlhqk37faiwma49/HLMdata.csv?dl=0.
5 If there are more than two hospitals, I will need to create k-1 different variables, where k represents the number of hospitals. Aguinis (2004) and Cohen et al. (2003) provide excellent discussions of how to create coded variables to represent groups.
6 The noncentrality parameter for the Chi-square distribution, $\lambda$, is given by $\lambda = N*w^2$, where $w$ is the effect size measure most widely used for the Chi-square test.

# 10 The Implications of Power Analyses

The power of a statistical test is the probability that you will reject the null hypothesis being tested, given that this null hypothesis is wrong. As we have noted throughout this book, the traditional null hypothesis that treatments have no effect whatsoever (or that the correlation between two variables is precisely zero, or any other hypothesis of "no effect") is very often wrong, and in this context, the statistical power of a test is essentially the probability that the test will lead to the correct conclusion. When testing the traditional null hypothesis, it is obvious that power should always be as high as possible. When testing a minimum-effect hypothesis (e.g., that the effect of treatments is negligibly small, but not necessarily precisely zero), the implications of varying levels of statistical power are potentially more complex, and a wider range of issues needs to be considered in determining how to use and interpret statistical power analysis.

This chapter begins with a discussion of the implications of statistical power analysis for tests of both traditional and minimum-effect null hypotheses. Next, we discuss the benefits of taking statistical power seriously. Some of these are direct and obvious (e.g., if you do a power analysis, you are less likely to conduct a study in which the odds of failure substantially outweigh the odds of success), but there are also many indirect benefits to doing power analyses that may, in the long run, be even more important than the direct benefits. Finally, we consider the question of whether power analysis renders the whole exercise of testing the traditional null hypothesis moot. If power is very high (or very low), the outcome of most statistical tests is not really in doubt, and the information from these tests might be severely limited as power reaches either extreme.

## Tests of the Traditional Null Hypothesis

In Chapter 1, we noted that two types of errors might be possible in testing a statistical hypothesis. First, you might reject the null hypothesis when it is, in fact, true (a Type I error). Second, you might fail to reject the null hypothesis when it is in fact wrong (a Type II error). Textbooks invariably stress the need to balance one type of error against the other (e.g., procedures that minimize

DOI: 10.4324/9781003296225-10

Type I errors also lead to low levels of power), but because the null hypothesis is almost certainly wrong there may be little to be gained and much to be lost by attempting to maintain such a "balance" (Murphy, 1990).

The fact that the traditional null hypothesis is so often wrong leads to three conclusions about statistical power: (1) you cannot have too much power, (2) you should take the simplest and most painless route to maximize power, and (3) tests with insufficient power should never be done.

### *You Cannot Have Too Much Power*

If the null hypothesis is very likely to be wrong, it is very unlikely you will ever make a Type I error, and the only way you are likely to make an error in statistical hypothesis testing is by failing to reject $H_0$. Our reason for repeating this point so many times is that it flies in the face of convention, where substantial attention is often devoted to the unlikely possibility that a Type I error might occur. In tests of the traditional null hypothesis, power is essentially the probability that the test will reach the right conclusion (because the traditional null is usually wrong), and there is little coherent statistical rationale for arguing that low levels of power should be acceptable. There are many practical problems with attaining high levels of power, as we note below. However, it is always to your advantage to maximize power in tests of the traditional null hypothesis.

### *Maximizing Power: The Hard Way and the Easy Way*

There are two practical and eminently sensible ways to attain high levels of power. The easy way is to change the alpha level. As we showed in Chapter 2, power is higher when a relatively lenient alpha level is used. Traditionally, the choice of criteria for defining statistical significance has been between alpha levels of .05 and .01. When testing the traditional null, there is little scientific or statistical advantage to using a stringent test, and you should generally set the alpha level for your tests as high as possible. Unfortunately, you are likely to meet resistance if you use alpha levels of .10, .20, or anything other than .05 or some other conventional figure, but this resistance is misplaced. Higher alpha rates yield more power, often with no meaningful increase in the likelihood of a Type I error.

The second strategy for maximizing power is to increase the sensitivity of your study, which generally implies using larger samples. Even though this strategy is more demanding than simply changing the alpha level, we strongly recommend it. Large, carefully constructed samples increase the generalizability and stability of your findings, and they decrease the possibility that sampling error will lead to meaningless or misleading results. We will say more about this in the sections that follow.

While it is harder to increase power by increasing $N$ than by increasing $\alpha$, this strategy has the immense benefit of improving your study. Simply changing the alpha level does nothing to enhance the meaningfulness or interpretability of your research, but the use of large samples helps to minimize one of the recurring problems in social science research – the over-reliance on the unstable results obtained in small samples (Schmidt, 1992).

*Tests with Insufficient Power Should Never Be Done*

Suppose you were diagnosed with an ulcer, and your doctor told you about a new treatment. Your doctor tells you that this treatment is more likely to make things worse than to make things better, and there are alternative treatments available that do not have this problem. Would you try the new treatment? Our answer is "no", and we believe this analogy applies exactly to statistical tests of the traditional null hypothesis. If power is low, you should not carry out a test of the traditional null hypothesis.

When power is less than .50 and you are virtually certain that $H_0$ is wrong, the test is more likely to yield a wrong answer than a right one. More to the point, the test is unlikely to produce new and useful knowledge; it is more likely to mislead you. If you are virtually certain before the test that $H_0$ is false, a test that rejects $H_0$ doesn't tell you much that you don't already know. A test that fails to reject $H_0$ shouldn't change your mind either (if $H_0$ is wrong virtually by definition, the results of your test shouldn't change this), but people will sometimes be misled by their data. Low-power tests are unlikely to have any effect except to mislead and confuse researchers and readers.

## Tests of Minimum-Effect Hypotheses

The alternative to testing the traditional null hypothesis that treatments have no effect is to test the minimum-effect null hypothesis that the effect of treatments is so small that it could be safely ignored. Different disciplines or research areas might require substantially different operational definitions of a "negligibly small" effect, and the standards suggested in previous chapters and in our One-Stop *F* Table (i.e., treatments accounting for less than 1%, or in some cases less than 5% of the variance in outcomes have negligibly small effects) will not always apply. Nevertheless, we believe that tests of minimum-effect null hypotheses are necessary if the whole enterprise of statistical hypothesis testing is to prove useful.

Statistical power analysis can be used to its fullest advantage in tests of minimum-effect null hypotheses. Because the hypothesis being tested may very well be true (i.e., although treatments are very unlikely to have absolutely no effect, the hypothesis that they have trivial effects is often a realistic one), it becomes important to develop specific procedures and criteria for "accepting the null", or determining when the evidence is consistent or inconsistent with the proposition that the effects of a particular treatment *are* indeed negligible; power analysis is extremely useful for this purpose. It also becomes important to give serious consideration to an issue that is usually (and incorrectly) presented in the context of traditional null hypothesis tests — i.e., the appropriate balance between Type I and Type II errors.

## Accepting the Null

In traditional null hypothesis testing, the idea of accepting the null hypothesis is sometimes treated as a sort of heresy. Rather than allowing one to accept the hypothesis that treatments have no effect, the traditional framework usually leaves you with two options: (1) deciding that there is sufficient evidence to reject the null (i.e., a significant outcome), and (2) deciding that there is not yet enough evidence to reject the null (i.e., a nonsignificant result). Because you already know that the traditional null is almost certain to be false, the fact that you have not yet accumulated enough evidence to confirm this fact tells you more about your study than about the substantive phenomenon you are studying.

As we have noted throughout, power is substantially affected by the size of one's sample. If $N$ is very small, you will not reject the null, no matter what research question you are pursuing. If $N$ is large enough, you will reject the traditional null, again no matter what research question you are pursuing. It is hard to resist the conclusion that tests of the traditional null hypothesis are little more than indirect measures of your sample size! In tests of the traditional null, the most logical interpretation of a nonsignificant result is that your sample is too small.

Occasionally, the door is left open for treating nonsignificant results as meaningful. For example, some journals allow for the possibility of publishing nonsignificant results, at least under some conditions (e.g., the inaugural issue of the journal *Human Performance* included an editorial suggesting that nonsignificant results would be treated as meaningful if specific research design criteria, including a demonstration of adequate statistical power, were met). The argument that is sometimes offered is that if well-designed studies fail to detect an effect, this might provide some evidence that that effect is likely to be a very small one, and that the null hypothesis might be very close to being true, even if the effect of treatments is not *precisely* zero. Bayesian approaches have been applied to the problem of statistically demonstrating that an effect is so small that it should be effectively ignored (Rouanet, 1996). Nevertheless, the bias against "accepting the null" runs so strong in tests of the traditional null hypothesis that this framework simply doesn't leave you with any appealing alternative when the effect of treatments *is* negligibly small. You will either collect a very large sample and reject the null (which may mislead you into thinking that the effect of treatments is something other than trivial) or you will fail to reject it and perhaps go out and collect more data about an essentially meaningless question.

In tests of minimum-effect hypotheses, there is a realistic possibility that the hypothesis being tested (i.e., that the effect of treatments is at best negligible) is indeed true, and there is a real need to develop procedures or conventions for deciding when to "accept the null". We believe that power analysis plays a critical role in determining and defining those procedures or conventions.

Suppose the hypothesis being tested is that the effect of treatments is at best negligible (e.g., treatments account for 1% or less of the variance in outcomes). A powerful study could provide strong evidence that this hypothesis is true. For example, if power is .80, this translates into odds of 4 to 1 that a statistical test will reject this hypothesis *if it is in fact false*. Failure to reject the null under these conditions can mean only one of two things: (1) the null really is true, or (2) the null is false, and this is that one test in five that yields the wrong result. The most logical conclusion to reach in this study is that the effects *are* negligibly small.

As we noted in Chapter 3, a complete evaluation of the meaning of the outcomes of statistical tests requires some knowledge about the probability that the null hypothesis being tested is true (i.e., the prior probability of $H_0$). The central weakness of traditional null hypothesis testing is that this prior probability is thought to be vanishingly small, and perhaps zero (Murphy, 1990). If this prior probability is zero, tests of the null hypothesis cannot provide much useful information.

---

### Type I Errors in Minimum-Effect Tests Revisited

In tests of the traditional null hypothesis, the likelihood if making a Type I error is defined by the alpha level of your statistical test. That is, if you define "statistically significant" in terms of a $p$ level of .05, this means that if the null hypothesis is true, the probability of falsely rejecting this hypothesis is .05. As we have noted many times in this book, the real risk of making a Type I error is much smaller than .05 because the traditional null hypothesis is rarely, if ever true.

What if you are testing a minimum effect hypothesis, for example, the hypothesis that treatments will explain 1% or less of the variance in your dependent variable? There is a meaningful possibility that this hypothesis *will* be true, and therefore, Type I errors need to be taken seriously. Unlike the traditional null hypothesis, the alpha level does not define the expected frequency of Type I errors. Rather, it defines the *maximum* probability of a Type I error. The actual probability of rejecting a minimum-effect null hypothesis may be substantially lower than the alpha level of this test.

Consider two different scenarios. In both scenarios, the null hypothesis is that treatments account for 1% of the variance of less. In the first case, treatments account for 9/10ths of one percent of the variance in outcomes in the population, but in the second, treatments for only 1/10th of one percent of variance (i.e., the true effects in the two scenarios are equivalent to $PV = .009$ vs. $PV = .001$). In each case, you set the alpha level for your test to be .05. As you might guess, the likelihood of making a Type I error is higher in the first scenario, where the null hypothesis is very close to being true than in the second, where it is not close at all. For example, consider a study examining the correlation between two variables that tests the null hypothesis that this correlation is .10 of less (i.e., that $PV$ is .01 or less). With a sample of $N = 100$, the probability of making a Type I error is the true effect is $PV = .001$ is .0012. If the true effect is $PV = .009$, the probability of making a Type I error is .045, close to the nominal alpha level. In general, the smaller the population effect, the less likely it is that you will make a Type I error.

Suppose you use a larger sample, say $N = 1,000$. The Type I error rates for the two scenarios studied here will be .0007 and .035, respectively. That is, as you increase your sample size you will increase the likelihood of correctly rejecting the minimum effect null *and* decrease the likelihood of making a Type I error. In tests of the traditional null hypothesis, the likelihood of a Type I error stays at .05 no matter what the sample size, *as long as this null hypothesis is true.* Minimal effect tests allow you to simultaneously increase power *and* decrease the likelihood of a Type I error.

---

The central weakness of the alternative approach described in this book, in which minimum-effect hypotheses are framed and tested, is that the prior probability that the null hypothesis is true is generally unknown. In Chapter 3, we noted that the prior probability of the traditional null hypothesis would necessarily be very low, and by most definitions is zero. It is likely that the prior probability of a minimum-effect null will also be somewhat low, especially if you are testing the hypothesis that the effect of treatments is at best negligible. Our rationale for believing that this prior will be small is that most treatments in the social and behavioral sciences do indeed have an effect. In Chapter 1, we noted that Lipsey and Wilson's (1993) review of over 300 meta-analyses of research studying the efficacy of psychological, educational, and behavioral treatments. These meta-analyses, in turn, summarize the results of thousands of studies in areas ranging from smoking cessation success rates to the effectiveness of computer-aided instruction. Over 85% of the meta-analyses they summarized reported effects that exceeded conventional criteria for "small effects" (i.e., $d = .20$ or $PV = .01$ or less). It is clear from the massive body of research summarized in that paper that a broad range of treatments and applications in the social and behavioral sciences have at least some effect, and the likelihood that a new treatment will have a negligible effect strikes us as relatively small, especially if the new treatment is solidly based in theory and research.

If you are working with well-established treatments that have been shown to have meaningful effects in a range of settings, the likelihood of a Type I error might be so small that they are no more likely when testing a minimum-effect null hypothesis than when testing the null hypothesis that there is no effect at all. In the end, questions about whether the null hypothesis is likely to be true is an empirical question, and the more certain you are that treatments have substantial effects, the less value there is in testing any type of null hypothesis. In these conditions, you should focus on effect size estimates, not on the outcomes of significance tests.

## Balancing Errors in Testing Minimum-Effect Hypotheses

In tests of the traditional null hypothesis, Type I errors are practically impossible and there is virtually nothing to be lost by setting alpha as high as possible. In tests of the minimum-effect hypothesis, the strategy of setting alpha as high as possible is no longer appropriate. The whole distinction between traditional and minimum-effect null hypotheses is that there is some realistic possibility that a minimum-effect null is true, and it is, therefore, possible to make a Type I error.

Although Type I errors are a real possibility in tests of minimum-effect null hypotheses, this does not mean that power should be ignored in carrying out these tests. The *possibility* of Type I errors should not blind you to the substantial likelihood that you will make Type II errors if you choose an unduly stringent alpha level. Choose any cell of the One-Stop $F$ Table in Appendix B, and you will see that: (a) a larger $F$ value is needed to reject a minimum-effect null than to reject the traditional null, given the same alpha level, and (b) a larger $F$ value is needed to reject the hypothesis that effects are small to moderate (e.g., they account for 5% or less of the variance) than to reject the hypothesis that these effects are negligibly small (i.e., they account for 1% or less of the variance). That is, all other things being equal, it is harder to reject a minimum-effect hypothesis than to reject the hypothesis that treatments have no effect whatsoever. The more demanding the hypothesis (e.g., 5% vs. 1% as the upper bound to be tested), the harder it is to reject $H_0$. In our view, there is usually no good reason to make things even more difficult than they already are by choosing an unrealistically stringent alpha level. Earlier, we suggested that when testing the traditional null hypothesis, the .01 alpha level should usually be avoided. We believe that the same advice holds for tests of minimum-effect hypotheses; the .01 alpha level should still be avoided in most cases.

We suggest a two-part test for determining whether should you use an alpha level of .01 rather than .05. First, .01 makes sense only if the consequences of a Type I error are relatively serious. In Chapter 4, we discussed concrete ways of comparing the perceived seriousness of Type I and Type II errors and noted that researchers often act as if falsely rejecting the null is much more serious than failing to reject the null when you should.

This may be true in some settings, but before deciding to set a stringent alpha level (thus markedly decreasing your power), we believe you should explicitly consider the relative costs of the two errors. Choose .01 only if there is a good reason to believe that a Type I error is substantially more serious than a Type II error.

Second, .01 makes sense only if the prior probability that the null hypothesis is true is reasonably high. That is, if the hypothesis to be tested is that a treatment has at most a negligible effect (e.g., it accounts for 1% or less of the variance in outcomes), you should be concerned with Type I errors *only* if there is some realistic possibility that the effects of treatments are indeed trivial. Finally, keep in mind that this is a two-part test. Use the .01 level rather than the .05 level only if the consequences of a Type I error are large, *and* the possibility that one might occur are substantial. In all other cases, we think you should use .05 as an alpha level, and that you should use an even more lenient alpha level whenever possible.

### Statistical Power and the Replication Crisis

In a widely cited article in *Nature*, the Open Science Collaboration (2015) carefully replicated 100 studies published in top psychology journals. They worked with the authors of these papers, used identical methods and materials, and sampled similar subject publications wherever possible. They found that key results from these papers often failed to replicate. Applying criteria that included both the size of the effect and whether results were statistically significant, they judged that only 39% of the published results held up in replication. The frequent failure of important findings to replicate had been recognized well before this paper (e.g., in 2012, the journal *Perspectives on Psychological Science* published a special issue discussing the replication crisis), but this paper served as an important wake-up call for psychologists and for scholars in other fields (e.g., Bettis, Ethiraj, Gambardella, Helfat & Mitchell, 2016).

There are many possible explanations for the replication crisis, including authors' use of questionable research practices (Banks, Rogelberg, Woznyj, Landis & Rupp, 2016), such as HARKing (creating hypotheses after the results are known) and *p*-hacking (conducting multiple tests and different types of data screening in search of results where $p$ is .05 or lower). However, it seems clear that statistical power is a major explanation for the frequent failure of research findings to replicate. As we have documented in several chapters in this book, the power of studies in the behavioral and social sciences is typically small. Because significance tests are almost always used as a gatekeeper for publication (e.g., virtually all of the original findings the authors participating in the Open Science

Consortium attempted to replicate were statistically significant), a combination of low power and a strong bias against publishing results that are not statistically significant means that the results that *are* published will often represent unusually strong effects and the likelihood that these findings will replicate will necessarily be small. For example, in the Open Science Consortium (2015) study, the average effects in the published journal articles that were the target for replication (expressed in terms of the correlation coefficient, *r*) were relatively strong, (the mean *r* equaled .403). In careful replications of these 100 studies, the effects were less than half this strong (the mean *r* in replications equaled .197).

One way to appreciate the role of significance tests in the replication crisis is to carry out simulation studies. For example, it is relatively simple to write **R** code that creates a normally distributed set of 10,000 effect size (ES) estimates that matched the distribution of effect sizes in the replication studies (i.e., a mean of $r = .197$ and a standard deviation of .257). If you simply screen out nonsignificant values (i.e., $p > .05$), this will dramatically bias the sorts of findings that *are* eligible for publication. For example, if $N = 60$ (the approximate average sample size in the original studies), screening out the nonsignificant correlations creates a distribution of findings with a mean effect size of .373, quite close to the mean effect size in the set of published studies (i.e., $r = .403$). In other words, if the true effect size is .197, simply screening out the nonsignificant values would create an upward bias nearly as large as the upward bias seen in published studies.

The bottom line is that if you carefully replicate studies that, despite their low power have survived publication review, you can be confident that your replications will report smaller effects than the original published studies. Further simulation studies have suggested that other factors, including the high rejection rates of some top journals and reviewers' preference for stronger rather than weaker effects, are also a factor in the tendency of good journals to publish papers that systematically overestimate the effects of treatments or interventions, but that the combination of low power and a strong reliance on significance testing are enough to create a substantial barrier to reproducibility in the social and behavioral sciences. Of course, many of these problems could have been avoided if more authors paid close attention to the power of their studies. The average sample size in the studies chosen for replication (all published in top journals) was approximately 60, and it is hard to believe that authors would have made this choice if they had conducted a power analysis. Based on the findings from the replications (which were no screened to reject nonsignificant results), the average power of these studies was probably about .32 and samples closer to $N = 200$ would have been needed to achieve power of .80.

# Power Analysis: Benefits, Costs, and Implications for Hypothesis Testing

If power analysis is taken seriously, there will be fundamental changes in the design, execution, and interpretation of research in the social and behavioral sciences. We believe that most of these changes will be beneficial, and we are enthusiastic advocates of power analysis. As we will note below, the indirect benefits of power analysis may prove, in the long run, even more important than the direct benefits of adopting this approach. There are, of course, some costs associated with incorporating power analysis in the design and interpretation of research; however, we believe the benefits still substantially outweigh the costs. Finally, it is useful to consider the implications of having extreme levels of power (either extremely high or extremely low) when conducting statistical hypothesis tests.

## Direct Benefits of Power Analysis

As we noted in Chapter 1, power analysis can be used as both a planning tool (e.g., determining how many subjects should be included in a study) and a diagnostic tool (e.g., making sense of previous studies that have either reported or failed to report "significant" results). Individuals who incorporate statistical power analysis into their research repertoire are better equipped to both plan and diagnose research studies, and they directly benefit from the information provided by power analyses.

### Planning Research

Statistical power analysis provides a rational framework for making important decisions about the design and scope of one's study. To be sure, many subjective decisions must be made in applying power analysis (e.g., what effect size do you anticipate, what alpha level is best), and the techniques described in this book do not represent a foolproof formula for making decisions about research design (e.g., choosing between repeated-measures or between-subjects designs) or sample size. The advantage of power analysis over other methods of making these important decisions, which are often made based on the force of habit, or by following the lead of other researchers, is that it makes explicit the consequences of these design choices for your study. If you are seriously interested in rejecting the null hypothesis, we think a power analysis is necessary in making good choices about study design and sample size.

Power analysis also highlights the importance of a decision that is usually ignored or made solely based on conventions in one's field – i.e., the alpha level that defines "statistical significance". The choice of stringent criteria (e.g., $\alpha = .01$) is sometimes interpreted as scientifically rigorous, whereas the choice of less rigorous criteria (e.g., $\alpha = .10$) is sometimes derided as "soft science". Nothing could be farther from the truth. Any decision about alpha levels implies some wish to balance Type I and Type II errors, and power analysis is necessary if you wish to make any kind of sense out of that balance. Once you appreciate the implications of choosing different alpha levels for the statistical power of your studies, you are more likely to make sensible choices about this critical parameter.

If power analysis is taken seriously, fewer studies with small samples or insufficient sensitivity will be done. In our view, researchers benefit substantially by knowing whether the study they have in mind has any real likelihood of detecting treatment effects. As we will note below, the indirect benefits to the field that might come with a decline in small-sample research are even greater than the benefits to the individual researcher.

## *Interpreting Research*

One criticism of tests of the traditional null hypothesis is that they routinely can mislead researchers and readers. Researchers who uncover a "significant" result are likely to confuse that with an important or meaningful result. This is hardly surprising; most dictionary definitions of "significant" include "important", "weighty", or "noteworthy" as synonyms. Similarly, "nonsignificant" is easily confused with "not important" or "non-meaningful". As power analysis clearly shows, very meaningful and important treatment effects are likely to be "nonsignificant" if the study lacks power, whereas completely trivial effects are likely to be "significant" if enough data are collected. It is impossible to sensibly interpret "significant" or "nonsignificant" results without considering the level of statistical power in the study that produced those results.

To give a concrete illustration, suppose you reviewed a dozen studies, all of which reported a "nonsignificant" correlation between attitudes toward drug use and subsequent drug consumption. What does this mean? If the studies are all based on small samples, it is entirely possible that there *is* a real and meaningful correlation between attitudes and subsequent behavior (e.g., if $N = 30$ and $\alpha = .05$, power for detecting a correlation as large as .30 is only .50), and that the studies simply did not have enough power to detect it. On the other hand, if all the studies included very large samples (e.g., $N = 2,500$), you could probably conclude that there is essentially no relationship between present attitudes and future behavior. Although the traditional null hypothesis might not be literally true in this instance, it would have to be very nearly true. With this much power, the studies you reviewed would have almost certainly detected any consistent relationships between attitudes and behavior.

## Is HARKing a Serious Problem?

HARKing (forming hypotheses after results are known) appears to be widespread, and it is thought to be a serious problem. Different surveys of researchers have suggested that between 30% and 90% of researchers have engaged in HARKing at some point in their careers (Bedian, Taylor & Miller, 2010; Fanelli, 2009; John, Loewenstein, & Prelec, 2012). At its worst, HARKing involves analyzing your data, seeing which relationships or effects are statistically significant and then writing an introduction section "predicting" whatever you found, no matter how unlikely the finding or how convoluted the explanation is. Less extreme versions of HARKing might involve using insights obtained by examining your data to fine-tune hypotheses to better fit what you have found.

HARKing is not only an ethical problem; it could also be a practical one. HARKing involves chasing your data, and if you do this, you are more likely to pay attention to and tailor your thinking in the direction of strong effects (which might represent unusual or unexpected results) rather than weaker ones. This practice could bias the types of effects that are reported in the literature and contribute to the replication crisis.

While HARKing has been widely discussed as a *potentially* serious problem, there have been few studies that have documented how much harm HARKing might lead to. To address this issue, Murphy and Aguinis (2019) conducted a series of studies that simulated different types of HARKing. For example, they looked at the implications of researchers looking at a set of statistical analyses that all examine the same question (e.g., the correlation between two specific variables) and cherry-picking the strongest finding for their publications. Their analyses suggest that this form of HARKing can bias findings, but that the effects of this bias are generally small. They also looked at a more extreme form of HARKing (which they call *question trolling*) in which researchers examine all sorts of relationships among different sets of variables and choose the strongest relationships as the focus of their study. This type of HARKing, in which the research question itself is not chosen until after the results are examined can have substantial effects.

Finally, Murphy and Aguinis (2019) remind us that HARKing is not always the result of author misconduct. Reviewers and journal editors sometimes suggest (some and sometimes even insist) that authors modify their introduction sections and their hypotheses in the light of the findings they report. No matter who is responsible for HARKing, it should be avoided. It undermines the credibility of research and whatever benefits it conveys to authors are likely to be short-lived.

## Indirect Benefits of Power Analysis

The widespread use of power analysis is likely to confer many indirect benefits. Most notably, studies that are designed with statistical power in mind are likely to use large samples and sensitive procedures. Perhaps even more important, power analysis directs the researcher's attention toward the most important parameter of all – i.e., the effect size. The ultimate benefit of statistical power analysis may be that it forces researchers to think about the strength of the effects they study, rather than thinking only about whether a particular effect is "significant".

### *Large Samples, Sensitive Procedures*

Small samples are the bane of social science research (Hunter & Schmidt, 1990; Schmidt, 1994). These studies produce unstable results, which in turn produce attempts to develop theories to "explain" what may be little more than sampling error. If power analyses were routinely included in the process of designing and planning studies, large samples would be the norm and sampling error would not loom so large as a barrier to cumulative progress in research.

Proponents of meta-analysis (e.g., Schmidt, 1994) note that by combining the outcomes of multiple-small sample studies, it is possible to draw sensible conclusions about effect sizes, even if the individual study samples are too small to provide either sufficient power or stable results. There is merit to this position, but there are also two problems with this solution to the problem of small samples. First, it creates a two-tiered structure in which the primary researchers do all the work, with little possibility of rewards (i.e., they do studies that cannot be published because of insufficient power and sensitivity) and the meta-analyst gets all the credit for amassing this material into an interpretable whole. Second, it leaves the meta-analyst at the mercy of a pool of primary researchers. Unless there are many studies examining *exactly* the question the meta-analyst wants to answer, the only alternatives are to change the question or to aggregate together studies that in fact differ in important ways. Neither alternative seems attractive, and if power analysis becomes routine, neither will be strictly necessary. If future studies include large samples and sensitive procedures, the need for meta-analyses will become less pressing than it is today.

The decision to use large samples is itself likely to improve other aspects of the research. For example, if you know that you will have to devote considerable time and resources to data collection, you will probably take more care to pretest, use reliable measures, follow well-laid-out procedures, etc. In contrast, if running a study amounts to little more than rounding up 25 undergraduates and herding them to your lab, the need for careful planning, precise measurement, etc. might not be pressing. In large-sample research, you may only have one chance to get things right, and you are less likely to rely on shoddy measures, incomplete procedures, etc. The net result of all this is that studies carried out with careful attention to statistical power are likely to be better and more useful than studies carried out with little regard for power.

*Focus on Effect Size*

If you scan most social science journals, you will find that the outcomes of significance tests are routinely reported, but effect size information is sometimes nowhere to be found. Our statistical training tends to focus our attention on *p* values and significance levels, and not on the substantive question of how well our treatments, interventions, tests, etc. work (See Cowles, 1989, for a historical analysis of why social scientists focus on significance tests). One of the most important advantages of statistical power analysis is that it makes it virtually impossible to ignore effect sizes.

The whole point of statistical analysis is to help you understand your data, and it has become increasingly clear over the years that an exclusive focus on significance testing is an impediment to understanding what the data mean (Cohen, 1994; Murphy, 2021; Schmidt, 1994; Wilkinson et al., 1999). Statistical power analysis forces you to think about the sort of effect you expect, or at least about the sort of effect you want to be able to detect, and once you start thinking along these lines, it is unlikely you will forget to think about the sort of effects you find. If power analysis did nothing more than direct researchers' attention to the size of their effects, it would be well worth the effort.

## Costs Associated with Power Analysis

Statistical power analysis brings many benefits, but there are also costs. Most notably, researchers who pay attention to statistical power will find it harder to carry out studies than researchers who do not think about power when planning or evaluating studies. Most researchers (the authors included) have done studies with small samples and insufficient power, and have "gotten away with it", in the sense that they reported significant results. Even when power is low, there is always some chance that you will reject $H_0$, and a clever researcher can make a career out of "getting lucky". Power analysis will lead you to do fewer small-sample studies, which in the long run might mean fewer studies period. It is relatively easy to do a dozen small-sample studies, with the knowledge that some will work, and some will not. It is not so easy to do a dozen large-sample studies, and one long-term result of applying power analysis is that the sheer number of studies performed in a field might go down. We don't see this as a bad thing, at least if many low-quality, small-sample studies are replaced with a few higher-quality, large-sample studies. Nevertheless, the prospects for building a lengthy vita by doing dozens of studies might be diminished if you pay serious attention to power analysis.

184  *The Implications of Power Analyses*

The most serious cost that might be associated with the widespread use of power analysis is an overemphasis on scientific conservatism. If studies are hard to carry out, and require significant resources (time, money, energy), there may be less willingness to try new ideas and approaches, or to test creative hypotheses. The long-term prospects for scientific progress are not good if researchers are unwilling or unable to take risks or try new ideas.

## Implications of Power Analysis: Can Power Be Too High?

Throughout this book, we have advocated paying attention to the probability that you will be able to reject a null hypothesis you believe to be wrong (i.e., power). The pitfalls of low power are reasonably obvious, but it is worth considering whether power can be too high. Suppose you follow the advice laid out in this book and design a study with a very high level of power (e.g., power = .95). One implication is there is little real doubt about the outcomes of your statistical tests; with few exceptions, your tests will yield "significant" outcomes.

When power is extreme (either high or low), you are not likely to learn much by conducting a formal hypothesis test. This might imply that power can be too high. We don't think so. Even when the outcome of a formal statistical hypothesis test is virtually known in advance, statistical analysis still has clear and obvious value. First, the statistics used in hypothesis testing usually provide an effect size estimate, or the information needed to make this estimate. Even if the statistical test itself provided little new information (with very high or very low power, you know how things will turn out), the *process* of carrying out a statistical test usually provides information that can be used to evaluate the stability and potential replicability of your results.

Consider, for example, the familiar $t$-test. The $t$ statistic is a ratio of the difference between sample means to the standard error of the difference.[1] If the level of power in your study is very high, tests of the significance of $t$ might not be all that informative; high power means that it will exceed the threshold for "significance" in virtually all cases. However, the *value* of the $t$ is still informative because it gives you an easy way of determining the standard error term, which in turn can be used in forming confidence intervals. For example, if the $M_1 - M_2 = 10.0$ and $t = 2.50$, it follows that the standard error of the difference between the means is 4.0 (i.e., 10.0/2.5), and a 95% confidence interval for the difference between means would be 7.84 units wide (i.e., $1.96 \times 4.0$). This confidence interval gives a very concrete indication of how much variation one might expect from study to study when comparing $M_1$ to $M_2$.

In general, the standard error terms for test statistics tend to become smaller as samples get larger. This is a concrete illustration of the general principle that large samples provide stable and consistent statistical estimates, whereas small samples provide unstable estimates. Even in settings where the significance of a particular statistical test is not in doubt, confidence intervals provide very useful information. Obtaining a confidence interval allows you to determine just how much sampling error you might expect in your statistical estimates.

The paragraph above illustrates a distinction that is sometimes blurred by researchers – i.e., the distinction between statistical analysis and null hypothesis testing. Researchers in the behavioral and social sciences have tended to over-emphasize formal hypothesis tests, and have paid too little attention to critical questions such as "How large is the effect of the treatments studied here?" (Cohen, 1994; Cowles, 1989; Wilkinson et al., 1999). Ironically, serious attention to the topic of power analysis is likely to *reduce* researchers' dependence on significance testing. The more you know about power, the more likely you are to take steps to maximize statistical power, which means that rejecting the null should be nearly a foregone conclusion. Once you understand that the null hypothesis test is *not* the most important facet of your statistical analysis, you are likely to turn your attention to the aspects of your analysis that are more important, such as estimating effect sizes.

Does all this mean that null hypothesis tests should be abandoned? Probably not. First, as we have noted throughout this book, many of the outstanding criticisms of the null hypothesis can be easily addressed by shifting from tests of point hypotheses (e.g., that treatments have no effect whatsoever) to tests of range or interval hypotheses (e.g., that the effects of treatments fall within some range of values denoting small effects). Second, the prospects for fundamental changes in research strategies seem poor, judging from the historical record (Cowles, 1989). Statisticians have been arguing for decades that the use of confidence intervals is preferable to the use of null hypothesis tests, with little apparent effect on actual research practice. Critics of null hypothesis tests have not suggested an alternative that is both viable *and* that is likely to be widely adopted. There is every indication that null hypothesis tests are here to stay, and that careful attention should be given to methods of making the process of hypothesis testing as useful and informative as possible. Careful attention to the principles of power analysis is likely to lead to better research and better statistical analyses.

## Note

1  The formula for $t$ is

$$t = \frac{M_1 - M_2}{SE_{M1-M2}} \quad \text{or} \quad t = \frac{M_1 - M_2}{\sqrt{\frac{\sigma_1^2}{n1} + \frac{\sigma_2^2}{n2}}}$$

# Appendix A
## Translating Common Statistics into F-Equivalent and PV Values

|  |  |  | Degrees of Freedom | |
|---|---|---|---|---|
| Statistic | F-equivalent | PV | $df_{hyp}$ | $df_{err}$ |
| t-test for difference between means[a] | $F(1, df_{err}) = t^2$ | $PV = \dfrac{t^2}{t^2 + df_{err}}$ | — | $N-2$ |
| Correlation coefficient[b] | $F(1, df_{err}) = \dfrac{r^2 df_{err}}{1 - r^2}$ | $PV = r^2$ | — | $N-2$ |
| Multiple $R^2$ | $F(df_{hyp}, df_{err}) = \dfrac{R^2 df_{err}}{(1 - R^2) df_{hyp}}$ | $PV = R^2$ | p | $N - p - 1$ |
| Change in $R^2$ | $F(df_{hyp}, df_{err}) = \dfrac{(R_F^2 - R_R^2)/df_{hyp}}{(1 - R_F^2)/df_{err}}$ | $PV = R^2$ | k | $N - p - 1$ |
| Chi square ($\chi 2$) | $F(df_{hyp}, df_{err}) = \chi 2 / df_{hyp}$ | c | $df_{hyp}$ | $\infty$ |
| Standardized mean Difference (d) | $F(1, df_{err}) = \dfrac{d^2 df_{err}}{4}$ | $PV = \dfrac{d^2}{d^2 + 4}$ | — | $N-2$ |
| d (repeated measures) | $F(1, df_{err}) = \dfrac{d^2 df_{err}}{4\sqrt{1 - r_{ab}}}$ | c | — | $N-2$ |
| Converting F to PV in a simple design |  | $PV = \dfrac{(df_{hyp} * F)}{[(df_{hyp} * F) + d_{err}]}$ | $df_{hyp}$ | $d_{err}$ |

Note: p = number of X variables in multiple regression equation. $R^2_F$ and $R^2_R$ represent the full and reduced model $R^2$ values, respectively, where the k represents the difference in the number of X variables in the two models. A $\chi 2$ variable with df = $df_{hyp}$ is distributed as F with degrees of freedom of $df_{hyp}$ and infinity. When d is used to describe the difference between two measures obtained from the same sample, $r_{ab}$ refers to the correlation between the two measures being compared.

a — Note that $d = \dfrac{2t}{\sqrt{df_{err}}}$

b — Note that $d = \dfrac{2r}{\sqrt{1 - r^2}}$

c — No simple transformation exists

# Appendix B
One-Stop *F* Table

# One-Stop F Table

| df Err | | | | | | | | | dfHyp | | | | | | | | |
|---|---|---|---|---|---|---|---|---|---|---|---|---|---|---|---|---|---|
| | | 1 | 2 | 3 | 4 | 5 | 6 | 7 | 8 | 9 | 10 | 12 | 15 | 20 | 30 | 40 | 60 | 120 |
| 3 | nil .05 | 10.13 | 9.55 | 9.28 | 9.12 | 9.01 | 8.94 | 8.89 | 8.85 | 8.81 | 8.79 | 8.74 | 8.70 | 8.66 | 8.62 | 8.59 | 8.57 | 8.55 |
| | nil .01 | 34.12 | 30.82 | 29.46 | 28.71 | 28.24 | 27.91 | 27.67 | 27.49 | 27.35 | 27.23 | 27.05 | 26.87 | 26.69 | 26.50 | 26.41 | 26.32 | 26.22 |
| | pow .50 | 8.26 | 7.21 | 6.78 | 6.54 | 6.39 | 6.29 | 6.22 | 6.16 | 6.12 | 6.08 | 6.02 | 5.97 | 5.91 | 5.86 | 5.82 | 5.80 | 5.76 |
| | pow .80 | 18.17 | 15.70 | 14.83 | 14.42 | 14.19 | 13.93 | 13.85 | 13.69 | 13.66 | 13.64 | 13.55 | 13.48 | 13.42 | 13.30 | 13.26 | 13.17 | 12.66 |
| | 1% .05 | 10.43 | 9.70 | 9.37 | 9.19 | 9.07 | 8.99 | 8.93 | 8.88 | 8.84 | 8.81 | 8.77 | 8.72 | 8.67 | 8.63 | 8.60 | 8.58 | 8.55 |
| | 1% .01 | 35.15 | 31.28 | 29.75 | 28.93 | 28.41 | 28.05 | 27.79 | 27.59 | 27.44 | 27.31 | 27.12 | 26.93 | 26.73 | 26.53 | 26.43 | 26.33 | 26.23 |
| | 1% .50 | 8.53 | 7.33 | 6.85 | 6.60 | 6.44 | 6.33 | 6.25 | 6.19 | 6.14 | 6.10 | 6.04 | 5.98 | 5.92 | 5.86 | 5.83 | 5.81 | 5.79 |
| | pow .80 | 18.72 | 16.04 | 14.96 | 14.50 | 14.25 | 13.98 | 13.89 | 13.82 | 13.69 | 13.66 | 13.56 | 13.48 | 13.42 | 13.34 | 13.31 | 13.25 | 13.07 |
| | 5% .05 | 11.72 | 10.30 | 9.76 | 9.48 | 9.30 | 9.18 | 9.09 | 9.02 | 8.97 | 8.92 | 8.86 | 8.79 | 8.73 | 8.66 | 8.63 | 8.59 | 8.56 |
| | 5% .01 | 39.41 | 33.23 | 31.00 | 29.84 | 29.13 | 28.64 | 28.30 | 28.03 | 27.82 | 27.66 | 27.41 | 27.15 | 26.90 | 26.64 | 26.52 | 26.39 | 26.26 |
| | pow .50 | 9.57 | 7.82 | 7.17 | 6.83 | 6.62 | 6.48 | 6.38 | 6.30 | 6.24 | 6.19 | 6.12 | 6.04 | 5.97 | 5.89 | 5.86 | 5.82 | 5.80 |
| | pow .80 | 20.77 | 17.02 | 15.60 | 14.98 | 14.63 | 14.30 | 14.16 | 14.07 | 13.90 | 13.86 | 13.72 | 13.62 | 13.51 | 13.40 | 13.38 | 13.33 | 13.18 |
| 4 | nil .05 | 7.71 | 6.94 | 6.59 | 6.39 | 6.26 | 6.16 | 6.09 | 6.04 | 6.00 | 5.96 | 5.91 | 5.86 | 5.80 | 5.75 | 5.72 | 5.69 | 5.66 |
| | nil .01 | 21.20 | 18.00 | 16.69 | 15.98 | 15.52 | 15.21 | 14.98 | 14.80 | 14.66 | 14.55 | 14.37 | 14.20 | 14.02 | 13.84 | 13.75 | 13.65 | 13.56 |
| | pow .50 | 6.68 | 5.48 | 5.00 | 4.73 | 4.55 | 4.43 | 4.34 | 4.27 | 4.22 | 4.17 | 4.10 | 4.03 | 3.96 | 3.89 | 3.86 | 3.82 | 3.78 |
| | pow .80 | 14.17 | 11.30 | 10.22 | 9.66 | 9.24 | 9.02 | 8.86 | 8.75 | 8.60 | 8.53 | 8.44 | 8.30 | 8.19 | 8.04 | 7.95 | 7.87 | 7.80 |
| | 1% .05 | 8.02 | 7.08 | 6.68 | 6.45 | 6.31 | 6.20 | 6.13 | 6.07 | 6.03 | 5.99 | 5.93 | 5.87 | 5.81 | 5.75 | 5.72 | 5.69 | 5.66 |
| | 1% .01 | 22.05 | 18.36 | 16.92 | 16.14 | 15.65 | 15.31 | 15.06 | 14.87 | 14.72 | 14.60 | 14.42 | 14.24 | 14.05 | 13.86 | 13.76 | 13.66 | 13.56 |
| | pow .50 | 6.94 | 5.60 | 5.07 | 4.78 | 4.60 | 4.46 | 4.37 | 4.30 | 4.24 | 4.19 | 4.12 | 4.05 | 3.97 | 3.90 | 3.86 | 3.82 | 3.79 |
| | pow .80 | 14.64 | 11.50 | 10.34 | 9.75 | 9.39 | 9.07 | 8.91 | 8.78 | 8.69 | 8.56 | 8.46 | 6.32 | 8.20 | 8.08 | 7.96 | 7.88 | 7.88 |
| | 5% .05 | 9.31 | 7.67 | 7.05 | 6.72 | 6.52 | 6.38 | 6.28 | 6.20 | 6.14 | 6.09 | 6.02 | 5.94 | 5.86 | 5.79 | 5.75 | 5.71 | 5.67 |
| | 5% .01 | 25.49 | 19.86 | 17.85 | 16.81 | 16.17 | 15.74 | 15.42 | 15.19 | 15.00 | 14.85 | 14.63 | 14.40 | 14.17 | 13.93 | 13.82 | 13.70 | 13.58 |
| | pow .50 | 8.05 | 6.10 | 5.39 | 5.01 | 4.77 | 4.61 | 4.50 | 4.40 | 4.33 | 4.28 | 4.19 | 4.10 | 4.02 | 3.93 | 3.88 | 3.84 | 3.80 |
| | pow .80 | 16.63 | 12.42 | 10.93 | 10.18 | 9.65 | 9.36 | 9.15 | 9.00 | 8.82 | 8.73 | 8.61 | 8.43 | 8.25 | 8.12 | 8.00 | 7.91 | 7.78 |

One-Stop F Table    189

| df2 | | α | | | | | | | | | | | | | | | | | |
|---|---|---|---|---|---|---|---|---|---|---|---|---|---|---|---|---|---|---|---|
| 5 | nil | .05 | 6.61 | 5.79 | 5.41 | 5.19 | 5.05 | 4.95 | 4.88 | 4.82 | 4.77 | 4.73 | 4.68 | 4.62 | 4.56 | 4.50 | 4.46 | 4.43 | 4.40 |
|   | nil | .01 | 16.26 | 13.27 | 12.06 | 11.39 | 10.97 | 10.67 | 10.46 | 10.29 | 10.16 | 10.08 | 9.89 | 9.72 | 9.55 | 9.38 | 9.29 | 9.20 | 9.11 |
|   | pow | .50 | 5.91 | 4.66 | 4.14 | 3.87 | 3.70 | 3.57 | 3.48 | 3.41 | 3.35 | 3.31 | 3.23 | 3.15 | 3.08 | 3.00 | 2.96 | 2.91 | 2.87 |
|   | pow | .80 | 12.35 | 9.38 | 8.19 | 7.60 | 7.24 | 6.94 | 6.77 | 6.59 | 6.50 | 6.43 | 6.28 | 6.11 | 5.97 | 5.83 | 5.74 | 5.65 | 5.51 |
|   | 1% | .05 | 6.94 | 5.93 | 5.50 | 5.26 | 5.10 | 4.99 | 4.91 | 4.85 | 4.80 | 4.76 | 4.70 | 4.63 | 4.57 | 4.50 | 4.47 | 4.44 | 4.40 |
|   | 1% | .01 | 17.07 | 13.61 | 12.26 | 11.54 | 11.08 | 10.76 | 10.53 | 10.35 | 10.21 | 10.10 | 9.93 | 9.75 | 9.58 | 9.39 | 9.30 | 9.21 | 9.12 |
|   | pow | .50 | 6.20 | 4.78 | 4.24 | 3.93 | 3.75 | 3.61 | 3.51 | 3.44 | 3.37 | 3.33 | 3.25 | 3.17 | 3.09 | 3.00 | 2.96 | 2.92 | 2.88 |
|   | pow | .80 | 12.85 | 9.50 | 8.38 | 7.68 | 7.30 | 6.99 | 6.81 | 6.68 | 6.53 | 6.46 | 6.30 | 6.17 | 5.98 | 5.84 | 5.76 | 5.66 | 5.54 |
|   | 5% | .05 | 8.31 | 6.54 | 5.88 | 5.53 | 5.32 | 5.17 | 5.06 | 4.98 | 4.91 | 4.86 | 4.78 | 4.70 | 4.62 | 4.54 | 4.49 | 4.45 | 4.41 |
|   | 5% | .01 | 20.28 | 14.97 | 13.10 | 12.13 | 11.54 | 11.14 | 10.85 | 10.63 | 10.45 | 10.31 | 10.10 | 9.89 | 9.68 | 9.46 | 9.35 | 9.24 | 9.13 |
|   | pow | .50 | 7.42 | 5.33 | 4.56 | 4.18 | 3.94 | 3.77 | 3.65 | 3.55 | 3.48 | 3.42 | 3.32 | 3.23 | 3.13 | 3.03 | 2.98 | 2.93 | 2.88 |
|   | pow | .80 | 14.92 | 10.5 | 8.90 | 8.11 | 7.64 | 7.27 | 7.05 | 6.84 | 6.72 | 6.62 | 6.44 | 6.24 | 6.07 | 5.89 | 5.78 | 5.67 | 5.55 |
| 6 | nil | .05 | 5.99 | 5.14 | 4.76 | 4.53 | 4.39 | 4.28 | 4.21 | 4.15 | 4.10 | 4.06 | 4.00 | 3.94 | 3.87 | 3.81 | 3.77 | 3.74 | 3.70 |
|   | nil | .01 | 13.74 | 10.92 | 9.78 | 9.15 | 8.75 | 8.47 | 8.26 | 8.10 | 7.98 | 7.87 | 7.72 | 7.56 | 7.40 | 7.23 | 7.14 | 7.06 | 6.96 |
|   | pow | .50 | 5.45 | 4.15 | 3.67 | 3.39 | 3.21 | 3.08 | 2.99 | 2.92 | 2.86 | 2.82 | 2.74 | 2.66 | 2.58 | 2.49 | 2.45 | 2.40 | 2.31 |
|   | pow | .80 | 11.33 | 8.29 | 7.15 | 6.51 | 6.12 | 5.84 | 5.65 | 5.50 | 5.38 | 5.28 | 5.10 | 4.95 | 4.80 | 4.65 | 4.56 | 4.46 | 4.31 |
|   | 1% | .05 | 6.35 | 5.30 | 4.85 | 4.60 | 4.44 | 4.33 | 4.24 | 4.18 | 4.13 | 4.08 | 4.02 | 3.95 | 3.89 | 3.82 | 3.78 | 3.74 | 3.70 |
|   | 1% | .01 | 14.56 | 11.25 | 9.93 | 9.29 | 8.85 | 8.55 | 8.33 | 8.16 | 8.03 | 7.92 | 7.76 | 7.59 | 7.42 | 7.24 | 7.15 | 7.06 | 6.97 |
|   | pow | .50 | 5.77 | 4.32 | 3.75 | 3.45 | 3.25 | 3.12 | 3.02 | 2.95 | 2.89 | 2.84 | 2.76 | 2.68 | 2.59 | 2.50 | 2.45 | 2.41 | 2.31 |
|   | pow | .80 | 11.86 | 8.54 | 7.28 | 6.60 | 6.19 | 5.90 | 5.69 | 5.53 | 5.41 | 5.31 | 5.16 | 5.00 | 4.82 | 4.66 | 4.56 | 4.46 | 4.32 |
|   | 5% | .05 | 7.82 | 5.94 | 5.25 | 4.89 | 4.66 | 4.51 | 4.40 | 4.31 | 4.24 | 4.19 | 4.11 | 4.02 | 3.94 | 3.85 | 3.80 | 3.76 | 3.71 |
|   | 5% | .01 | 17.73 | 12.58 | 10.78 | 9.86 | 9.29 | 8.91 | 8.63 | 8.42 | 8.25 | 8.12 | 7.92 | 7.72 | 7.51 | 7.30 | 7.20 | 7.09 | 6.98 |
|   | pow | .50 | 7.11 | 4.88 | 4.13 | 3.72 | 3.47 | 3.29 | 3.17 | 3.06 | 2.98 | 2.93 | 2.83 | 2.74 | 2.64 | 2.53 | 2.48 | 2.42 | 2.32 |
|   | pow | .80 | 14.05 | 9.45 | 7.88 | 7.04 | 6.53 | 6.18 | 5.93 | 5.70 | 5.55 | 5.43 | 5.26 | 5.08 | 4.90 | 4.69 | 4.60 | 4.49 | 4.33 |
| 8 | nil | .05 | 5.32 | 4.46 | 4.07 | 3.84 | 3.69 | 3.58 | 3.50 | 3.44 | 3.39 | 3.35 | 3.28 | 3.22 | 3.15 | 3.08 | 3.04 | 3.00 | 2.96 |
|   | nil | .01 | 11.26 | 8.65 | 7.59 | 7.01 | 6.63 | 6.37 | 6.18 | 6.03 | 5.91 | 5.81 | 5.67 | 5.52 | 5.36 | 5.20 | 5.12 | 5.03 | 4.94 |
|   | pow | .50 | 4.94 | 3.63 | 3.12 | 2.82 | 2.66 | 2.52 | 2.41 | 2.36 | 2.30 | 2.24 | 2.18 | 2.11 | 2.04 | 1.95 | 1.91 | 1.86 | 1.73 |
|   | pow | .80 | 10.22 | 7.17 | 5.99 | 5.33 | 4.95 | 4.65 | 4.44 | 4.30 | 4.16 | 4.05 | 3.91 | 3.75 | 3.59 | 3.41 | 3.33 | 3.23 | 3.12 |
|   | 1% | .05 | 5.74 | 4.64 | 4.18 | 3.92 | 3.75 | 3.63 | 3.54 | 3.47 | 3.42 | 3.37 | 3.31 | 3.24 | 3.16 | 3.09 | 3.05 | 3.01 | 2.96 |
|   | 1% | .01 | 12.14 | 8.99 | 7.79 | 7.15 | 6.74 | 6.46 | 6.25 | 6.09 | 5.96 | 5.86 | 5.70 | 5.54 | 5.38 | 5.21 | 5.13 | 5.04 | 4.94 |
|   | pow | .50 | 5.36 | 3.79 | 3.22 | 2.88 | 2.71 | 2.56 | 2.45 | 2.39 | 2.32 | 2.29 | 2.20 | 2.13 | 2.05 | 1.96 | 1.91 | 1.86 | 1.73 |

*(Continued)*

# 190  One-Stop F Table

|    |     |       | 1     | 2     | 3    | 4    | 5    | 6    | 7    | 8    | 9    | 10   | 12   | 15   | 20   | 30   | 40   | 60   | 120  |
|----|-----|-------|-------|-------|------|------|------|------|------|------|------|------|------|------|------|------|------|------|------|
|    | pow | .80   | 10.86 | 7.42  | 6.14 | 5.44 | 5.03 | 4.72 | 4.49 | 4.34 | 4.20 | 4.12 | 3.93 | 3.77 | 3.60 | 3.42 | 3.34 | 3.23 | 3.13 |
|    | 5%  | .05   | 7.44  | 5.37  | 4.62 | 4.24 | 3.99 | 3.93 | 3.71 | 3.62 | 3.55 | 3.49 | 3.40 | 3.31 | 3.22 | 3.12 | 3.07 | 3.03 | 2.97 |
|    | 5%  | .01   | 15.41 | 10.35 | 8.61 | 7.72 | 7.18 | 6.81 | 6.54 | 6.34 | 6.19 | 6.06 | 5.86 | 5.67 | 5.47 | 5.27 | 5.17 | 5.07 | 4.96 |
|    | pow | .50   | 6.94  | 4.48  | 3.65 | 3.20 | 2.92 | 2.76 | 2.62 | 2.54 | 2.45 | 2.38 | 2.30 | 2.20 | 2.10 | 2.00 | 1.94 | 1.88 | 1.74 |
|    | pow | .80   | 13.34 | 8.47  | 6.79 | 5.90 | 5.35 | 5.01 | 4.74 | 4.56 | 4.39 | 4.25 | 4.07 | 3.88 | 3.69 | 3.48 | 3.37 | 3.26 | 3.14 |
| 10 | nil | .05   | 4.96  | 4.10  | 3.71 | 3.48 | 3.33 | 3.22 | 3.14 | 3.07 | 3.02 | 2.98 | 2.91 | 2.84 | 2.77 | 2.70 | 2.66 | 2.62 | 2.58 |
|    | nil | .01   | 10.04 | 7.56  | 6.55 | 5.99 | 5.64 | 5.39 | 5.20 | 5.06 | 4.94 | 4.85 | 4.71 | 4.56 | 4.11 | 4.25 | 4.17 | 4.08 | 3.99 |
|    | pow | .50   | 4.68  | 3.33  | 2.83 | 2.53 | 2.34 | 2.21 | 2.11 | 2.04 | 1.98 | 1.93 | 1.86 | 1.81 | 1.74 | 1.66 | 1.62 | 1.58 | 1.41 |
|    | pow | .80   | 9.65  | 6.58  | 5.40 | 4.75 | 4.34 | 4.05 | 3.84 | 3.68 | 3.55 | 3.44 | 3.28 | 3.14 | 2.97 | 2.75 | 2.69 | 2.59 | 2.50 |
|    | 1%  | .05   | 5.46  | 4.31  | 3.83 | 3.57 | 3.39 | 3.27 | 3.18 | 3.11 | 3.05 | 3.01 | 2.94 | 2.86 | 2.79 | 2.71 | 2.67 | 2.63 | 2.58 |
|    | 1%  | .01   | 11.02 | 7.93  | 6.77 | 6.14 | 5.75 | 5.48 | 5.28 | 5.12 | 5.00 | 4.90 | 4.75 | 4.59 | 4.43 | 4.26 | 4.18 | 4.09 | 3.99 |
|    | pow | .50   | 5.14  | 3.56  | 2.94 | 2.60 | 2.40 | 2.25 | 2.15 | 2.07 | 2.01 | 1.99 | 1.91 | 1.83 | 1.75 | 1.67 | 1.63 | 1.58 | 1.42 |
|    | pow | .80   | 10.37 | 6.89  | 5.57 | 4.87 | 4.42 | 4.12 | 3.90 | 3.73 | 3.59 | 3.51 | 3.34 | 3.16 | 2.98 | 2.79 | 2.71 | 2.60 | 2.50 |
|    | 5%  | .05   | 7.39  | 5.13  | 4.34 | 3.93 | 3.67 | 3.50 | 3.37 | 3.27 | 3.20 | 3.13 | 3.04 | 2.94 | 2.85 | 2.75 | 2.70 | 2.64 | 2.59 |
|    | 5%  | .01   | 14.48 | 9.38  | 7.64 | 6.75 | 6.21 | 5.85 | 5.58 | 5.38 | 5.23 | 5.10 | 4.91 | 4.72 | 4.52 | 4.32 | 4.22 | 4.12 | 4.01 |
|    | pow | .50   | 6.94  | 4.35  | 3.43 | 2.96 | 2.67 | 2.48 | 2.34 | 2.23 | 2.15 | 2.09 | 1.99 | 1.89 | 1.81 | 1.71 | 1.65 | 1.60 | 1.42 |
|    | pow | .80   | 13.14 | 8.06  | 6.28 | 5.37 | 4.81 | 4.44 | 4.16 | 3.95 | 3.79 | 3.66 | 3.46 | 3.26 | 3.07 | 2.85 | 2.74 | 2.63 | 2.52 |
| 12 | nil | .05   | 4.74  | 3.89  | 3.49 | 3.26 | 3.11 | 3.00 | 2.91 | 2.85 | 2.80 | 2.75 | 2.69 | 2.62 | 2.54 | 2.47 | 2.43 | 2.38 | 2.34 |
|    | nil | .01   | 9.33  | 6.93  | 5.95 | 5.41 | 5.06 | 4.82 | 4.64 | 4.50 | 4.39 | 4.30 | 4.16 | 4.01 | 3.86 | 3.70 | 3.62 | 3.54 | 3.45 |
|    | pow | .50   | 4.52  | 3.17  | 2.63 | 2.33 | 2.15 | 2.02 | 1.93 | 1.86 | 1.80 | 1.76 | 1.66 | 1.61 | 1.53 | 1.47 | 1.43 | 1.39 | 1.35 |
|    | pow | .80   | 9.30  | 6.23  | 5.03 | 4.38 | 3.97 | 3.69 | 3.48 | 3.32 | 3.20 | 3.09 | 2.91 | 2.76 | 2.58 | 2.41 | 2.31 | 2.21 | 2.09 |
|    | 1%  | .05   | 5.31  | 4.12  | 3.63 | 3.36 | 3.18 | 3.06 | 2.96 | 2.89 | 2.83 | 2.79 | 2.71 | 2.64 | 2.56 | 2.48 | 2.43 | 2.39 | 2.34 |
|    | 1%  | .01   | 10.40 | 7.34  | 6.19 | 5.57 | 5.19 | 4.92 | 4.72 | 4.57 | 4.45 | 4.35 | 4.20 | 4.04 | 3.88 | 3.72 | 3.63 | 3.54 | 3.45 |
|    | pow | .50   | 5.05  | 3.38  | 2.75 | 2.42 | 2.21 | 2.07 | 1.97 | 1.89 | 1.83 | 1.79 | 1.68 | 1.62 | 1.54 | 1.47 | 1.43 | 1.40 | 1.36 |
|    | pow | .80   | 10.12 | 6.55  | 5.22 | 4.51 | 4.07 | 3.77 | 3.54 | 3.37 | 3.24 | 3.13 | 2.95 | 2.78 | 2.60 | 2.42 | 2.31 | 2.21 | 2.10 |
|    | 5%  | .05   | 7.47  | 5.0:  | 4.20 | 3.76 | 3.49 | 3.31 | 3.17 | 3.07 | 2.99 | 2.93 | 2.83 | 2.73 | 2.62 | 2.52 | 2.46 | 2.41 | 2.35 |

dfHyp

## One-Stop F Table

| | | | | | | | | | | | | | | | | |
|---|---|---|---|---|---|---|---|---|---|---|---|---|---|---|---|---|
| | 5% | .01 | 14.07 | 8.88 | 7.12 | 6.23 | 5.68 | 5.31 | 5.05 | 4.85 | 4.69 | 4.56 | 4.37 | 4.18 | 3.98 | 3.78 | 3.68 | 3.57 | 3.47 |
| | pow | .50 | 7.08 | 4.26 | 3.30 | 2.81 | 2.51 | 2.32 | 2.18 | 2.08 | 2.00 | 1.93 | 1.80 | 1.72 | 1.61 | 1.52 | 1.47 | 1.42 | 1.36 |
| | pow | .80 | 13.18 | 7.84 | 6.00 | 5.07 | 4.49 | 4.11 | 3.83 | 3.62 | 3.46 | 3.33 | 3.10 | 2.91 | 2.69 | 2.48 | 2.36 | 2.24 | 2.11 |
| 14 | nil | .05 | 4.60 | 3.74 | 3.34 | 3.11 | 2.96 | 2.85 | 2.76 | 2.70 | 2.65 | 2.60 | 2.53 | 2.46 | 2.39 | 2.31 | 2.27 | 2.22 | 2.18 |
| | nil | .01 | 8.86 | 6.51 | 5.56 | 5.04 | 4.69 | 4.46 | 4.28 | 4.14 | 4.03 | 3.94 | 3.80 | 3.66 | 3.50 | 3.35 | 3.27 | 3.18 | 3.09 |
| | pow | .50 | 4.41 | 3.06 | 2.52 | 2.23 | 2.00 | 1.88 | 1.79 | 1.72 | 1.67 | 1.59 | 1.54 | 1.46 | 1.39 | 1.32 | 1.28 | 1.26 | 1.23 |
| | pow | .80 | 9.06 | 5.99 | 4.80 | 4.16 | 3.72 | 3.44 | 3.23 | 3.07 | 2.95 | 2.82 | 2.67 | 2.50 | 2.33 | 2.14 | 2.04 | 1.95 | 1.83 |
| | 1% | .05 | 5.24 | 4.00 | 3.50 | 3.22 | 3.04 | 2.91 | 2.82 | 2.75 | 2.69 | 2.64 | 2.56 | 2.49 | 2.40 | 2.32 | 2.27 | 2.23 | 2.18 |
| | 1% | .01 | 10.04 | 6.96 | 5.82 | 5.21 | 4.83 | 4.56 | 4.36 | 4.21 | 4.09 | 3.99 | 3.84 | 3.69 | 3.53 | 3.36 | 3.28 | 3.19 | 3.10 |
| | pow | .50 | 5.01 | 3.30 | 2.66 | 2.32 | 2.12 | 1.93 | 1.83 | 1.76 | 1.70 | 1.66 | 1.56 | 1.47 | 1.41 | 1.32 | 1.28 | 1.26 | 1.23 |
| | pow | .80 | 9.98 | 6.35 | 5.01 | 4.30 | 3.85 | 3.52 | 3.30 | 3.13 | 3.00 | 2.89 | 2.71 | 2.53 | 2.35 | 2.15 | 2.05 | 1.95 | 1.84 |
| | 5% | .05 | 7.62 | 5.02 | 4.13 | 3.66 | 3.38 | 3.19 | 3.05 | 2.94 | 2.86 | 2.79 | 2.69 | 2.58 | 2.47 | 2.36 | 2.31 | 2.25 | 2.19 |
| | 5% | .01 | 13.93 | 8.61 | 6.82 | 5.91 | 5.36 | 4.98 | 4.72 | 4.51 | 4.35 | 4.22 | 4.03 | 3.83 | 3.63 | 3.43 | 3.33 | 3.22 | 3.11 |
| | pow | .50 | 7.25 | 4.27 | 3.26 | 2.76 | 2.45 | 2.20 | 2.06 | 1.96 | 1.88 | 1.78 | 1.69 | 1.58 | 1.48 | 1.37 | 1.32 | 1.29 | 1.25 |
| | pow | .80 | 13.32 | 7.76 | 5.86 | 4.90 | 4.32 | 3.89 | 3.61 | 3.40 | 3.23 | 3.08 | 2.88 | 2.66 | 2.45 | 2.22 | 2.10 | 1.98 | 1.85 |
| 16 | nil | .05 | 4.49 | 3.63 | 3.24 | 3.01 | 2.85 | 2.74 | 2.66 | 2.59 | 2.54 | 2.49 | 2.42 | 2.35 | 2.27 | 2.19 | 2.15 | 2.10 | 2.06 |
| | nil | .01 | 8.53 | 6.23 | 5.29 | 4.77 | 4.44 | 4.20 | 4.03 | 3.89 | 3.78 | 3.69 | 3.55 | 3.41 | 3.26 | 3.10 | 3.02 | 2.93 | 2.84 |
| | pow | .50 | 4.33 | 2.98 | 2.39 | 2.10 | 1.92 | 1.80 | 1.67 | 1.61 | 1.56 | 1.53 | 1.44 | 1.36 | 1.28 | 1.22 | 1.18 | 1.15 | 1.12 |
| | pow | .80 | 8.88 | 5.83 | 4.61 | 3.97 | 3.56 | 3.28 | 3.05 | 2.89 | 2.77 | 2.67 | 2.49 | 2.32 | 2.14 | 1.96 | 1.86 | 1.76 | 1.64 |
| | 1% | .05 | 5.20 | 3.92 | 3.41 | 3.13 | 2.94 | 2.81 | 2.72 | 2.64 | 2.58 | 2.53 | 2.46 | 2.38 | 2.29 | 2.20 | 2.16 | 2.11 | 2.06 |
| | 1% | .01 | 9.81 | 6.71 | 5.57 | 4.96 | 4.58 | 4.31 | 4.12 | 3.97 | 3.85 | 3.75 | 3.60 | 3.45 | 3.29 | 3.12 | 3.03 | 2.94 | 2.85 |
| | pow | .50 | 5.00 | 3.25 | 2.60 | 2.26 | 2.00 | 1.86 | 1.77 | 1.65 | 1.60 | 1.56 | 1.46 | 1.38 | 1.29 | 1.23 | 1.18 | 1.15 | 1.13 |
| | pow | .80 | 9.90 | 6.22 | 4.87 | 4.15 | 3.68 | 3.37 | 3.15 | 2.95 | 2.82 | 2.71 | 2.53 | 2.35 | 2.16 | 1.98 | 1.87 | 1.76 | 1.65 |
| | 5% | .05 | 7.81 | 5.04 | 4.10 | 3.61 | 3.32 | 3.12 | 2.97 | 2.86 | 2.77 | 2.70 | 2.59 | 2.48 | 2.37 | 2.25 | 2.19 | 2.13 | 2.07 |
| | 5% | .01 | 13.91 | 8.47 | 6.63 | 5.71 | 5.15 | 4.77 | 4.49 | 4.29 | 4.13 | 4.00 | 3.80 | 3.60 | 3.39 | 3.19 | 3.08 | 2.97 | 2.86 |
| | pow | .50 | 7.45 | 4.31 | 3.26 | 2.73 | 2.36 | 2.16 | 2.02 | 1.86 | 1.79 | 1.73 | 1.60 | 1.49 | 1.37 | 1.28 | 1.22 | 1.18 | 1.14 |
| | pow | .80 | 13.52 | 7.75 | 5.79 | 4.80 | 4.17 | 3.77 | 3.48 | 3.24 | 3.07 | 2.94 | 2.71 | 2.49 | 2.26 | 2.04 | 1.92 | 1.80 | 1.66 |
| 18 | nil | .05 | 4.41 | 3.55 | 3.16 | 2.93 | 2.77 | 2.66 | 2.58 | 2.51 | 2.46 | 2.41 | 2.34 | 2.27 | 2.19 | 2.11 | 2.06 | 2.02 | 1.97 |
| | nil | .01 | 8.28 | 6.01 | 5.09 | 4.58 | 4.25 | 4.01 | 3.84 | 3.71 | 3.60 | 3.51 | 3.37 | 3.23 | 3.08 | 2.92 | 2.84 | 2.75 | 2.66 |

*(Continued)*

192  *One-Stop* F *Table*

$dfHyp$

| | | 1 | 2 | 3 | 4 | 5 | 6 | 7 | 8 | 9 | 10 | 12 | 15 | 20 | 30 | 40 | 60 | 120 |
|---|---|---|---|---|---|---|---|---|---|---|---|---|---|---|---|---|---|---|
| pow | .50 | 4.21 | 2.87 | 2.34 | 2.05 | 1.87 | 1.70 | 1.62 | 1.56 | 1.47 | 1.44 | 1.35 | 1.28 | 1.21 | 1.13 | 1.08 | 1.06 | 1.04 |
| pow | .80 | 8.73 | 5.68 | 4.49 | 3.84 | 3.44 | 3.13 | 2.93 | 2.77 | 2.62 | 2.52 | 2.35 | 2.19 | 2.01 | 1.82 | 1.71 | 1.61 | 1.50 |
| 1% | .05 | 5.19 | 3.87 | 3.35 | 3.06 | 2.87 | 2.74 | 2.64 | 2.57 | 2.50 | 2.45 | 2.38 | 2.30 | 2.21 | 2.12 | 2.07 | 2.02 | 1.97 |
| 1% | .01 | 9.67 | 6.54 | 5.39 | 4.78 | 4.40 | 4.13 | 3.94 | 3.79 | 3.67 | 3.57 | 3.42 | 3.27 | 3.11 | 2.94 | 2.85 | 2.76 | 2.66 |
| pow | .50 | 5.01 | 3.21 | 2.56 | 2.16 | 1.95 | 1.82 | 1.67 | 1.60 | 1.55 | 1.47 | 1.38 | 1.30 | 1.22 | 1.14 | 1.11 | 1.06 | 1.04 |
| pow | .80 | 9.87 | 6.13 | 4.76 | 4.01 | 3.56 | 3.26 | 3.01 | 2.84 | 2.71 | 2.58 | 2.39 | 2.22 | 2.03 | 1.84 | 1.73 | 1.62 | 1.50 |
| 5% | .05 | 8.02 | 5.09 | 4.09 | 3.59 | 3.28 | 3.07 | 2.92 | 2.80 | 2.71 | 2.64 | 2.52 | 2.41 | 2.29 | 2.17 | 2.11 | 2.05 | 1.98 |
| 5% | .01 | 13.99 | 8.40 | 6.52 | 5.58 | 5.00 | 4.62 | 4.34 | 4.13 | 3.97 | 3.83 | 3.63 | 3.43 | 3.22 | 3.01 | 2.90 | 2.79 | 2.68 |
| pow | .50 | 7.65 | 4.37 | 3.27 | 2.66 | 2.34 | 2.14 | 1.94 | 1.83 | 1.76 | 1.65 | 1.52 | 1.42 | 1.31 | 1.20 | 1.13 | 1.09 | 1.06 |
| pow | .80 | 13.75 | 7.77 | 5.76 | 4.71 | 4.10 | 3.69 | 3.37 | 3.14 | 2.98 | 2.81 | 2.59 | 2.37 | 2.14 | 1.91 | 1.77 | 1.65 | 1.52 |
| 20 nil | .05 | 4.34 | 3.49 | 3.10 | 2.87 | 2.71 | 2.60 | 2.51 | 2.45 | 2.39 | 2.35 | 2.28 | 2.20 | 2.12 | 2.04 | 1.99 | 1.95 | 1.90 |
| nil | .01 | 8.09 | 5.85 | 4.94 | 4.43 | 4.10 | 3.87 | 3.70 | 3.56 | 3.46 | 3.37 | 3.23 | 3.09 | 2.94 | 2.78 | 2.69 | 2.61 | 2.52 |
| pow | .50 | 4.17 | 2.82 | 2.29 | 2.00 | 1.77 | 1.66 | 1.58 | 1.47 | 1.43 | 1.35 | 1.32 | 1.21 | 1.14 | 1.05 | 1.01 | 0.99 | 0.97 |
| pow | .80 | 8.63 | 5.58 | 4.39 | 3.75 | 3.32 | 3.04 | 2.83 | 2.65 | 2.53 | 2.41 | 2.26 | 2.07 | 1.90 | 1.70 | 1.60 | 1.50 | 1.18 |
| 1% | .05 | 5.20 | 3.84 | 3.30 | 3.01 | 2.82 | 2.69 | 2.59 | 2.51 | 2.45 | 2.39 | 2.32 | 2.23 | 2.14 | 2.05 | 2.00 | 1.95 | 1.90 |
| 1% | .01 | 9.58 | 6.41 | 5.26 | 4.65 | 4.27 | 4.00 | 3.80 | 3.65 | 3.53 | 3.44 | 3.29 | 3.13 | 2.97 | 2.80 | 2.71 | 2.62 | 2.52 |
| pow | .50 | 5.03 | 3.19 | 2.47 | 2.13 | 1.92 | 1.73 | 1.63 | 1.57 | 1.47 | 1.43 | 1.34 | 1.23 | 1.15 | 1.06 | 1.04 | 0.99 | 0.97 |
| pow | .80 | 9.87 | 6.07 | 4.66 | 3.93 | 3.48 | 3.14 | 2.92 | 2.75 | 2.59 | 2.49 | 2.31 | 2.11 | 1.92 | 1.72 | 1.62 | 1.51 | 1.39 |
| 5% | .05 | 8.20 | 5.15 | 4.11 | 3.58 | 3.26 | 3.04 | 2.88 | 2.76 | 2.67 | 2.59 | 2.47 | 2.36 | 2.23 | 2.11 | 2.05 | 1.98 | 1.91 |
| 5% | .01 | 14.11 | 8.38 | 6.46 | 5.49 | 4.91 | 4.51 | 4.23 | 4.02 | 3.85 | 3.71 | 3.51 | 3.30 | 3.09 | 2.88 | 2.77 | 2.65 | 2.54 |
| pow | .50 | 7.95 | 4.43 | 3.30 | 2.67 | 2.34 | 2.12 | 1.92 | 1.81 | 1.68 | 1.63 | 1.50 | 1.35 | 1.24 | 1.12 | 1.06 | 1.02 | 0.98 |
| pow | .80 | 14.02 | 7.82 | 5.75 | 4.68 | 4.05 | 3.63 | 3.30 | 3.08 | 2.88 | 2.74 | 2.51 | 2.27 | 2.04 | 1.79 | 1.66 | 1.54 | 1.40 |
| 22 nil | .05 | 4.29 | 3.44 | 3.05 | 2.82 | 2.66 | 2.55 | 2.46 | 2.40 | 2.34 | 2.30 | 2.22 | 2.15 | 2.07 | 1.98 | 1.94 | 1.89 | 1.84 |
| nil | .01 | 7.94 | 5.72 | 4.82 | 4.31 | 3.99 | 3.76 | 3.59 | 3.45 | 3.35 | 3.26 | 3.12 | 2.98 | 2.83 | 2.67 | 2.58 | 2.49 | 2.40 |
| pow | .50 | 4.13 | 2.79 | 2.25 | 1.97 | 1.74 | 1.62 | 1.49 | 1.44 | 1.35 | 1.32 | 1.24 | 1.18 | 1.08 | 1.00 | 0.97 | 0.92 | 0.90 |
| pow | .80 | 8.55 | 5.50 | 4.32 | 3.67 | 3.24 | 2.96 | 2.73 | 2.58 | 2.43 | 2.33 | 2.16 | 2.00 | 1.81 | 1.62 | 1.52 | 1.40 | 1.29 |

One-Stop F Table    193

| | | | | | | | | | | | | | | | | | | |
|---|---|---|---|---|---|---|---|---|---|---|---|---|---|---|---|---|---|---|
| | 1% | .05 | 5.23 | 3.82 | 3.27 | 2.97 | 2.78 | 2.64 | 2.54 | 2.46 | 2.40 | 2.35 | 2.27 | 2.18 | 2.09 | 2.00 | 1.95 | 1.90 | 1.84 |
| | 1% | .01 | 9.53 | 6.32 | 5.16 | 4.55 | 4.16 | 3.90 | 3.70 | 3.55 | 3.43 | 3.33 | 3.18 | 3.02 | 2.86 | 2.69 | 2.60 | 2.50 | 2.41 |
| | pow | .50 | 5.06 | 3.18 | 2.45 | 2.10 | 1.89 | 1.70 | 1.61 | 1.49 | 1.44 | 1.36 | 1.27 | 1.20 | 1.09 | 1.01 | 0.97 | 0.94 | 0.90 |
| | pow | .80 | 9.88 | 6.03 | 4.60 | 3.87 | 3.41 | 3.08 | 2.85 | 2.66 | 2.52 | 2.39 | 2.21 | 2.04 | 1.83 | 1.63 | 1.53 | 1.42 | 1.29 |
| | 5% | .05 | 8.41 | 5.22 | 4.13 | 3.58 | 3.25 | 3.02 | 2.86 | 2.73 | 2.63 | 2.56 | 2.44 | 2.31 | 2.19 | 2.06 | 1.99 | 1.92 | 1.85 |
| | 5% | .01 | 14.26 | 8.40 | 6.43 | 5.44 | 4.84 | 4.44 | 4.15 | 3.93 | 3.76 | 3.62 | 3.41 | 3.20 | 2.99 | 2.77 | 2.66 | 2.54 | 2.43 |
| | pow | .50 | 8.17 | 4.51 | 3.33 | 2.68 | 2.34 | 2.06 | 1.91 | 1.75 | 1.67 | 1.56 | 1.43 | 1.33 | 1.19 | 1.07 | 1.02 | 0.95 | 0.92 |
| | pow | .80 | 14.27 | 7.89 | 5.77 | 4.66 | 4.02 | 3.56 | 3.26 | 3.00 | 2.82 | 2.66 | 2.43 | 2.20 | 1.95 | 1.71 | 1.58 | 1.45 | 1.31 |
| 24 | nil | .05 | 4.25 | 3.40 | 3.01 | 2.78 | 2.62 | 2.51 | 2.42 | 2.35 | 2.30 | 2.25 | 2.18 | 2.11 | 2.03 | 1.94 | 1.89 | 1.84 | 1.79 |
| | nil | .01 | 7.82 | 5.61 | 4.72 | 4.22 | 3.90 | 3.67 | 3.50 | 3.36 | 3.26 | 3.17 | 3.03 | 2.89 | 2.74 | 2.58 | 2.49 | 2.40 | 2.31 |
| | pow | .50 | 4.10 | 2.76 | 2.22 | 1.94 | 1.71 | 1.60 | 1.47 | 1.41 | 1.32 | 1.30 | 1.22 | 1.12 | 1.02 | 0.96 | 0.90 | 0.87 | 0.85 |
| | pow | .80 | 8.48 | 5.44 | 4.25 | 3.61 | 3.18 | 2.90 | 2.67 | 2.52 | 2.37 | 2.27 | 2.10 | 1.92 | 1.73 | 1.54 | 1.44 | 1.33 | 1.21 |
| | 1% | .05 | 5.26 | 3.80 | 3.25 | 2.94 | 2.75 | 2.61 | 2.50 | 2.42 | 2.36 | 2.31 | 2.23 | 2.14 | 2.05 | 1.95 | 1.90 | 1.85 | 1.79 |
| | 1% | .01 | 9.51 | 6.25 | 5.08 | 4.47 | 4.08 | 3.81 | 3.62 | 3.46 | 3.34 | 3.24 | 3.09 | 2.93 | 2.77 | 2.60 | 2.51 | 2.41 | 2.31 |
| | pow | .50 | 5.10 | 3.18 | 2.43 | 2.08 | 1.81 | 1.68 | 1.53 | 1.47 | 1.42 | 1.34 | 1.25 | 1.14 | 1.08 | 0.96 | 0.93 | 0.89 | 0.85 |
| | pow | .80 | 9.91 | 6.00 | 4.56 | 3.82 | 3.33 | 3.02 | 2.77 | 2.60 | 2.47 | 2.34 | 2.15 | 1.96 | 1.78 | 1.56 | 1.46 | 1.34 | 1.22 |
| | 5% | .05 | 8.63 | 5.30 | 4.16 | 3.59 | 3.25 | 3.01 | 2.84 | 2.71 | 2.61 | 2.53 | 2.41 | 2.28 | 2.15 | 2.02 | 1.95 | 1.88 | 1.81 |
| | 5% | .01 | 14.43 | 8.43 | 6.42 | 5.41 | 4.80 | 4.39 | 4.09 | 3.87 | 3.69 | 3.55 | 3.34 | 3.13 | 2.91 | 2.68 | 2.57 | 2.45 | 2.33 |
| | pow | .50 | 8.38 | 4.58 | 3.37 | 2.G9 | 2.35 | 2.06 | 1.91 | 1.74 | 1.66 | 1.55 | 1.42 | 1.32 | 1.17 | 1.03 | 0.98 | 0.92 | 0.87 |
| | pow | .80 | 14.52 | 7.97 | 5.79 | 4.66 | 4.01 | 3.54 | 3.23 | 2.96 | 2.79 | 2.62 | 2.38 | 2.16 | 1.90 | 1.64 | 1.52 | 1.38 | 1.24 |
| 26 | nil | .05 | 4.22 | 3.37 | 2.98 | 2.74 | 2.59 | 2.47 | 2.39 | 2.32 | 2.26 | 2.22 | 2.15 | 2.07 | 1.99 | 1.90 | 1.85 | 1.80 | 1.75 |
| | nil | .01 | 7.72 | 5.53 | 4.64 | 4.14 | 3.82 | 3.59 | 3.42 | 3.29 | 3.18 | 3.09 | 2.96 | 2.81 | 2.66 | 2.50 | 2.42 | 2.33 | 2.23 |
| | pow | .50 | 4.07 | 2.73 | 2.20 | 1.85 | 1.69 | 1.52 | 1.44 | 1.34 | 1.30 | 1.22 | 1.15 | 1.10 | 1.01 | 0.91 | 0.86 | 0.84 | 0.80 |
| | pow | .80 | 8.42 | 5.38 | 4.20 | 3.53 | 3.13 | 2.82 | 2.62 | 2.44 | 2.32 | 2.20 | 2.03 | 1.87 | 1.68 | 1.48 | 1.37 | 1.27 | 1.15 |
| | 1% | .05 | 5.29 | 3.80 | 3.23 | 2.92 | 2.72 | 2.58 | 2.48 | 2.40 | 2.33 | 2.28 | 2.19 | 2.11 | 2.01 | 1.92 | 1.86 | 1.81 | 1.75 |
| | 1% | .01 | 9.51 | 6.21 | 5.03 | 4.40 | 4.01 | 3.75 | 3.55 | 3.39 | 3.27 | 3.17 | 3.02 | 2.86 | 2.70 | 2.52 | 2.43 | 2.34 | 2.24 |
| | pow | .50 | 5.14 | 3.18 | 2.42 | 2.07 | 1.80 | 1.66 | 1.51 | 1.45 | 1.35 | 1.32 | 1.23 | 1.12 | 1.02 | 0.92 | 0.90 | 0.84 | 0.81 |
| | pow | .80 | 9.95 | 5.98 | 4.53 | 3.78 | 3.29 | 2.98 | 2.73 | 2.55 | 2.40 | 2.29 | 2.11 | 1.91 | 1.71 | 1.50 | 1.40 | 1.28 | 1.15 |
| | 5% | .05 | 8.85 | 5.38 | 4.20 | 3.61 | 3.25 | 3.01 | 2.84 | 2.70 | 2.60 | 2.51 | 2.39 | 2.26 | 2.12 | 1.99 | 1.92 | 1.84 | 1.77 |
| | 5% | .01 | 14.63 | 9.48 | 6.43 | 5.39 | 4.77 | 4.35 | 4.05 | 3.82 | 3.G4 | 3.50 | 3.29 | 3.07 | 2.84 | 2.62 | 2.50 | 2.38 | 2.26 |

(*Continued*)

194   One-Stop F Table

|  |  |  | 1 | 2 | 3 | 4 | 5 | 6 | 7 | 8 | 9 | 10 | 12 | 15 | 20 | 30 | 40 | 60 | 120 |
|---|---|---|---|---|---|---|---|---|---|---|---|---|---|---|---|---|---|---|---|
|  | pow | .50 | 8.59 | 4.66 | 3.41 | 2.71 | 2.36 | 2.06 | 1.91 | 1.74 | 1.66 | 1.54 | 1.41 | 1.26 | 1.12 | 1.02 | 0.95 | 0.88 | 0.83 |
|  | pow | .80 | 14.78 | 8.06 | 5.83 | 4.67 | 4.00 | 3.52 | 3.21 | 2.94 | 2.76 | 2.58 | 2.35 | 2.09 | 1.84 | 1.60 | 1.46 | 1.32 | 1.17 |
| 28 | nil | .05 | 4.19 | 3.34 | 2.95 | 2.71 | 2.56 | 2.44 | 2.36 | 2.29 | 2.23 | 2.19 | 2.12 | 2.04 | 1.96 | 1.87 | 1.82 | 1.77 | 1.71 |
|  | nil | .01 | 7.63 | 5.45 | 4.57 | 4.07 | 3.75 | 3.53 | 3.36 | 3.23 | 3.12 | 3.03 | 2.90 | 2.75 | 2.60 | 2.44 | 2.35 | 2.26 | 2.17 |
|  | pow | .50 | 4.05 | 2.71 | 2.18 | 1.83 | 1.67 | 1.50 | 1.42 | 1.32 | 1.28 | 1.21 | 1.13 | 1.04 | 0.95 | 0.90 | 0.83 | 0.79 | 0.77 |
|  | pow | .80 | 8.38 | 5.34 | 4.15 | 3.49 | 3.08 | 2.78 | 2.58 | 2.40 | 2.28 | 2.16 | 1.99 | 1.81 | 1.62 | 1.44 | 1.32 | 1.21 | 1.09 |
|  | 1% | .05 | 5.33 | 3.80 | 3.22 | 2.90 | 2.70 | 2.56 | 2.45 | 2.37 | 2.30 | 2.25 | 2.17 | 2.08 | 1.98 | 1.89 | 1.83 | 1.78 | 1.72 |
|  | 1% | .01 | 9.52 | 6.17 | 4.98 | 4.35 | 3.96 | 3.69 | 3.49 | 3.34 | 3.22 | 3.12 | 2.96 | 2.80 | 2.64 | 2.46 | 2.37 | 2.27 | 2.17 |
|  | pow | .50 | 5.19 | 3.19 | 2.42 | 2.06 | 1.78 | 1.65 | 1.50 | 1.43 | 1.33 | 1.30 | 1.17 | 1.11 | 1.01 | 0.91 | 0.86 | 0.82 | 0.77 |
|  | pow | .80 | 10.00 | 5.97 | 4.50 | 3.75 | 3.26 | 2.94 | 2.69 | 2.52 | 2.36 | 2.25 | 2.04 | 1.87 | 1.67 | 1.46 | 1.34 | 1.23 | 1.10 |
|  | 5% | .05 | 9.07 | 5.47 | 4.25 | 3.64 | 3.26 | 3.01 | 2.83 | 2.70 | 2.59 | 2.50 | 2.37 | 2.24 | 2.10 | 1.96 | 1.89 | 1.81 | 1.73 |
|  | 5% | .01 | 14.83 | 8.55 | 6.45 | 5.39 | 4.75 | 4.33 | 4.02 | 3.79 | 3.61 | 3.46 | 3.24 | 3.02 | 2.79 | 2.56 | 2.44 | 2.32 | 2.19 |
|  | pow | .50 | 8.80 | 4.74 | 3.45 | 2.74 | 2.37 | 2.07 | 1.91 | 1.74 | 1.59 | 1.54 | 1.41 | 1.25 | 1.11 | 0.98 | 0.91 | 0.85 | 0.78 |
|  | pow | .80 | 15.04 | 8.16 | 5.87 | 4.69 | 4.00 | 3.52 | 3.19 | 2.92 | 2.71 | 2.56 | 2.32 | 2.00 | 1.81 | 1.54 | 1.41 | 1.27 | 1.12 |
| 30 | nil | .05 | 4.16 | 3.32 | 2.92 | 2.69 | 2.53 | 2.42 | 2.33 | 2.27 | 2.21 | 2.16 | 2.09 | 2.01 | 1.93 | 1.84 | 1.79 | 1.74 | 1.68 |
|  | nil | .01 | 7.56 | 5.39 | 4.51 | 4.02 | 3.70 | 3.47 | 3.30 | 3.17 | 3.07 | 2.98 | 2.84 | 2.70 | 2.55 | 2.39 | 2.30 | 2.21 | 2.11 |
|  | pow | .50 | 4.03 | 2.69 | 2.16 | 1.82 | 1.65 | 1.48 | 1.41 | 1.30 | 1.27 | 1.19 | 1.12 | 1.02 | 0.94 | 0.86 | 0.82 | 0.76 | 0.73 |
|  | pow | .80 | 8.33 | 5.30 | 4.12 | 3.45 | 3.05 | 2.74 | 2.54 | 2.36 | 2.24 | 2.12 | 1.95 | 1.77 | 1.58 | 1.39 | 1.28 | 1.17 | 1.04 |
|  | 1% | .05 | 5.38 | 3.80 | 3.21 | 2.89 | 2.68 | 2.54 | 2.43 | 2.35 | 2.28 | 2.23 | 2.14 | 2.05 | 1.96 | 1.86 | 1.80 | 1.75 | 1.69 |
|  | 1% | .01 | 9.54 | 6.15 | 4.94 | 4.31 | 3.92 | 3.64 | 3.44 | 3.29 | 3.17 | 3.07 | 2.91 | 2.75 | 2.59 | 2.41 | 2.32 | 2.22 | 2.12 |
|  | pow | .50 | 5.23 | 3.19 | 2.41 | 2.05 | 1.77 | 1.64 | 1.49 | 1.42 | 1.32 | 1.24 | 1.15 | 1.05 | 0.96 | 0.87 | 0.82 | 0.77 | 0.73 |
|  | pow | .80 | 10.06 | 5.97 | 4.49 | 3.73 | 3.23 | 2.91 | 2.66 | 2.49 | 2.32 | 2.19 | 2.01 | 1.81 | 1.61 | 1.41 | 1.30 | 1.17 | 1.05 |
|  | 5% | .05 | 9.29 | 5.57 | 4.29 | 3.66 | 3.28 | 3.02 | 2.83 | 2.69 | 2.58 | 2.49 | 2.36 | 2.22 | 2.08 | 1.94 | 1.86 | 1.78 | 1.70 |
|  | 5% | .01 | 15.04 | 8.62 | 6.47 | 5.40 | 4.75 | 4.31 | 4.00 | 3.76 | 3.58 | 3.43 | 3.20 | 2.98 | 2.75 | 2.51 | 2.39 | 2.27 | 2.14 |
|  | pow | .50 | 9.01 | 4.82 | 3.50 | 2.76 | 2.39 | 2.08 | 1.92 | 1.74 | 1.59 | 1.54 | 1.40 | 1.25 | 1.11 | 0.94 | 0.88 | 0.82 | 0.76 |
|  | pow | .80 | 15.31 | 8.26 | 5.91 | 4.71 | 4.01 | 3.51 | 3.19 | 2.91 | 2.69 | 2.54 | 2.30 | 2.04 | 1.78 | 1.50 | 1.36 | 1.23 | 1.07 |

dfHyp

| df | | α | | | | | | | | | | | | | | | | | | |
|----|---|---|---|---|---|---|---|---|---|---|---|---|---|---|---|---|---|---|---|---|
| 40 | nil | .05 | 4.08 | 3.23 | 2.84 | 2.61 | 2.45 | 2.34 | 2.25 | 2.18 | 2.12 | 2.08 | 2.00 | 1.92 | 1.84 | 1.74 | 1.69 | 1.64 | 1.58 |
|    | nil | .01 | 7.31 | 5.18 | 4.31 | 3.83 | 3.51 | 3.29 | 3.12 | 2.99 | 2.99 | 2.80 | 2.66 | 2.52 | 2.37 | 2.20 | 2.11 | 2.02 | 1.92 |
|    | pow | .50 | 3.97 | 2.63 | 2.02 | 1.76 | 1.52 | 1.42 | 1.29 | 1.25 | 1.16 | 1.08 | 1.02 | 0.93 | 0.86 | 0.76 | 0.70 | 0.65 | 0.61 |
|    | pow | .80 | 8.20 | 5.16 | 3.96 | 3.32 | 2.89 | 2.61 | 2.39 | 2.23 | 2.09 | 1.97 | 1.80 | 1.62 | 1.44 | 1.23 | 1.12 | 1.00 | 0.88 |
|    | 1%  | .05 | 5.64 | 3.85 | 3.21 | 2.86 | 2.64 | 2.49 | 2.38 | 2.29 | 2.22 | 2.16 | 2.07 | 1.97 | 1.87 | 1.77 | 1.71 | 1.65 | 1.58 |
|    | 1%  | .01 | 9.75 | 6.12 | 4.86 | 4.20 | 3.79 | 3.51 | 3.30 | 3.14 | 3.01 | 2.91 | 2.75 | 2.59 | 2.42 | 2.23 | 2.14 | 2.03 | 1.92 |
|    | pow | .50 | 5.49 | 3.26 | 2.42 | 2.04 | 1.75 | 1.54 | 1.45 | 1.33 | 1.22 | 1.20 | 1.12 | 1.01 | 0.88 | 0.77 | 0.71 | 0.66 | 0.61 |
|    | pow | .80 | 10.38 | 6.01 | 4.46 | 3.67 | 3.15 | 2.80 | 2.56 | 2.36 | 2.19 | 2.08 | 1.90 | 1.70 | 1.43 | 1.25 | 1.14 | 1.01 | 0.88 |
|    | 5%  | .05 | 10.44 | 6.01 | 4.56 | 3.83 | 3.39 | 3.09 | 2.88 | 2.72 | 2.59 | 2.49 | 2.34 | 2.19 | 2.03 | 1.86 | 1.7b | 1.69 | 1.60 |
|    | 5%  | .01 | 16.14 | 9.06 | 6.70 | 5.51 | 4.80 | 4.32 | 3.98 | 3.72 | 3.52 | 3.35 | 3.11 | 2.86 | 2.61 | 2.36 | 2.22 | 2.09 | 1.95 |
|    | pow | .50 | 10.00 | 5.35 | 3.73 | 3.00 | 2.49 | 2.15 | 1.97 | 1.77 | 1.61 | 1.55 | 1.35 | 1.19 | 1.05 | 0.88 | 0.80 | 0.72 | 0.63 |
|    | pow | .80 | 16.64 | 8.82 | 6.19 | 4.91 | 4.10 | 3.56 | 3.20 | 2.90 | 2.67 | 2.51 | 2.22 | 1.94 | 1.67 | 1.38 | 1.23 | 1.07 | 0.91 |
| 50 | nil | .05 | 4.03 | 3.18 | 2.79 | 2.56 | 2.40 | 2.29 | 2.20 | 2.13 | 2.07 | 2.02 | 1.95 | 1.87 | 1.78 | 1.68 | 1.63 | 1.57 | 1.51 |
|    | nil | .01 | 7.17 | 5.06 | 4.20 | 3.72 | 3.41 | 3.19 | 3.02 | 2.89 | 2.78 | 2.70 | 2.56 | 2.42 | 2.26 | 2.10 | 2.01 | 1.91 | 1.80 |
|    | pow | .50 | 3.93 | 2.59 | 1.99 | 1.72 | 1.49 | 1.39 | 1.26 | 1.15 | 1.13 | 1.06 | 0.99 | 0.91 | 0.79 | 0.70 | 0.63 | 0.60 | 0.54 |
|    | pow | .80 | 8.11 | 5.09 | 3.88 | 3.25 | 2.82 | 2.54 | 2.31 | 2.13 | 2.01 | 1.89 | 1.73 | 1.55 | 1.35 | 1.14 | 1.02 | 0.91 | 0.78 |
|    | 1%  | .05 | 5.93 | 3.94 | 3.24 | 2.87 | 2.63 | 2.47 | 2.35 | 2.26 | 2.19 | 2.12 | 2.03 | 1.93 | 1.83 | 1.71 | 1.65 | 1.59 | 1.52 |
|    | 1%  | .01 | 10.04 | 6.18 | 4.85 | 4.16 | 3.74 | 3.44 | 3.23 | 3.07 | 2.94 | 2.83 | 2.67 | 2.50 | 2.32 | 2.13 | 2.03 | 1.92 | 1.81 |
|    | pow | .50 | 5.76 | 3.35 | 2.46 | 1.97 | 1.75 | 1.54 | 1.44 | 1.32 | 1.21 | 1.18 | 1.04 | 0.94 | 0.81 | 0.71 | 0.67 | 0.60 | 0.54 |
|    | pow | .80 | 10.74 | 6.11 | 4.48 | 3.63 | 3.13 | 2.76 | 2.52 | 2.31 | 2.14 | 2.03 | 1.82 | 1.61 | 1.39 | 1.17 | 1.05 | 0.92 | 0.78 |
|    | 5%  | .05 | 11.39 | 6.50 | 4.84 | 4.02 | 3.53 | 3.20 | 2.96 | 2.78 | 2.64 | 2.53 | 2.36 | 2.19 | 2.01 | 1.83 | 1.74 | 1.64 | 1.54 |
|    | 5%  | .01 | 17.26 | 9.56 | 6.98 | 5.69 | 4.92 | 4.40 | 4.03 | 3.75 | 3.53 | 3.35 | 3.09 | 2.82 | 2.55 | 2.28 | 2.14 | 1.99 | 1.84 |
|    | pow | .50 | 11.09 | 5.76 | 4.08 | 3.16 | 2.61 | 2.23 | 2.03 | 1.82 | 1.65 | 1.58 | 1.37 | 1.20 | 1.00 | 0.84 | 0.76 | 0.66 | 0.57 |
|    | pow | .80 | 17.86 | 9.36 | 6.55 | 5.11 | 4.24 | 3.66 | 3.27 | 2.95 | 2.70 | 2.53 | 2.22 | 1.92 | 1.61 | 1.31 | 1.16 | 0.99 | 0.81 |
| 60 | nil | .05 | 3.99 | 3.15 | 2.76 | 2.53 | 2.37 | 2.25 | 2.17 | 2.10 | 2.04 | 1.99 | 1.92 | 1.83 | 1.75 | 1.65 | 1.59 | 1.53 | 1.47 |
|    | nil | .01 | 7.07 | 4.98 | 4.13 | 3.65 | 3.34 | 3.12 | 2.95 | 2.82 | 2.72 | 2.63 | 2.50 | 2.35 | 2.20 | 2.03 | 1.94 | 1.84 | 1.73 |
|    | pow | .50 | 3.90 | 2.57 | 1.97 | 1.70 | 1.47 | 1.30 | 1.24 | 1.13 | 1.04 | 1.04 | 0.91 | 0.84 | 0.78 | 0.65 | 0.58 | 0.54 | 0.48 |
|    | pow | .80 | 8.06 | 5.04 | 3.83 | 3.20 | 2.77 | 2.46 | 2.26 | 2.08 | 1.94 | 1.85 | 1.66 | 1.48 | 1.30 | 1.08 | 0.96 | 0.84 | 0.70 |
|    | 1%  | .05 | 6.24 | 4.04 | 3.29 | 2.90 | 2.64 | 2.47 | 2.35 | 2.25 | 2.17 | 2.11 | 2.01 | 1.91 | 1.80 | 1.68 | 1.62 | 1.55 | 1.47 |
|    | 1%  | .01 | 10.35 | 6.28 | 4.88 | 4.16 | 3.72 | 3.42 | 3.20 | 3.03 | 2.90 | 2.79 | 2.62 | 2.45 | 2.26 | 2.07 | 1.96 | 1.85 | 1.73 |

*(Continued)*

196  *One-Stop* F *Table*

|  |  |  | | | | | | dfHyp | | | | | | | | | |
|---|---|---|---|---|---|---|---|---|---|---|---|---|---|---|---|---|---|
|  |  | 1 | 2 | 3 | 4 | 5 | 6 | 7 | 8 | 9 | 10 | 12 | 15 | 20 | 30 | 40 | 60 | 120 |
| 70 | pow .50 | 6.04 | 3.44 | 2.50 | 1.99 | 1.76 | 1.54 | 1.37 | 1.31 | 1.20 | 1.11 | 1.03 | 0.93 | 0.80 | 0.70 | 0.63 | 0.55 | 0.50 |
|  | pow .80 | 11.13 | 6.22 | 4.53 | 3.65 | 3.13 | 2.75 | 2.47 | 2.29 | 2.11 | 1.97 | 1.78 | 1.58 | 1.35 | 1.13 | 0.99 | 0.86 | 0.72 |
|  | 5% .05 | 12.49 | 6.94 | 5.14 | 4.23 | 3.68 | 3.31 | 3.05 | 2.86 | 2.70 | 2.58 | 2.39 | 2.20 | 2.01 | 1.82 | 1.72 | 1.61 | 1.50 |
|  | 5% .01 | 18.38 | 10.06 | 7.29 | 5.90 | 5.07 | 4.51 | 4.11 | 3.81 | 3.58 | 3.39 | 3.11 | 2.82 | 2.53 | 2.24 | 2.09 | 1.93 | 1.77 |
|  | pow .50 | 11.97 | 6.30 | 4.33 | 3.33 | 2.82 | 2.41 | 2.11 | 1.88 | 1.77 | 1.62 | 1.39 | 1.21 | 1.01 | 0.84 | 0.72 | 0.61 | 0.53 |
|  | pow .80 | 19.10 | 9.93 | 6.86 | 5.33 | 4.44 | 3.81 | 3.36 | 3.02 | 2.78 | 2.57 | 2.24 | 1.93 | 1.60 | 1.29 | 1.11 | 0.93 | 0.75 |
|  | nil .05 | 3.97 | 3.13 | 2.74 | 2.50 | 2.35 | 2.23 | 2.14 | 2.07 | 2.01 | 1.97 | 1.89 | 1.81 | 1.72 | 1.62 | 1.56 | 1.50 | 1.43 |
|  | nil .01 | 7.01 | 4.92 | 4.07 | 3.60 | 3.29 | 3.07 | 2.91 | 2.78 | 2.67 | 2.58 | 2.45 | 2.30 | 2.15 | 1.98 | 1.89 | 1.78 | 1.67 |
|  | pow .50 | 3.88 | 2.55 | 1.95 | 1.68 | 1.46 | 1.28 | 1.23 | 1.12 | 1.03 | 1.02 | 0.90 | 0.82 | 0.72 | 0.60 | 0.58 | 0.51 | 0.45 |
|  | pow .80 | 8.02 | 5.00 | 3.80 | 3.16 | 2.73 | 2.43 | 2.23 | 2.05 | 1.91 | 1.81 | 1.62 | 1.44 | 1.24 | 1.03 | 0.92 | 0.80 | 0.66 |
|  | 1% .05 | 6.57 | 4.14 | 3.35 | 2.92 | 2.66 | 2.48 | 2.35 | 2.25 | 2.17 | 2.10 | 2.00 | 1.89 | 1.78 | 1.66 | 1.59 | 1.52 | 1.44 |
|  | 1% .01 | 10.67 | 6.39 | 4.93 | 4.18 | 3.73 | 3.41 | 3.19 | 3.01 | 2.87 | 2.76 | 2.59 | 2.41 | 2.23 | 2.03 | 1.92 | 1.80 | 1.68 |
|  | pow .50 | 6.32 | 3.54 | 2.55 | 2.11 | 1.78 | 1.55 | 1.37 | 1.32 | 1.20 | 1.11 | 1.03 | 0.93 | 0.80 | 0.66 | 0.59 | 0.52 | 0.45 |
|  | pow .80 | 11.55 | 6.35 | 4.59 | 3.71 | 3.14 | 2.75 | 2.47 | 2.28 | 2.10 | 1.96 | 1.76 | 1.55 | 1.32 | 1.08 | 0.95 | 0.81 | 0.66 |
|  | 5% .05 | 13.34 | 7.42 | 5.45 | 4.43 | 3.84 | 3.44 | 3.16 | 2.94 | 2.77 | 2.64 | 2.44 | 2.23 | 2.03 | 1.81 | 1.71 | 1.59 | 1.48 |
|  | 5% .01 | 19.46 | 10.58 | 7.61 | 6.13 | 5.23 | 4.64 | 4.21 | 3.89 | 3.64 | 3.44 | 3.14 | 2.84 | 2.53 | 2.22 | 2.06 | 1.89 | 1.72 |
|  | pow .50 | 13.03 | 6.69 | 4.56 | 3.60 | 2.94 | 2.50 | 2.18 | 2.02 | 1.82 | 1.66 | 1.49 | 1.23 | 1.02 | 0.80 | 0.72 | 0.61 | 0.50 |
|  | pow .80 | 20.22 | 10.46 | 7.20 | 5.59 | 4.60 | 3.93 | 3.45 | 3.13 | 2.84 | 2.62 | 2.30 | 1.94 | 1.60 | 1.26 | 1.09 | 0.90 | 0.71 |
| 80 | nil .05 | 3.95 | 3.11 | 2.72 | 2.49 | 2.33 | 2.21 | 2.12 | 2.05 | 2.00 | 1.95 | 1.87 | 1.79 | 1.70 | 1.60 | 1.54 | 1.48 | 1.41 |
|  | nil .01 | 6.96 | 4.88 | 4.04 | 3.56 | 3.26 | 3.04 | 2.87 | 2.74 | 2.64 | 2.55 | 2.41 | 2.27 | 2.11 | 1.94 | 1.85 | 1.75 | 1.63 |
|  | pow .50 | 3.87 | 2.54 | 1.94 | 1.67 | 1.44 | 1.27 | 1.21 | 1.11 | 1.02 | 0.94 | 0.89 | 0.82 | 0.71 | 0.60 | 0.54 | 0.48 | 0.42 |
|  | pow .80 | 7.99 | 4.97 | 3.77 | 3.14 | 2.71 | 2.40 | 2.20 | 2.02 | 1.88 | 1.76 | 1.60 | 1.42 | 1.22 | 1.00 | 0.88 | 0.76 | 0.62 |
|  | 1% .05 | 6.83 | 4.26 | 3.41 | 2.96 | 2.69 | 2.50 | 2.36 | 2.26 | 2.17 | 2.10 | 2.00 | 1.89 | 1.77 | 1.64 | 1.57 | 1.50 | 1.42 |
|  | 1% .01 | 10.98 | 6.51 | 4.99 | 4.22 | 3.74 | 3.42 | 3.19 | 3.01 | 2.86 | 2.75 | 2.57 | 2.39 | 2.20 | 1.99 | 1.89 | 1.77 | 1.64 |
|  | pow .50 | 6.73 | 3.64 | 2.60 | 2.14 | 1.80 | 1.56 | 1.38 | 1.32 | 1.21 | 1.11 | 1.03 | 0.87 | 0.79 | 0.65 | 0.58 | 0.52 | 0.44 |
|  | pow .80 | 11.95 | 6.48 | 4.66 | 3.75 | 3.16 | 2.76 | 2.47 | 2.28 | 2.10 | 1.95 | 1.75 | 1.52 | 1.31 | 1.06 | 0.93 | 0.79 | 0.63 |

| | | | | | | | | | | | | | | | | | |
|---|---|---|---|---|---|---|---|---|---|---|---|---|---|---|---|---|---|
| 90 | 5% | .05 | 14.39 | 7.84 | 5.71 | 4.65 | 4.01 | 3.58 | 3.26 | 3.03 | 2.85 | 2.71 | 2.49 | 2.27 | 2.04 | 1.82 | 1.70 | 1.58 | 1.46 |
| | 5% | .01 | 20.52 | 11.08 | 7.93 | 6.35 | 5.41 | 4.77 | 4.32 | 3.98 | 3.72 | 3.50 | 3.18 | 2.86 | 2.54 | 2.21 | 2.04 | 1.87 | 1.69 |
| | pow | .50 | 13.83 | 7.22 | 4.92 | 3.76 | 3.06 | 2.59 | 2.34 | 2.08 | 1.87 | 1.70 | 1.52 | 1.25 | 1.03 | 0.80 | 0.69 | 0.58 | 0.47 |
| | pow | .80 | 21.36 | 11.02 | 7.55 | 5. | | | | | | | | | | | | | |
| | nil | .05 | 3.94 | 3.10 | 2.71 | 2.47 | 2.32 | 2.20 | 2.11 | 2.04 | 1.98 | 1.94 | 1.86 | 1.78 | 1.69 | 1.58 | 1.53 | 1.46 | 1.39 |
| | nil | .01 | 6.92 | 4.85 | 4.01 | 3.53 | 3.23 | 3.01 | 2.84 | 2.72 | 2.61 | 2.52 | 2.39 | 2.24 | 2.09 | 1.91 | 1.82 | 1.72 | 1.60 |
| | pow | .50 | 3.86 | 2.53 | 1.93 | 1.66 | 1.43 | 1.26 | 1.21 | 1.10 | 1.01 | 0.94 | 0.89 | 0.81 | 0.70 | 0.59 | 0.54 | 0.46 | 0.40 |
| | pow | .80 | 7.97 | 4.95 | 3.75 | 3.12 | 2.69 | 2.38 | 2.18 | 2.00 | 1.86 | 1.74 | 1.58 | 1.40 | 1.20 | 0.98 | 0.86 | 0.73 | 0.59 |
| | 1% | .05 | 6.97 | 4.37 | 3.48 | 3.00 | 2.71 | 2.52 | 2.38 | 2.26 | 2.18 | 2.11 | 2.00 | 1.88 | 1.76 | 1.63 | 1.56 | 1.48 | 1.40 |
| | 1% | .01 | 11.29 | 6.64 | 5.06 | 4.26 | 3.77 | 3.43 | 3.19 | 3.01 | 2.86 | 2.74 | 2.56 | 2.38 | 2.18 | 1.97 | 1.86 | 1.74 | 1.61 |
| | pow | .50 | 6.86 | 3.74 | 2.66 | 2.18 | 1.83 | 1.58 | 1.47 | 1.33 | 1.21 | 1.12 | 1.03 | 0.87 | 0.79 | 0.65 | 0.55 | 0.49 | 0.40 |
| | pow | .80 | 12.12 | 6.62 | 4.74 | 3.19 | 3.19 | 2.78 | 2.52 | 2.28 | 2.10 | 1.95 | 1.75 | 1.51 | 1.29 | 1.04 | 0.89 | 0.76 | 0.60 |
| | 5% | .05 | 15.17 | 8.31 | 6.02 | 4.88 | 4.15 | 3.70 | 3.37 | 3.12 | 2.93 | 2.77 | 2.54 | 2.30 | 2.07 | 1.83 | 1.70 | 1.58 | 1.44 |
| | 5% | .01 | 21.57 | 11.59 | 8.25 | 6.59 | 5.58 | 4.92 | 4.44 | 4.08 | 3.80 | 3.57 | 3.24 | 2.90 | 2.55 | 2.21 | 2.03 | 1.85 | 1.66 |
| | pow | .50 | 14.87 | 7.58 | 5.14 | 3.92 | 3.29 | 2.78 | 2.41 | 2.14 | 1.92 | 1.82 | 1.55 | 1.34 | 1.09 | 0.85 | 0.69 | 0.58 | 0.45 |
| | pow | .80 | 22.43 | 11.51 | 7.87 | 6.04 | 4.97 | 4.23 | 3.70 | 3.30 | 2.99 | 2.77 | 2.39 | 2.03 | 1.65 | 1.27 | 1.06 | 0.87 | 0.65 |
| 100 | nil | .05 | 3.93 | 3.09 | 2.70 | 2.46 | 2.30 | 2.19 | 2.10 | 2.03 | 1.97 | 1.92 | 1.85 | 1.77 | 1.67 | 1.57 | 1.51 | 1.45 | 1.37 |
| | nil | .01 | 6.89 | 4.82 | 3.98 | 3.51 | 3.21 | 2.99 | 2.82 | 2.69 | 2.59 | 2.50 | 2.37 | 2.22 | 2.07 | 1.89 | 1.80 | 1.69 | 1.57 |
| | pow | .50 | 3.85 | 2.52 | 1.92 | 1.66 | 1.43 | 1.26 | 1.20 | 1.10 | 1.01 | 0.93 | 0.88 | 0.80 | 0.70 | 0.59 | 0.50 | 0.46 | 0.39 |
| | pow | .80 | 7.95 | 4.94 | 3.73 | 3.10 | 2.67 | 2.37 | 2.17 | 1.99 | 1.84 | 1.72 | 1.56 | 1.38 | 1.18 | 0.97 | 0.83 | 0.71 | O.57 |
| | 1% | .05 | 7.24 | 4.49 | 3.55 | 3.04 | 2.74 | 2.54 | 2.39 | 2.28 | 2.19 | 2.11 | 2.00 | 1.88 | 1.76 | 1.62 | 1.55 | 1.47 | 1.38 |
| | 1% | .01 | 11.60 | 6.76 | 5.13 | 4.30 | 3.80 | 3.45 | 3.21 | 3.02 | 2.87 | 2.75 | 2.56 | 2.37 | 2.17 | 1.96 | 1.84 | 1.72 | 1.58 |
| | pow | .50 | 7.11 | 3.84 | 2.71 | 2.22 | 1.85 | 1.59 | 1.49 | 1.34 | 1.22 | 1.12 | 1.04 | 0.87 | 0.74 | 0.61 | 0.55 | 0.49 | 0.39 |
| | pow | .80 | 12.45 | 6.76 | 4.82 | 3.83 | 3.22 | 2.80 | 2.53 | 2.29 | 2.11 | 1.95 | 1.75 | 1.50 | 1.26 | 1.01 | 0.88 | 0.74 | 0.58 |
| | 5% | .05 | 16.18 | 8.81 | 6.27 | 5.05 | 4.32 | 3.83 | 3.49 | 3.21 | 3.00 | 2.84 | 2.60 | 2.34 | 2.09 | 1.84 | 1.71 | 1.57 | 1.43 |
| | 5% | .01 | 22.59 | 12.08 | 8.57 | 6.82 | 5.76 | 5.06 | 4.56 | 4.18 | 3.88 | 3.65 | 3.29 | 2.94 | 2.58 | 2.21 | 2.03 | 1.84 | 1.64 |
| | pow | .50 | 15.62 | 7.93 | 5.51 | 4.19 | 3.40 | 2.87 | 2.49 | 2.29 | 2.06 | 1.87 | 1.58 | 1.36 | 1.11 | 0.86 | 0.70 | 0.59 | 0.44 |
| | pow | .80 | 23.49 | 12.03 | 8.22 | 6.30 | 5.14 | 4.36 | 3.81 | 3.43 | 3.10 | 2.83 | 2.43 | 2.06 | 1.67 | 1.28 | 1.06 | 0.86 | 0.63 |

(*Continued*)

198   One-Stop F Table

| | | | | | | | | | | dfHyp | | | | | | | |
|---|---|---|---|---|---|---|---|---|---|---|---|---|---|---|---|---|---|
| | | | 1 | 2 | 3 | 4 | 5 | 6 | 7 | 8 | 9 | 10 | 12 | 15 | 20 | 30 | 40 | 60 | 120 |
| 120 | nil | .05 | 3.91 | 3.07 | 2.68 | 2.45 | 2.29 | 2.17 | 2.09 | 2.01 | 1.96 | 1.91 | 1.83 | 1.75 | 1.66 | 1.55 | 1.49 | 1.43 | 1.35 |
| | nil | .01 | 6.85 | 4.79 | 3.95 | 3.48 | 3.17 | 2.96 | 2.79 | 2.66 | 2.56 | 2.47 | 2.34 | 2.19 | 2.03 | 1.86 | 1.76 | 1.65 | 1.53 |
| | pow | .50 | 3.84 | 2.51 | 1.91 | 1.56 | 1.42 | 1.25 | 1.11 | 1.09 | 1.00 | 0.92 | 0.87 | 0.74 | 0.64 | 0.54 | 0.50 | 0.43 | 0.36 |
| | pow | .80 | 7.93 | 4.91 | 3.71 | 3.05 | 2.65 | 2.34 | 2.12 | 1.97 | 1.82 | 1.70 | 1.54 | 1.34 | 1.14 | 0.92 | 0.81 | 0.68 | 0.53 |
| | 1% | .05 | 7.76 | 4.74 | 3.66 | 3.13 | 2.81 | 2.59 | 2.43 | 2.31 | 2.21 | 2.13 | 2.01 | 1.89 | 1.75 | 1.61 | 1.54 | 1.45 | 1.36 |
| | 1% | .01 | 12.20 | 7.02 | 5.28 | 4.40 | 3.86 | 3.50 | 3.24 | 3.04 | 2.88 | 2.76 | 2.56 | 2.36 | 2.15 | 1.93 | 1.81 | 1.69 | 1.55 |
| | pow | .50 | 7.58 | 4.04 | 2.92 | 2.29 | 1.90 | 1.63 | 1.52 | 1.36 | 1.24 | 1.13 | 1.05 | 0.87 | 0.74 | 0.61 | 0.54 | 0.46 | 0.37 |
| | pow | .80 | 13.10 | 7.05 | 4.98 | 3.93 | 3.29 | 2.85 | 2.56 | 2.32 | 2.12 | 1.97 | 1.75 | 1.50 | 1.25 | 1.00 | 0.86 | 0.71 | 0.55 |
| | 5% | .05 | 17.88 | 9.64 | 6.89 | 5.45 | 4.64 | 4.09 | 3.70 | 3.41 | 3.17 | 2.98 | 2.71 | 2.43 | 2.15 | 1.87 | 1.72 | 1.57 | 1.42 |
| | 5% | .01 | 24.59 | 13.05 | 9.20 | 7.28 | 6.12 | 5.35 | 4.80 | 4.38 | 4.06 | 3.80 | 3.41 | 3.02 | 2.63 | 2.23 | 2.03 | 1.83 | 1.61 |
| | pow | .50 | 17.37 | 8.79 | 5.92 | 4.63 | 3.74 | 3.15 | 2.73 | 2.41 | 2.25 | 2.04 | 1.72 | 1.47 | 1.13 | 0.87 | 0.74 | 0.59 | 0.42 |
| | pow | .80 | 25.54 | 13.02 | 8.83 | 6.78 | 5.51 | 4.67 | 4.06 | 3.61 | 3.30 | 3.00 | 2.57 | 2.16 | 1.71 | 1.29 | 1.08 | 0.85 | 0.61 |
| 150 | nil | .05 | 3.89 | 3.06 | 2.67 | 2.43 | 2.27 | 2.16 | 2.07 | 2.00 | 1.94 | 1.89 | 1.81 | 1.73 | 1.64 | 1.53 | 1.47 | 1.40 | 1.32 |
| | nil | .01 | 6.80 | 4.75 | 3.92 | 3.45 | 3.14 | 2.92 | 2.76 | 2.63 | 2.53 | 2.44 | 2.30 | 2.16 | 2.00 | 1.83 | 1.73 | 1.62 | 1.49 |
| | pow | .50 | 3.83 | 2.50 | 1.90 | 1.55 | 1.41 | 1.24 | 1.10 | 1.08 | 0.99 | 0.92 | 0.86 | 0.73 | 0.63 | 0.54 | 0.45 | 0.40 | 0.33 |
| | pow | .80 | 7.90 | 4.89 | 3.09 | 3.02 | 2.63 | 2.32 | 2.09 | 1.94 | 1.80 | 1.68 | 1.52 | 1.31 | 1.11 | 0.90 | 0.77 | 0.64 | 0.49 |
| | 1% | .05 | 8.61 | 5.01 | 3.86 | 3.28 | 2.92 | 2.66 | 2.49 | 2.36 | 2.25 | 2.17 | 2.03 | 1.90 | 1.76 | 1.61 | 1.53 | 1.44 | 1.34 |
| | 1% | .01 | 13.04 | 7.40 | 5.51 | 4.56 | 3.98 | 3.59 | 3.31 | 3.09 | 2.93 | 2.79 | 2.58 | 2.37 | 2.15 | 1.92 | 1.79 | 1.66 | 1.51 |
| | pow | .50 | 8.26 | 4.42 | 3.09 | 2.40 | 1.98 | 1.78 | 1.56 | 1.40 | 1.27 | 1.15 | 1.06 | 0.88 | 0.74 | 0.61 | 0.51 | 0.43 | 0.34 |
| | pow | .80 | 14.11 | 7.43 | 5.21 | 4.09 | 3.40 | 2.96 | 2.62 | 2.37 | 2.16 | 2.00 | 1.77 | 1.51 | 1.25 | 0.98 | 0.83 | 0.68 | 0.51 |
| | 5% | .05 | 20.52 | 10.86 | 7.64 | 6.06 | 5.11 | 4.48 | 4.03 | 3.69 | 3.41 | 3.20 | 2.88 | 2.57 | 2.24 | 1.92 | 1.75 | 1.59 | 1.41 |
| | 5% | .01 | 27.47 | 14.46 | 10.12 | 7.95 | 6.65 | 5.78 | 5.10 | 4.69 | 4.33 | 4.04 | 3.60 | 3.17 | 2.73 | 2.28 | 2.06 | 1.83 | 1.59 |
| | pow | .50 | 19.73 | 10.24 | 6.86 | 5.19 | 4.19 | 3.52 | 3.04 | 2.67 | 2.48 | 2.25 | 1.90 | 1.61 | 1.23 | 0.93 | 0.75 | 0.59 | 0.41 |
| | pow | .80 | 28.49 | 14.57 | 9.81 | 7.46 | 6.05 | 5.10 | 4.43 | 3.92 | 3.56 | 3.24 | 2.76 | 2.31 | 1.81 | 1.35 | 1.09 | 0.85 | 0.58 |
| 200 | nil | .05 | 3.88 | 3.04 | 2.65 | 2.42 | 2.26 | 2.14 | 2.05 | 1.98 | 1.93 | 1.88 | 1.80 | 1.71 | 1.62 | 1.51 | 1.45 | 1.38 | 1.30 |
| | nil | .01 | 6.76 | 4.71 | 3.88 | 3.41 | 3.11 | 2.89 | 2.73 | 2.60 | 2.50 | 2.41 | 2.27 | 2.13 | 1.97 | 1.79 | 1.69 | 1.58 | 1.45 |

One-Stop F Table  199

| | | | | | | | | | | | | | | | | | |
|---|---|---|---|---|---|---|---|---|---|---|---|---|---|---|---|---|---|
| | pow | .50 | 3.82 | 2.48 | 1.89 | 1.54 | 1.40 | 1.23 | 1.09 | 1.07 | 0.98 | 0.91 | 0.79 | 0.72 | 0.63 | 0.49 | 0.45 | 0.37 | 0.30 |
| | pow | .80 | 7.88 | 4.86 | 3.66 | 3.00 | 2.60 | 2.30 | 2.07 | 1.92 | 1.78 | 1.65 | 1.47 | 1.29 | 1.09 | 0.86 | 0.75 | 0.60 | 0.45 |
| | 1% | .05 | 9.58 | 5.57 | 4.22 | 3.49 | 3.08 | 2.81 | 2.61 | 2.46 | 2.33 | 2.23 | 2.09 | 1.93 | 1.78 | 1.61 | 1.52 | 1.43 | 1.32 |
| | 1% | .01 | 14.39 | 8.02 | 5.90 | 4.83 | 4.18 | 3.75 | 3.43 | 3.20 | 3.01 | 2.86 | 2.63 | 2.40 | 2.16 | 1.91 | 1.75 | 1.64 | 1.48 |
| | pow | .50 | 9.37 | 4.88 | 3.36 | 2.68 | 2.20 | 1.88 | 1.64 | 1.46 | 1.40 | 1.28 | 1.09 | 0.96 | 0.75 | 0.61 | 0.51 | 0.43 | 0.33 |
| | pow | .80 | 15.40 | 8.08 | 5.62 | 4.38 | 3.62 | 3.11 | 2.74 | 2.46 | 2.27 | 2.09 | 1.81 | 1.56 | 1.26 | 0.98 | 0.82 | 0.66 | 0.48 |
| | 5% | .05 | 24.55 | 12.87 | 8.94 | 7.02 | 5.88 | 5.11 | 4.56 | 4.15 | 3.84 | 3.56 | 3.18 | 2.79 | 2.40 | 2.01 | 1.82 | 1.62 | 1.41 |
| | 5% | .01 | 32.04 | 16.72 | 11.60 | 9.04 | 7.51 | 6.49 | 5.76 | 5.21 | 4.78 | 4.44 | 3.93 | 3.42 | 2.90 | 2.38 | 2.12 | 1.85 | 1.58 |
| | pow | .50 | 23.65 | 11.85 | 8.17 | 6.17 | 4.96 | 4.16 | 3.58 | 3.15 | 2.81 | 2.53 | 2.21 | 1.79 | 1.42 | 1.01 | 0.81 | 0.63 | 0.41 |
| | pow | .80 | 33.10 | 16.67 | 11.31 | 8.57 | 6.92 | 5.82 | 5.04 | 4.45 | 3.99 | 3.61 | 3.10 | 2.54 | 2.00 | 1.44 | 1.15 | 0.88 | 0.58 |
| 300 | nil | .05 | 3.86 | 3.03 | 2.63 | 2.40 | 2.24 | 2.13 | 2.04 | 1.97 | 1.91 | 1.86 | 1.78 | 1.70 | 1.60 | 1.49 | 1.43 | 1.36 | 1.27 |
| | nil | .01 | 6.72 | 4.68 | 3.85 | 3.38 | 3.08 | 2.86 | 2.70 | 2.57 | 2.47 | 2.38 | 2.24 | 2.10 | 1.94 | 1.76 | 1.66 | 1.55 | 1.41 |
| | pow | .50 | 3.80 | 2.47 | 1.88 | 1.53 | 1.39 | 1.22 | 1.09 | 0.98 | 0.97 | 0.90 | 0.78 | 0.72 | 0.62 | 0.48 | 0.41 | 0.37 | 0.26 |
| | pow | .80 | 7.85 | 4.84 | 3.64 | 2.98 | 2.58 | 2.28 | 2.05 | 1.87 | 1.75 | 1.63 | 1.44 | 1.27 | 1.07 | 0.83 | 0.71 | 0.58 | 0.41 |
| | 1% | .05 | 11.62 | 6.54 | 4.85 | 3.97 | 3.43 | 3.08 | 2.84 | 2.65 | 2.51 | 2.38 | 2.20 | 2.02 | 1.83 | 1.64 | 1.54 | 1.43 | 1.31 |
| | 1% | .01 | 16.85 | 9.18 | 6.63 | 5.36 | 4.59 | 4.07 | 3.70 | 3.42 | 3.20 | 3.03 | 2.76 | 2.49 | 2.22 | 1.93 | 1.78 | 1.62 | 1.45 |
| | pow | .50 | 11.36 | 5.83 | 3.97 | 3.17 | 2.56 | 2.17 | 1.89 | 1.67 | 1.50 | 1.45 | 1.23 | 1.01 | 0.84 | 0.62 | 0.51 | 0.43 | 0.30 |
| | pow | .80 | 17.91 | 9.26 | 6.37 | 4.96 | 4.04 | 3.44 | 3.02 | 2.69 | 2.44 | 2.27 | 1.95 | 1.63 | 1.33 | 1.00 | 0.82 | 0.66 | 0.45 |
| | 5% | .05 | 32.04 | 16.65 | 11.52 | 8.94 | 7.35 | 6.32 | 5.59 | 5.05 | 4.62 | 4.28 | 3.77 | 3.25 | 2.74 | 2.22 | 1.97 | 1.70 | 1.44 |
| | 5% | .01 | 40.62 | 20.94 | 14.40 | 11.13 | 9.16 | 7.86 | 6.92 | 6.21 | 5.67 | 5.23 | 4.58 | 3.92 | 3.26 | 2.60 | 2.27 | 1.94 | 1.59 |
| | pow | .50 | 31.22 | 15.66 | 10.47 | 7.87 | 6.49 | 5.42 | 4.66 | 4.09 | 3.64 | 3.29 | 2.75 | 2.29 | 1.73 | 1.22 | 0.96 | 0.70 | 0.43 |
| | pow | .80 | 41.71 | 20.99 | 14.09 | 10.62 | 8.63 | 7.22 | 6.23 | 5.48 | 4.90 | 4.43 | 3.73 | 3.07 | 2.35 | 1.66 | 1.31 | 0.96 | 0.59 |
| 400 | nil | .05 | 3.85 | 3.02 | 2.63 | 2.39 | 2.24 | 2.12 | 2.03 | 1.96 | 1.90 | 1.85 | 1.77 | 1.69 | 1.59 | 1.48 | 1.42 | 1.35 | 1.26 |
| | nil | .01 | 6.70 | 4.66 | 3.83 | 3.37 | 3.06 | 2.85 | 2.68 | 2.56 | 2.45 | 2.36 | 2.23 | 2.08 | 1.92 | 1.74 | 1.64 | 1.53 | 1.39 |
| | pow | .50 | 3.80 | 2.47 | 1.88 | 1.52 | 1.38 | 1.21 | 1.08 | 0.97 | 0.97 | 0.90 | 0.78 | 0.71 | 0.62 | 0.48 | 0.41 | 0.34 | 0.25 |
| | pow | .80 | 7.84 | 4.83 | 3.63 | 2.96 | 2.57 | 2.26 | 2.04 | 1.86 | 1.74 | 1.62 | 1.43 | 1.25 | 1.06 | 0.82 | 0.69 | 0.55 | 0.39 |
| | 1% | .05 | 13.49 | 7.44 | 5.43 | 4.42 | 3.78 | 3.35 | 3.07 | 2.85 | 2.68 | 2.54 | 2.32 | 2.11 | 1.90 | 1.68 | 1.56 | 1.44 | 1.30 |
| | 1% | .01 | 19.02 | 10.26 | 7.33 | 5.87 | 4.98 | 4.39 | 3.97 | 3.65 | 3.40 | 3.20 | 2.90 | 2.60 | 2.29 | 1.97 | 1.80 | 1.63 | 1.44 |
| | pow | .50 | 13.21 | 6.73 | 4.56 | 3.47 | 2.79 | 2.46 | 2.14 | 1.89 | 1.70 | 1.54 | 1.38 | 1.12 | 0.92 | 0.68 | 0.56 | 0.43 | 0.30 |

(*Continued*)

200   One-Stop F Table

| | | | | | | | | | dfHyp | | | | | | | | | |
|---|---|---|---|---|---|---|---|---|---|---|---|---|---|---|---|---|---|---|
| | | | 1 | 2 | 3 | 4 | 5 | 6 | 7 | 8 | 9 | 10 | 12 | 15 | 20 | 30 | 40 | 60 | 120 |
| | pow | .80 | 20.18 | 10.35 | 7.07 | 5.42 | 4.40 | 3.77 | 3.29 | 2.93 | 2.65 | 2.42 | 2.10 | 1.75 | 1.41 | 1.04 | 0.86 | 0.66 | 0.44 |
| | 5% | .05 | 39.07 | 20.15 | 13.84 | 10.68 | 8.79 | 7.53 | 6.62 | 5.95 | 5.42 | 4.98 | 4.33 | 3.71 | 3.08 | 2.44 | 2.12 | 1.80 | 1.48 |
| | 5% | .01 | 48.68 | 24.94 | 17.05 | 13.11 | 10.74 | 9.16 | 8.04 | 7.19 | 6.53 | 6.00 | 5.21 | 4.42 | 3.63 | 2.84 | 2.44 | 2.04 | 1.63 |
| | pow | .50 | 38.66 | 19.38 | 12.95 | 9.73 | 7.80 | 6.51 | 5.59 | 4.90 | 4.36 | 4.04 | 3.37 | 2.71 | 2.04 | 1.42 | 1.11 | 0.80 | 0.48 |
| | pow | .80 | 49.91 | 25.07 | 16.78 | 12.64 | 10.16 | 8.50 | 7.32 | 6.43 | 5.73 | 5.24 | 4.39 | 3.55 | 2.71 | 1.88 | 1.47 | 1.06 | 0.63 |
| 500 | nil | .05 | 3.85 | 3.01 | 2.62 | 2.39 | 2.23 | 2.12 | 2.03 | 1.95 | 1.90 | 1.85 | 1.77 | 1.68 | 1.59 | 1.48 | 1.41 | 1.34 | 1.25 |
| | nil | .01 | 6.68 | 4.65 | 3.82 | 3.36 | 3.05 | 2.84 | 2.67 | 2.55 | 2.44 | 2.36 | 2.22 | 2.07 | 1.91 | 1.73 | 1.63 | 1.52 | 1.38 |
| | pow | .50 | 3.79 | 2.46 | 1.87 | 1.52 | 1.38 | 1.21 | 1.08 | 0.97 | 0.97 | 0.89 | 0.77 | 0.71 | 0.56 | 0.48 | 0.41 | 0.34 | 0.25 |
| | pow | .80 | 7.83 | 4.82 | 3.62 | 2.96 | 2.56 | 2.26 | 2.03 | 1.85 | 1.73 | 1.61 | 1.42 | 1.25 | 1.03 | 0.82 | 0.69 | 0.55 | 0.38 |
| | 1% | .05 | 15.23 | 8.29 | 5.98 | 4.82 | 4.13 | 3.65 | 3.29 | 3.04 | 2.84 | 2.69 | 2.45 | 2.21 | 1.97 | 1.72 | 1.59 | 1.45 | 1.30 |
| | 1% | .01 | 21.10 | 11.28 | 8.00 | 6.36 | 5.37 | 4.70 | 4.23 | 3.89 | 3.60 | 3.38 | 3.04 | 2.71 | 2.36 | 2.01 | 1.83 | 1.64 | 1.43 |
| | pow | .50 | 14.99 | 7.61 | 5.14 | 3.90 | 3.15 | 2.64 | 2.38 | 2.10 | 1.89 | 1.71 | 1.44 | 1.24 | 0.95 | 0.69 | 0.56 | 0.44 | 0.30 |
| | pow | .80 | 22.31 | 11.39 | 7.74 | 5.91 | 4.81 | 4.06 | 3.57 | 3.16 | 2.85 | 2.60 | 2.22 | 1.86 | 1.46 | 1.07 | 0.87 | 0.67 | 9.44 |
| | 5% | .05 | 46.31 | 23.76 | 16.24 | 12.45 | 10.16 | 8.63 | 7.57 | 6.77 | 6.15 | 5.65 | 4.91 | 4.16 | 3.40 | 2.66 | 2.28 | 1.90 | 1.52 |
| | 5% | .01 | 56.40 | 28.82 | 19.60 | 15.02 | 12.26 | 10.43 | 9.11 | 8.13 | 7.36 | 6.75 | 5.83 | 4.91 | 3.99 | 3.07 | 2.61 | 2.14 | 1.67 |
| | pow | .50 | 45.18 | 22.62 | 15.10 | 11.32 | 9.05 | 7.54 | 6.65 | 5.82 | 5.13 | 4.67 | 3.90 | 3.13 | 2.42 | 1.62 | 1.26 | 0.90 | 0.52 |
| | pow | .80 | 57.52 | 28.85 | 19.30 | 14.50 | 11.61 | 9.68 | 8.44 | 7.40 | 6.60 | 5.96 | 4.99 | 4.03 | 3.09 | 2.11 | 1.64 | 1.17 | 0.67 |
| 600 | nil | .05 | 3.85 | 3.01 | 2.62 | 2.39 | 2.23 | 2.11 | 2.02 | 1.95 | 1.89 | 1.84 | 1.77 | 1.68 | 1.58 | 1.47 | 1.41 | 1.34 | 1.25 |
| | nil | .01 | 6.67 | 4.64 | 3.81 | 3.35 | 3.05 | 2.83 | 2.67 | 2.54 | 2.44 | 2.35 | 2.21 | 2.07 | 1.91 | 1.73 | 1.63 | 1.51 | 1.37 |
| | pow | .50 | 3.79 | 2.46 | 1.87 | 1.52 | 1.38 | 1.21 | 1.08 | 0.97 | 0.96 | 0.89 | 0.77 | 0.71 | 0.56 | 0.48 | 0.41 | 0.34 | 0.25 |
| | pow | .80 | 7.82 | 4.82 | 3.62 | 2.95 | 2.56 | 2.25 | 2.02 | 1.85 | 1.73 | 1.61 | 1.42 | 1.24 | 1.02 | 0.81 | 0.68 | 0.54 | 0.38 |
| | 1% | .05 | 16.94 | 9.11 | 6.51 | 5.21 | 4.43 | 3.91 | 3.54 | 3.24 | 3.01 | 2.84 | 2.57 | 2.31 | 2.03 | 1.76 | 1.62 | 1.47 | 1.31 |
| | 1% | .01 | 23.08 | 12.25 | 8.64 | 6.83 | 5.74 | 5.01 | 4.49 | 4.10 | 3.80 | 3.55 | 3.18 | 2.81 | 2.44 | 2.06 | 1.86 | 1.66 | 1.44 |
| | pow | .50 | 16.14 | 8.46 | 5.71 | 4.32 | 3.49 | 2.93 | 2.52 | 2.32 | 2.07 | 1.88 | 1.59 | 1.28 | 1.04 | 0.75 | 0.61 | 0.47 | 0.30 |
| | pow | .80 | 24.06 | 12.38 | 8.38 | 6.38 | 5.18 | 4.38 | 3.80 | 3.41 | 3.05 | 2.78 | 2.37 | 1.95 | 1.55 | 1.13 | 0.91 | 0.69 | 0.44 |
| | 5% | .05 | 52.82 | 26.98 | 18.38 | 14.08 | 11.50 | 9.78 | 8.55 | 7.63 | 6.91 | 6.34 | 5.48 | 4.59 | 3.73 | 2.86 | 2.44 | 2.00 | 1.56 |

| | | | | | | | | | | | | | | | | | | | |
|---|---|---|---|---|---|---|---|---|---|---|---|---|---|---|---|---|---|---|---|
| | 5% | .01 | 63.87 | 32.55 | 22.11 | 16.89 | 13.75 | 11.65 | 10.16 | 9.05 | 8.18 | 7.48 | 6.44 | 5.39 | 4.35 | 3.30 | 2.78 | 2.25 | 1.72 |
| | pow | .50 | 52.51 | 26.28 | 17.54 | 13.17 | 10.55 | 8.80 | 7.55 | 6.61 | 5.88 | 5.30 | 4.42 | 3.63 | 2.73 | 1.88 | 1.41 | 1.00 | 0.57 |
| | pow | .80 | 65.29 | 32.72 | 21.87 | 16.44 | 13.19 | 11.02 | 9.47 | 8.30 | 7.40 | 6.67 | 5.59 | 4.54 | 3.44 | 2.37 | 1.80 | 1.27 | 0.72 |
| 1,000 | nil | .05 | 3.84 | 3.00 | 2.61 | 2.38 | 2.22 | 2.11 | 2.02 | 1.95 | 1.89 | 1.84 | 1.76 | 1.67 | 1.58 | 1.47 | 1.40 | 1.33 | 1.24 |
| | nil | .01 | 6.66 | 4.63 | 3.80 | 3.34 | 3.04 | 2.82 | 2.66 | 2.53 | 2.42 | 2.34 | 2.20 | 2.06 | 1.90 | 1.71 | 1.61 | 1.49 | 1.35 |
| | pow | .50 | 3.79 | 2.46 | 1.87 | 1.51 | 1.37 | 1.20 | 1.07 | 0.97 | 0.96 | 0.89 | 0.77 | 0.71 | 0.55 | 0.48 | 0.41 | 0.30 | 0.23 |
| | pow | .80 | 7.81 | 4.81 | 3.61 | 2.94 | 2.55 | 2.24 | 2.02 | 1.84 | 1.72 | 1.60 | 1.41 | 1.23 | 1.01 | 0.80 | 0.67 | 0.52 | 0.36 |
| | 1% | .05 | 23.25 | 12.26 | 8.59 | 6.76 | 5.66 | 4.93 | 4.40 | 3.99 | 3.66 | 3.42 | 3.05 | 2.68 | 2.30 | 1.93 | 1.74 | 1.54 | 1.34 |
| | 1% | .01 | 30.44 | 15.89 | 11.01 | 8.59 | 7.13 | 6.16 | 5.47 | 4.95 | 4.54 | 4.22 | 3.73 | 3.24 | 2.75 | 2.25 | 2.00 | 1.74 | 1.46 |
| | pow | .50 | 22.91 | 11.53 | 7.73 | 5.83 | 4.68 | 3.92 | 3.37 | 3.08 | 2.73 | 2.47 | 2.07 | 1.67 | 1.33 | 0.95 | 0.72 | 0.54 | 0.33 |
| | pow | .80 | 31.72 | 16.01 | 10.78 | 8.16 | 6.58 | 5.53 | 4.78 | 4.27 | 3.81 | 3.45 | 2.91 | 2.38 | 1.86 | 1.33 | 1.04 | 0.76 | 0.47 |
| | 5% | .05 | 78.99 | 40.07 | 27.09 | 20.61 | 16.71 | 14.12 | 12.27 | 10.88 | 9.79 | 8.93 | 7.63 | 6.32 | 5.00 | 3.71 | 3.06 | 2.41 | 1.75 |
| | 5% | .01 | 92.43 | 46.81 | 31.60 | 23.96 | 19.41 | 16.37 | 14.20 | 12.57 | 11.30 | 10.29 | 8.77 | 7.25 | 5.73 | 4.21 | 3.45 | 2.68 | 1.92 |
| | pow | .50 | 78.54 | 39.29 | 26.20 | 19.66 | 15.74 | 13.12 | 11.25 | 9.85 | 8.75 | 7.88 | 6.57 | 5.36 | 4.02 | 2.69 | 2.06 | 1.40 | 0.75 |
| | pow | .80 | 93.82 | 46.97 | 31.35 | 23.54 | 18.86 | 15.73 | 13.50 | 11.83 | 10.53 | 9.48 | 7.92 | 6.42 | 4.83 | 3.25 | 2.49 | 1.70 | 0.92 |
| 10,000 | nil | .05 | 3.84 | 3.00 | 2.61 | 2.37 | 2.21 | 2.10 | 2.01 | 1.94 | 1.88 | 1.83 | 1.75 | 1.66 | 1.57 | 1.46 | 1.39 | 1.32 | 1.22 |
| | nil | .01 | 6.64 | 4.61 | 3.78 | 3.32 | 3.02 | 2.80 | 2.64 | 2.51 | 2.41 | 2.32 | 2.19 | 2.04 | 1.88 | 1.70 | 1.59 | 1.47 | 1.33 |
| | pow | .50 | 3.79 | 2.34 | 1.86 | 1.51 | 1.37 | 1.20 | 1.07 | 0.96 | 0.87 | 0.84 | 0.77 | 0.63 | 0.55 | 0.43 | 0.36 | 0.30 | 0.22 |
| | pow | .80 | 7.81 | 4.76 | 3.60 | 2.93 | 2.54 | 2.23 | 2.00 | 1.83 | 1.68 | 1.59 | 1.40 | 1.20 | 1.00 | 0.77 | 0.64 | 0.51 | 0.34 |
| | 1% | .05 | 135.8 | 68.43 | 45.99 | 34.77 | 28.04 | 23.55 | 20.34 | 17.94 | 16.07 | 14.57 | 12.33 | 10.06 | 7.80 | 5.56 | 4.44 | 3.31 | 2.19 |
| | 1% | .01 | 152.7 | 76.89 | 51.63 | 38.99 | 31.39 | 26.35 | 22.74 | 20.04 | 17.94 | 16.26 | 13.73 | 11.21 | 8.69 | 6.16 | 4.90 | 3.63 | 2.36 |
| | pow | .50 | 134.7 | 67.36 | 44.90 | 33.68 | 26.95 | 22.46 | 19.25 | 16.85 | 14.98 | 13.48 | 11.23 | 8.99 | 6.74 | 4.55 | 3.41 | 2.31 | 1.19 |
| | pow | .80 | 154.1 | 77.06 | 51.39 | 38.56 | 30.86 | 25.73 | 22.06 | 19.31 | 17.17 | 15.46 | 12.90 | 10.32 | 7.75 | 5.22 | 3.93 | 2.66 | 1.37 |
| | 5% | .05 | 601.3 | 301.2 | 201.2 | 151.1 | 121.1 | 101.1 | 86.81 | 76.09 | 67.76 | 61.09 | 51.08 | 41.08 | 31.07 | 21.06 | 16.05 | 11.04 | 6.03 |
| | 5% | .01 | 637.8 | 319.4 | 213.3 | 160.3 | 128.4 | 107.2 | 92.03 | 80.66 | 71.81 | 64.74 | 54.12 | 43.51 | 32.89 | 22.28 | 16.97 | 11.67 | 6.36 |
| | pow | .50 | 600.6 | 300.3 | 201.1 | 150.1 | 120.1 | 100.1 | 85.79 | 75.05 | 66.72 | 60.06 | 50.04 | 40.02 | 30.01 | 20.01 | 15.01 | 10.04 | 5.03 |
| | pow | .80 | 639.9 | 319.5 | 213.1 | 159.9 | 127.7 | 106.6 | 91.23 | 79.99 | 71.07 | 64.00 | 53.31 | 42.68 | 31.94 | 21.33 | 16.00 | 10.68 | 5.38 |

## R Code Used to Generate One-Stop $F$ Table

```
# ONE-STOP F TABLE
# User defined functions
noncenfn<-function(df, pv) df * pv / (1 - pv)
noncenfn2<-function(df, FF) df * FF
critFn<-function(df1, df2, alph, effect) qf((1-alph), df1, df2, noncenfn(df2, effect))
findF<-function(FF) 1-pf(critF, dfhyp, dferr, FF * dfhyp)-powtarget

options(digits=3)

# Global variables
dfhyps<-c(1, 2, 3, 4, 5, 6, 7, 8, 9, 10, 12, 15, 20, 30, 40, 60, 120)
dferrs<-c(3, 4, 5, 6, 8, 10, 12, 14, 16, 18, 20, 22, 24, 26, 28, 30, 40, 50, 60, 70, 80, 90, 100, 120, 150, 200, 300, 400, 500,600, 1000, 10000)

onestop<-matrix(nrow=12, ncol=length(dfhyps))
rownames(onestop)<-c("nil .05", "nil .01", "pow .5", "pow .8", "1% .05", "1% .01", "pow .5", "pow .8", "5% .05", "5% .01", "pow .5", "pow .8")
colnames(onestop)<-dfhyps

for (dferr in dferrs) {
    cat("\ndfErr = ", dferr, "\t\t\t\t\tdfHyp\n")

    for (i in 1:length(dfhyps)) {
        dfhyp=dfhyps[i]

        # Nil hypotheses
        effect=0
        critF<-critFn(dfhyp, dferr, 0.05, effect)
        onestop[1, i]=critF
        onestop[2, i]=critFn(dfhyp, dferr, 0.01, effect)

        powtarget=0.5
        Fresult<-uniroot(findF,c(0,650))
        onestop[3, i]=Fresult$root
        powtarget=0.8
        Fresult<-uniroot(findF,c(0,650))
        onestop[4, i]=Fresult$root

        # 1% minimum effects
        effect=0.01
        critF<-critFn(dfhyp, dferr, 0.05, effect)
```

```
    onestop[5, i]=critF
    onestop[6, i]=critFn(dfhyp, dferr, 0.01, effect)

    powtarget=0.5
    Fresult<-uniroot(findF,c(0,650))
    onestop[7, i]=Fresult$root
    powtarget=0.8
    Fresult<-uniroot(findF,c(0,650))
    onestop[8, i]=Fresult$root

    # 5% minimum effects
    effect=0.05
    critF<-critFn(dfhyp, dferr, 0.05, effect)
    onestop[9, i]=critF
    onestop[10,i]=critFn(dfhyp, dferr, 0.01, effect)

    powtarget=0.5
    Fresult<-uniroot(findF,c(0,650))
    onestop[11, i]=Fresult$root
    powtarget=0.8
    Fresult<-uniroot(findF,c(0,650))
    onestop[12,i]=Fresult$root
  }
  print(onestop)
}
```

# Appendix C
One-Stop *PV* Table

## One-Stop PV Table

|  |  |  | | | | | | dfHyp | | | | | | | |
|---|---|---|---|---|---|---|---|---|---|---|---|---|---|---|---|
| dfErr | | | 1 | 2 | 3 | 4 | 5 | 6 | 7 | 8 | 9 | 10 | 12 | 15 | 20 | 30 | 40 |
| 3 | nil | .05 | 0.771 | 0.864 | 0.903 | 0.924 | 0.938 | 0.947 | 0.954 | 0.959 | 0.964 | 0.967 | 0.972 | 0.978 | 0.983 | 0.989 | 0.991 |
|  | nil | .01 | 0.919 | 0.954 | 0.967 | 0.975 | 0.979 | 0.982 | 0.985 | 0.987 | 0.988 | 0.989 | 0.991 | 0.993 | 0.994 | 0.996 | 0.997 |
|  | pow | .50 | 0.734 | 0.828 | 0.871 | 0.897 | 0.914 | 0.926 | 0.936 | 0.943 | 0.948 | 0.953 | 0.960 | 0.968 | 0.975 | 0.983 | 0.987 |
|  | pow | .80 | 0.858 | 0.913 | 0.937 | 0.951 | 0.959 | 0.965 | 0.970 | 0.973 | 0.976 | 0.978 | 0.982 | 0.985 | 0.989 | 0.993 | 0.994 |
|  | 1% | .05 | 0.777 | 0.866 | 0.904 | 0.925 | 0.938 | 0.947 | 0.954 | 0.959 | 0.964 | 0.967 | 0.972 | 0.978 | 0.983 | 0.989 | 0.991 |
|  | 1% | .01 | 0.921 | 0.954 | 0.967 | 0.975 | 0.979 | 0.982 | 0.985 | 0.987 | 0.988 | 0.989 | 0.991 | 0.993 | 0.994 | 0.996 | 0.997 |
|  | pow | .50 | 0.740 | 0.830 | 0.873 | 0.898 | 0.915 | 0.927 | 0.936 | 0.943 | 0.948 | 0.953 | 0.960 | 0.968 | 0.975 | 0.983 | 0.987 |
|  | pow | .80 | 0.862 | 0.914 | 0.937 | 0.951 | 0.960 | 0.965 | 0.970 | 0.974 | 0.976 | 0.979 | 0.982 | 0.985 | 0.989 | 0.993 | 0.994 |
|  | 5% | .05 | 0.796 | 0.873 | 0.907 | 0.927 | 0.939 | 0.948 | 0.955 | 0.960 | 0.964 | 0.967 | 0.973 | 0.978 | 0.983 | 0.989 | 0.991 |
|  | 5% | .01 | 0.929 | 0.957 | 0.969 | 0.975 | 0.980 | 0.983 | 0.985 | 0.987 | 0.988 | 0.989 | 0.991 | 0.993 | 0.994 | 0.996 | 0.997 |
|  | pow | .50 | 0.761 | 0.839 | 0.878 | 0.901 | 0.917 | 0.928 | 0.937 | 0.944 | 0.949 | 0.954 | 0.961 | 0.968 | 0.975 | 0.983 | 0.987 |
|  | pow | .80 | 0.874 | 0.919 | 0.940 | 0.952 | 0.961 | 0.966 | 0.971 | 0.974 | 0.977 | 0.979 | 0.982 | 0.986 | 0.989 | 0.993 | 0.994 |
| 4 | nil | .05 | 0.658 | 0.776 | 0.832 | 0.865 | 0.887 | 0.902 | 0.914 | 0.924 | 0.931 | 0.937 | 0.947 | 0.956 | 0.967 | 0.977 | 0.983 |
|  | nil | .01 | 0.841 | 0.900 | 0.926 | 0.941 | 0.951 | 0.958 | 0.963 | 0.967 | 0.971 | 0.973 | 0.977 | 0.982 | 0.986 | 0.990 | 0.993 |
|  | pow | .50 | 0.625 | 0.733 | 0.789 | 0.825 | 0.850 | 0.869 | 0.884 | 0.895 | 0.905 | 0.912 | 0.925 | 0.938 | 0.952 | 0.967 | 0.975 |
|  | pow | .80 | 0.780 | 0.850 | 0.885 | 0.906 | 0.920 | 0.931 | 0.939 | 0.946 | 0.951 | 0.955 | 0.962 | 0.969 | 0.976 | 0.984 | 0.988 |
|  | 1% | .05 | 0.667 | 0.780 | 0.834 | 0.866 | 0.887 | 0.903 | 0.915 | 0.924 | 0.931 | 0.937 | 0.947 | 0.957 | 0.967 | 0.977 | 0.983 |
|  | 1% | .01 | 0.846 | 0.902 | 0.927 | 0.942 | 0.951 | 0.958 | 0.963 | 0.967 | 0.971 | 0.973 | 0.977 | 0.982 | 0.986 | 0.990 | 0.993 |
|  | pow | .50 | 0.634 | 0.737 | 0.792 | 0.827 | 0.852 | 0.870 | 0.884 | 0.896 | 0.905 | 0.913 | 0.925 | 0.938 | 0.952 | 0.967 | 0.975 |
|  | pow | .80 | 0.785 | 0.852 | 0.886 | 0.907 | 0.921 | 0.932 | 0.940 | 0.946 | 0.951 | 0.955 | 0.962 | 0.969 | 0.976 | 0.984 | 0.988 |
|  | 5% | .05 | 0.699 | 0.793 | 0.841 | 0.871 | 0.891 | 0.905 | 0.917 | 0.925 | 0.932 | 0.938 | 0.947 | 0.957 | 0.967 | 0.977 | 0.983 |
|  | 5% | .01 | 0.864 | 0.909 | 0.931 | 0.944 | 0.953 | 0.959 | 0.964 | 0.968 | 0.971 | 0.974 | 0.978 | 0.982 | 0.986 | 0.991 | 0.993 |
|  | pow | .50 | 0.668 | 0.753 | 0.802 | 0.834 | 0.856 | 0.874 | 0.887 | 0.898 | 0.907 | 0.914 | 0.926 | 0.939 | 0.953 | 0.967 | 0.975 |
|  | pow | .80 | 0.806 | 0.861 | 0.891 | 0.911 | 0.923 | 0.933 | 0.941 | 0.947 | 0.952 | 0.956 | 0.963 | 0.969 | 0.976 | 0.984 | 0.988 |

*(Continued)*

206  *One-Stop PV Table*

| | | | | | | | | | *df Hyp* | | | | | | | |
|---|---|---|---|---|---|---|---|---|---|---|---|---|---|---|---|---|
| | | 1 | 2 | 3 | 4 | 5 | 6 | 7 | 8 | 9 | 10 | 12 | 15 | 20 | 30 | 40 |
| 5 | nil .05 | 0.569 | 0.698 | 0.764 | 0.806 | 0.835 | 0.856 | 0.872 | 0.885 | 0.896 | 0.904 | 0.918 | 0.933 | 0.948 | 0.964 | 0.973 |
| | nil .01 | 0.765 | 0.842 | 0.879 | 0.901 | 0.916 | 0.928 | 0.936 | 0.943 | 0.948 | 0.953 | 0.960 | 0.967 | 0.974 | 0.983 | 0.987 |
| | pow .50 | 0.542 | 0.651 | 0.713 | 0.756 | 0.787 | 0.811 | 0.830 | 0.845 | 0.858 | 0.869 | 0.886 | 0.904 | 0.925 | 0.947 | 0.959 |
| | pow .80 | 0.712 | 0.790 | 0.831 | 0.859 | 0.879 | 0.893 | 0.905 | 0.913 | 0.921 | 0.928 | 0.938 | 0.948 | 0.960 | 0.972 | 0.979 |
| | 1% .05 | 0.581 | 0.703 | 0.767 | 0.808 | 0.836 | 0.857 | 0.873 | 0.886 | 0.896 | 0.905 | 0.919 | 0.933 | 0.948 | 0.964 | 0.973 |
| | 1% .01 | 0.773 | 0.845 | 0.880 | 0.902 | 0.917 | 0.928 | 0.936 | 0.943 | 0.948 | 0.953 | 0.960 | 0.967 | 0.975 | 0.983 | 0.987 |
| | pow .50 | 0.553 | 0.657 | 0.718 | 0.759 | 0.789 | 0.812 | 0.831 | 0.846 | 0.859 | 0.869 | 0.886 | 0.905 | 0.925 | 0.947 | 0.959 |
| | pow .80 | 0.720 | 0.793 | 0.834 | 0.860 | 0.880 | 0.894 | 0.905 | 0.914 | 0.922 | 0.928 | 0.938 | 0.949 | 0.960 | 0.972 | 0.979 |
| | 5% .05 | 0.624 | 0.723 | 0.779 | 0.816 | 0.842 | 0.861 | 0.876 | 0.888 | 0.898 | 0.907 | 0.920 | 0.934 | 0.949 | 0.965 | 0.973 |
| | 5% .01 | 0.802 | 0.857 | 0.887 | 0.907 | 0.920 | 0.930 | 0.938 | 0.944 | 0.950 | 0.954 | 0.960 | 0.967 | 0.975 | 0.983 | 0.987 |
| | pow .50 | 0.597 | 0.681 | 0.732 | 0.770 | 0.798 | 0.819 | 0.836 | 0.850 | 0.862 | 0.872 | 0.889 | 0.906 | 0.926 | 0.948 | 0.960 |
| | pow .80 | 0.749 | 0.808 | 0.842 | 0.866 | 0.884 | 0.897 | 0.908 | 0.916 | 0.924 | 0.930 | 0.939 | 0.949 | 0.960 | 0.972 | 0.979 |
| 6 | nil .05 | 0.499 | 0.632 | 0.704 | 0.751 | 0.785 | 0.811 | 0.831 | 0.847 | 0.860 | 0.871 | 0.889 | 0.908 | 0.928 | 0.950 | 0.962 |
| | nil .01 | 0.696 | 0.785 | 0.830 | 0.859 | 0.879 | 0.894 | 0.906 | 0.915 | 0.923 | 0.929 | 0.939 | 0.950 | 0.961 | 0.973 | 0.979 |
| | pow .50 | 0.476 | 0.581 | 0.647 | 0.693 | 0.728 | 0.755 | 0.777 | 0.796 | 0.811 | 0.824 | 0.845 | 0.869 | 0.896 | 0.926 | 0.942 |
| | pow .80 | 0.654 | 0.734 | 0.781 | 0.813 | 0.836 | 0.854 | 0.868 | 0.880 | 0.890 | 0.898 | 0.911 | 0.925 | 0.941 | 0.959 | 0.968 |
| | 1% .05 | 0.514 | 0.638 | 0.708 | 0.754 | 0.787 | 0.812 | 0.832 | 0.848 | 0.861 | 0.872 | 0.889 | 0.908 | 0.928 | 0.950 | 0.962 |
| | 1% .01 | 0.708 | 0.790 | 0.833 | 0.861 | 0.881 | 0.895 | 0.907 | 0.916 | 0.923 | 0.930 | 0.939 | 0.950 | 0.961 | 0.973 | 0.979 |
| | pow .50 | 0.490 | 0.590 | 0.652 | 0.697 | 0.731 | 0.757 | 0.779 | 0.797 | 0.812 | 0.825 | 0.847 | 0.870 | 0.896 | 0.926 | 0.942 |
| | pow .80 | 0.664 | 0.740 | 0.784 | 0.815 | 0.838 | 0.855 | 0.869 | 0.881 | 0.890 | 0.898 | 0.912 | 0.926 | 0.941 | 0.959 | 0.968 |
| | 5% .05 | 0.566 | 0.664 | 0.724 | 0.765 | 0.795 | 0.818 | 0.837 | 0.852 | 0.864 | 0.875 | 0.891 | 0.910 | 0.929 | 0.951 | 0.962 |
| | 5% .01 | 0.747 | 0.807 | 0.844 | 0.868 | 0.886 | 0.899 | 0.910 | 0.918 | 0.925 | 0.931 | 0.941 | 0.951 | 0.962 | 0.973 | 0.980 |
| | pow .50 | 0.542 | 0.619 | 0.674 | 0.713 | 0.743 | 0.767 | 0.787 | 0.803 | 0.817 | 0.830 | 0.850 | 0.872 | 0.898 | 0.927 | 0.943 |
| | pow .80 | 0.701 | 0.759 | 0.798 | 0.824 | 0.845 | 0.861 | 0.874 | 0.884 | 0.893 | 0.901 | 0.913 | 0.927 | 0.942 | 0.959 | 0.968 |
| 8 | nil .05 | 0.399 | 0.527 | 0.604 | 0.657 | 0.647 | 0.729 | 0.754 | 0.775 | 0.792 | 0.807 | 0.831 | 0.858 | 0.887 | 0.920 | 0.938 |
| | nil .01 | 0.585 | 0.684 | 0.740 | 0.778 | 0.806 | 0.827 | 0.844 | 0.858 | 0.869 | 0.879 | 0.895 | 0.912 | 0.931 | 0.951 | 0.962 |

One-Stop PV Table 207

| | | | | | | | | | | | | | | |
|---|---|---|---|---|---|---|---|---|---|---|---|---|---|---|
| pow | .50 | 0.382 | 0.476 | 0.539 | 0.585 | 0.624 | 0.654 | 0.679 | 0.703 | 0.721 | 0.737 | 0.766 | 0.798 | 0.836 | 0.880 | 0.905 |
| pow | .80 | 0.561 | 0.642 | 0.692 | 0.727 | 0.756 | 0.777 | 0.795 | 0.811 | 0.824 | 0.835 | 0.854 | 0.875 | 0.900 | 0.928 | 0.943 |
| 1% | .05 | 0.418 | 0.537 | 0.610 | 0.662 | 0.701 | 0.731 | 0.756 | 0.776 | 0.794 | 0.808 | 0.832 | 0.858 | 0.888 | 0.920 | 0.938 |
| 1% | .01 | 0.603 | 0.692 | 0.745 | 0.781 | 0.808 | 0.829 | 0.845 | 0.859 | 0.870 | 0.880 | 0.895 | 0.912 | 0.931 | 0.951 | 0.962 |
| pow | .50 | 0.401 | 0.487 | 0.547 | 0.591 | 0.629 | 0.657 | 0.682 | 0.705 | 0.723 | 0.741 | 0.767 | 0.799 | 0.837 | 0.880 | 0.905 |
| pow | .80 | 0.576 | 0.650 | 0.697 | 0.731 | 0.759 | 0.780 | 0.797 | 0.813 | 0.825 | 0.837 | 0.855 | 0.876 | 0.900 | 0.928 | 0.943 |
| 5% | .05 | 0.482 | 0.573 | 0.634 | 0.679 | 0.714 | 0.742 | 0.764 | 0.783 | 0.800 | 0.813 | 0.836 | 0.861 | 0.889 | 0.921 | 0.939 |
| 5% | .01 | 0.658 | 0.721 | 0.764 | 0.794 | 0.818 | 0.836 | 0.851 | 0.864 | 0.874 | 0.883 | 0.898 | 0.914 | 0.932 | 0.952 | 0.963 |
| pow | .50 | 0.465 | 0.528 | 0.578 | 0.615 | 0.646 | 0.674 | 0.696 | 0.717 | 0.734 | 0.749 | 0.775 | 0.805 | 0.840 | 0.882 | 0.906 |
| pow | .80 | 0.625 | 0.679 | 0.718 | 0.747 | 0.770 | 0.790 | 0.806 | 0.820 | 0.832 | 0.842 | 0.859 | 0.879 | 0.902 | 0.929 | 0.944 |
| 10 | | | | | | | | | | | | | | | |
| nil | .05 | 0.332 | 0.451 | 0.527 | 0.582 | 0.624 | 0.659 | 0.687 | 0.711 | 0.731 | 0.749 | 0.778 | 0.810 | 0.847 | 0.890 | 0.914 |
| nil | .01 | 0.501 | 0.602 | 0.663 | 0.706 | 0.738 | 0.764 | 0.784 | 0.802 | 0.816 | 0.829 | 0.850 | 0.872 | 0.898 | 0.927 | 0.943 |
| pow | .50 | 0.319 | 0.400 | 0.459 | 0.503 | 0.539 | 0.570 | 0.596 | 0.620 | 0.640 | 0.659 | 0.691 | 0.731 | 0.776 | 0.833 | 0.866 |
| pow | .80 | 0.491 | 0.568 | 0.618 | 0.655 | 0.684 | 0.708 | 0.729 | 0.746 | 0.761 | 0.775 | 0.798 | 0.825 | 0.856 | 0.893 | 0.915 |
| 1% | .05 | 0.353 | 0.463 | 0.535 | 0.588 | 0.629 | 0.662 | 0.690 | 0.713 | 0.733 | 0.750 | 0.779 | 0.811 | 0.848 | 0.890 | 0.914 |
| 1% | .01 | 0.524 | 0.613 | 0.670 | 0.711 | 0.742 | 0.767 | 0.787 | 0.804 | 0.818 | 0.830 | 0.851 | 0.873 | 0.899 | 0.927 | 0.944 |
| pow | .50 | 0.339 | 0.416 | 0.468 | 0.510 | 0.545 | 0.575 | 0.600 | 0.623 | 0.643 | 0.665 | 0.696 | 0.733 | 0.778 | 0.833 | 0.867 |
| pow | .80 | 0.509 | 0.579 | 0.626 | 0.661 | 0.689 | 0.712 | 0.732 | 0.749 | 0.764 | 0.778 | 0.800 | 0.826 | 0.856 | 0.893 | 0.915 |
| 5% | .05 | 0.425 | 0.507 | 0.566 | 0.611 | 0.647 | 0.677 | 0.702 | 0.724 | 0.742 | 0.758 | 0.785 | 0.815 | 0.851 | 0.892 | 0.915 |
| 5% | .01 | 0.591 | 0.652 | 0.696 | 0.730 | 0.756 | 0.778 | 0.796 | 0.812 | 0.825 | 0.836 | 0.855 | 0.876 | 0.900 | 0.928 | 0.944 |
| pow | .50 | 0.410 | 0.465 | 0.507 | 0.542 | 0.572 | 0.598 | 0.621 | 0.641 | 0.659 | 0.676 | 0.705 | 0.739 | 0.784 | 0.837 | 0.869 |
| pow | .80 | 0.568 | 0.617 | 0.653 | 0.682 | 0.706 | 0.727 | 0.744 | 0.760 | 0.773 | 0.785 | 0.806 | 0.830 | 0.860 | 0.895 | 0.916 |
| 12 | | | | | | | | | | | | | | | |
| nil | .05 | 0.283 | 0.393 | 0.466 | 0.521 | 0.564 | 0.600 | 0.630 | 0.655 | 0.677 | 0.696 | 0.729 | 0.766 | 0.809 | 0.860 | 0.890 |
| nil | .01 | 0.437 | 0.536 | 0.598 | 0.643 | 0.678 | 0.707 | 0.730 | 0.750 | 0.767 | 0.782 | 0.806 | 0.834 | 0.865 | 0.902 | 0.923 |
| pow | .50 | 0.274 | 0.346 | 0.396 | 0.437 | 0.472 | 0.502 | 0.529 | 0.553 | 0.575 | 0.595 | 0.625 | 0.667 | 0.718 | 0.786 | 0.826 |
| pow | .80 | 0.436 | 0.509 | 0.557 | 0.594 | 0.623 | 0.648 | 0.670 | 0.689 | 0.706 | 0.721 | 0.744 | 0.775 | 0.811 | 0.858 | 0.885 |
| 1% | .05 | 0.307 | 0.407 | 0.476 | 0.528 | 0.570 | 0.604 | 0.634 | 0.658 | 0.680 | 0.699 | 0.731 | 0.767 | 0.810 | 0.861 | 0.890 |
| 1% | .01 | 0.464 | 0.550 | 0.607 | 0.650 | 0.684 | 0.711 | 0.734 | 0.753 | 0.769 | 0.784 | 0.808 | 0.835 | 0.866 | 0.903 | 0.924 |
| pow | .50 | 0.296 | 0.360 | 0.408 | 0.446 | 0.479 | 0.508 | 0.534 | 0.558 | 0.579 | 0.598 | 0.627 | 0.670 | 0.719 | 0.787 | 0.827 |

*(Continued)*

208   One-Stop PV Table

|    |      |     | \multicolumn{15}{c}{dfHyp} |
| --- | --- | --- | --- | --- | --- | --- | --- | --- | --- | --- | --- | --- | --- | --- | --- | --- | --- |
|    |      |     | 1 | 2 | 3 | 4 | 5 | 6 | 7 | 8 | 9 | 10 | 12 | 15 | 20 | 30 | 40 |
|    | pow  | .80 | 0.457 | 0.522 | 0.566 | 0.601 | 0.629 | 0.653 | 0.674 | 0.692 | 0.709 | 0.723 | 0.747 | 0.777 | 0.812 | 0.858 | 0.885 |
|    | 5%   | .05 | 0.384 | 0.457 | 0.512 | 0.556 | 0.592 | 0.623 | 0.649 | 0.672 | 0.692 | 0.709 | 0.739 | 0.773 | 0.814 | 0.863 | 0.891 |
|    | 5%   | .01 | 0.540 | 0.597 | 0.640 | 0.675 | 0.703 | 0.727 | 0.747 | 0.764 | 0.779 | 0.792 | 0.814 | 0.839 | 0.869 | 0.904 | 0.925 |
|    | pow  | .50 | 0.371 | 0.415 | 0.452 | 0.483 | 0.512 | 0.537 | 0.560 | 0.580 | 0.599 | 0.617 | 0.643 | 0.682 | 0.728 | 0.792 | 0.830 |
|    | pow  | .80 | 0.523 | 0.566 | 0.600 | 0.628 | 0.652 | 0.673 | 0.691 | 0.707 | 0.722 | 0.735 | 0.756 | 0.784 | 0.818 | 0.861 | 0.887 |
| 14 | nil  | .05 | 0.247 | 0.348 | 0.417 | 0.471 | 0.514 | 0.550 | 0.580 | 0.607 | 0.630 | 0.650 | 0.685 | 0.725 | 0.773 | 0.832 | 0.866 |
|    | nil  | .01 | 0.388 | 0.482 | 0.544 | 0.590 | 0.626 | 0.656 | 0.681 | 0.703 | 0.721 | 0.738 | 0.765 | 0.797 | 0.834 | 0.878 | 0.903 |
|    | pow  | .50 | 0.240 | 0.304 | 0.351 | 0.389 | 0.416 | 0.446 | 0.472 | 0.496 | 0.518 | 0.532 | 0.569 | 0.610 | 0.666 | 0.738 | 0.785 |
|    | pow  | .80 | 0.393 | 0.461 | 0.507 | 0.543 | 0.571 | 0.596 | 0.618 | 0.637 | 0.655 | 0.669 | 0.696 | 0.728 | 0.769 | 0.821 | 0.854 |
|    | 1%   | .05 | 0.272 | 0.364 | 0.429 | 0.479 | 0.521 | 0.555 | 0.585 | 0.611 | 0.633 | 0.653 | 0.687 | 0.727 | 0.774 | 0.832 | 0.867 |
|    | 1%   | .01 | 0.418 | 0.499 | 0.555 | 0.598 | 0.633 | 0.662 | 0.686 | 0.706 | 0.725 | 0.740 | 0.767 | 0.798 | 0.835 | 0.878 | 0.904 |
|    | pow  | .50 | 0.264 | 0.320 | 0.363 | 0.399 | 0.430 | 0.453 | 0.478 | 0.501 | 0.523 | 0.543 | 0.572 | 0.612 | 0.668 | 0.739 | 0.786 |
|    | pow  | .80 | 0.416 | 0.476 | 0.518 | 0.551 | 0.579 | 0.602 | 0.623 | 0.642 | 0.659 | 0.674 | 0.699 | 0.730 | 0.770 | 0.822 | 0.854 |
|    | 5%   | .05 | 0.353 | 0.418 | 0.469 | 0.511 | 0.547 | 0.577 | 0.604 | 0.627 | 0.648 | 0.666 | 0.697 | 0.735 | 0.780 | 0.835 | 0.868 |
|    | 5%   | .01 | 0.499 | 0.552 | 0.594 | 0.628 | 0.657 | 0.681 | 0.702 | 0.721 | 0.737 | 0.751 | 0.776 | 0.804 | 0.838 | 0.880 | 0.905 |
|    | pow  | .50 | 0.341 | 0.379 | 0.411 | 0.440 | 0.467 | 0.485 | 0.508 | 0.528 | 0.547 | 0.559 | 0.592 | 0.628 | 0.679 | 0.746 | 0.791 |
|    | pow  | .80 | 0.488 | 0.526 | 0.557 | 0.583 | 0.607 | 0.625 | 0.644 | 0.660 | 0.675 | 0.687 | 0.711 | 0.740 | 0.778 | 0.826 | 0.857 |
| 16 | nil  | .05 | 0.219 | 0.312 | 0.378 | 0.429 | 0.471 | 0.507 | 0.538 | 0.564 | 0.588 | 0.609 | 0.645 | 0.688 | 0.740 | 0.804 | 0.843 |
|    | nil  | .01 | 0.348 | 0.438 | 0.498 | 0.544 | 0.581 | 0.612 | 0.638 | 0.660 | 0.680 | 0.698 | 0.727 | 0.762 | 0.803 | 0.853 | 0.883 |
|    | pow  | .50 | 0.213 | 0.271 | 0.310 | 0.345 | 0.376 | 0.403 | 0.422 | 0.446 | 0.468 | 0.489 | 0.519 | 0.561 | 0.615 | 0.696 | 0.746 |
|    | pow  | .80 | 0.357 | 0.421 | 0.464 | 0.498 | 0.527 | 0.551 | 0.571 | 0.591 | 0.609 | 0.625 | 0.652 | 0.685 | 0.728 | 0.786 | 0.823 |
|    | 1%   | .05 | 0.245 | 0.329 | 0.390 | 0.439 | 0.479 | 0.513 | 0.543 | 0.569 | 0.592 | 0.613 | 0.648 | 0.690 | 0.741 | 0.805 | 0.844 |
|    | 1%   | .01 | 0.380 | 0.456 | 0.511 | 0.554 | 0.589 | 0.618 | 0.643 | 0.665 | 0.684 | 0.701 | 0.730 | 0.764 | 0.804 | 0.854 | 0.883 |
|    | pow  | .50 | 0.238 | 0.289 | 0.328 | 0.361 | 0.385 | 0.411 | 0.436 | 0.452 | 0.473 | 0.493 | 0.523 | 0.564 | 0.617 | 0.697 | 0.747 |
|    | pow  | .80 | 0.382 | 0.437 | 0.477 | 0.509 | 0.535 | 0.558 | 0.579 | 0.596 | 0.613 | 0.629 | 0.655 | 0.688 | 0.730 | 0.788 | 0.824 |
|    | 5%   | .05 | 0.328 | 0.387 | 0.434 | 0.475 | 0.509 | 0.539 | 0.565 | 0.588 | 0.609 | 0.628 | 0.660 | 0.699 | 0.748 | 0.809 | 0.846 |

## One-Stop PV Table

|   |   |   |   |   |   |   |   |   |   |   |   |   |   |   |   |   |   |
|---|---|---|---|---|---|---|---|---|---|---|---|---|---|---|---|---|---|
|   | 5% | .01 | 0.465 | 0.514 | 0.554 | 0.588 | 0.617 | 0.641 | 0.663 | 0.682 | 0.699 | 0.714 | 0.740 | 0.771 | 0.809 | 0.857 | 0.885 |
|   | pow | .50 | 0.318 | 0.350 | 0.379 | 0.406 | 0.425 | 0.447 | 0.469 | 0.482 | 0.501 | 0.519 | 0.546 | 0.583 | 0.632 | 0.706 | 0.753 |
|   | pow | .80 | 0.458 | 0.492 | 0.521 | 0.546 | 0.566 | 0.586 | 0.604 | 0.618 | 0.633 | 0.647 | 0.670 | 0.700 | 0.739 | 0.793 | 0.827 |
| 18 | nil | .05 | 0.197 | 0.283 | 0.345 | 0.394 | 0.435 | 0.470 | 0.500 | 0.527 | 0.551 | 0.573 | 0.610 | 0.654 | 0.709 | 0.778 | 0.821 |
|   | nil | .01 | 0.315 | 0.401 | 0.459 | 0.504 | 0.541 | 0.572 | 0.599 | 0.622 | 0.643 | 0.661 | 0.692 | 0.729 | 0.774 | 0.829 | 0.863 |
|   | pow | .50 | 0.190 | 0.242 | 0.280 | 0.313 | 0.342 | 0.362 | 0.386 | 0.409 | 0.423 | 0.444 | 0.474 | 0.517 | 0.573 | 0.654 | 0.706 |
|   | pow | .80 | 0.327 | 0.387 | 0.428 | 0.461 | 0.488 | 0.511 | 0.532 | 0.552 | 0.567 | 0.584 | 0.611 | 0.646 | 0.691 | 0.752 | 0.792 |
|   | 1% | .05 | 0.224 | 0.301 | 0.358 | 0.405 | 0.444 | 0.477 | 0.507 | 0.533 | 0.556 | 0.577 | 0.613 | 0.657 | 0.711 | 0.779 | 0.822 |
|   | 1% | .01 | 0.349 | 0.421 | 0.473 | 0.515 | 0.550 | 0.580 | 0.605 | 0.627 | 0.647 | 0.665 | 0.695 | 0.731 | 0.775 | 0.830 | 0.864 |
|   | pow | .50 | 0.218 | 0.263 | 0.299 | 0.324 | 0.352 | 0.377 | 0.394 | 0.416 | 0.437 | 0.449 | 0.478 | 0.520 | 0.575 | 0.656 | 0.711 |
|   | pow | .80 | 0.354 | 0.405 | 0.443 | 0.471 | 0.497 | 0.521 | 0.539 | 0.558 | 0.575 | 0.589 | 0.615 | 0.649 | 0.693 | 0.754 | 0.794 |
|   | 5% | .05 | 0.308 | 0.361 | 0.406 | 0.444 | 0.477 | 0.506 | 0.531 | 0.554 | 0.575 | 0.594 | 0.627 | 0.668 | 0.718 | 0.784 | 0.824 |
|   | 5% | .01 | 0.437 | 0.483 | 0.521 | 0.553 | 0.582 | 0.606 | 0.628 | 0.647 | 0.665 | 0.680 | 0.708 | 0.741 | 0.782 | 0.834 | 0.866 |
|   | pow | .50 | 0.298 | 0.327 | 0.353 | 0.372 | 0.394 | 0.416 | 0.430 | 0.449 | 0.467 | 0.478 | 0.504 | 0.542 | 0.592 | 0.667 | 0.715 |
|   | pow | .80 | 0.433 | 0.463 | 0.490 | 0.511 | 0.532 | 0.551 | 0.567 | 0.583 | 0.598 | 0.610 | 0.633 | 0.664 | 0.704 | 0.761 | 0.798 |
| 20 | nil | .05 | 0.178 | 0.259 | 0.317 | 0.364 | 0.404 | 0.438 | 0.468 | 0.495 | 0.518 | 0.540 | 0.577 | 0.623 | 0.680 | 0.754 | 0.799 |
|   | nil | .01 | 0.288 | 0.369 | 0.426 | 0.470 | 0.S06 | 0.537 | 0.564 | 0.588 | 0.609 | 0.627 | 0.660 | 0.698 | 0.746 | 0.806 | 0.843 |
|   | pow | .50 | 0.172 | 0.220 | 0.256 | 0.286 | 0.307 | 0.332 | 0.356 | 0.370 | 0.391 | 0.403 | 0.441 | 0.475 | 0.533 | 0.612 | 0.669 |
|   | pow | .80 | 0.301 | 0.358 | 0.397 | 0.428 | 0.453 | 0.477 | 0.498 | 0.515 | 0.532 | 0.546 | 0.576 | 0.609 | 0.655 | 0.719 | 0.762 |
|   | 1% | .05 | 0.206 | 0.277 | 0.331 | 0.376 | 0.413 | 0.446 | 0.475 | 0.501 | 0.524 | 0.545 | 0.581 | 0.626 | 0.682 | 0.755 | 0.800 |
|   | 1% | .01 | 0.324 | 0.391 | 0.441 | 0.482 | 0.516 | 0.545 | 0.571 | 0.594 | 0.614 | 0.632 | 0.663 | 0.701 | 0.748 | 0.808 | 0.844 |
|   | pow | .50 | 0.201 | 0.242 | 0.270 | 0.298 | 0.324 | 0.341 | 0.364 | 0.385 | 0.398 | 0.418 | 0.446 | 0.479 | 0.536 | 0.614 | 0.675 |
|   | pow | .80 | 0.330 | 0.378 | 0.411 | 0.440 | 0.465 | 0.485 | 0.505 | 0.524 | 0.538 | 0.554 | 0.580 | 0.613 | 0.658 | 0.720 | 0.764 |
|   | 5% | .05 | 0.291 | 0.340 | 0.381 | 0.417 | 0.449 | 0.477 | 0.502 | 0.525 | 0.545 | 0.564 | 0.597 | 0.639 | 0.691 | 0.760 | 0.804 |
|   | 5% | .01 | 0.414 | 0.456 | 0.492 | 0.524 | 0.551 | 0.575 | 0.597 | 0.616 | 0.634 | 0.650 | 0.678 | 0.712 | 0.756 | 0.812 | 0.847 |
|   | pow | .50 | 0.285 | 0.307 | 0.331 | 0.348 | 0.369 | 0.389 | 0.402 | 0.420 | 0.431 | 0.448 | 0.473 | 0.503 | 0.555 | 0.627 | 0.679 |
|   | pow | .80 | 0.412 | 0.439 | 0.463 | 0.483 | 0.503 | 0.521 | 0.536 | 0.552 | 0.564 | 0.578 | 0.601 | 0.630 | 0.671 | 0.729 | 0.769 |
| 22 | nil | .05 | 0.163 | 0.238 | 0.294 | 0.339 | 0.377 | 0.410 | 0.439 | 0.466 | 0.489 | 0.511 | 0.548 | 0.594 | 0.653 | 0.730 | 0.779 |
|   | nil | .01 | 0.265 | 0.342 | 0.396 | 0.440 | 0.475 | 0.506 | 0.533 | 0.557 | 0.578 | 0.597 | 0.630 | 0.670 | 0.720 | 0.784 | 0.824 |

(*Continued*)

210  One-Stop PV Table

| | | | | | | | | | dfHyp | | | | | | | |
|---|---|---|---|---|---|---|---|---|---|---|---|---|---|---|---|---|
| | | | 1 | 2 | 3 | 4 | 5 | 6 | 7 | 8 | 9 | 10 | 12 | 15 | 20 | 30 | 40 |
| | pow | .50 | 0.158 | 0.202 | 0.235 | 0.263 | 0.283 | 0.307 | 0.322 | 0.343 | 0.356 | 0.375 | 0.404 | 0.446 | 0.495 | 0.578 | 0.638 |
| | pow | .80 | 0.280 | 0.333 | 0.370 | 0.400 | 0.424 | 0.447 | 0.465 | 0.484 | 0.499 | 0.515 | 0.541 | 0.577 | 0.622 | 0.688 | 0.734 |
| | 1% | .05 | 0.192 | 0.258 | 0.308 | 0.351 | 0.387 | 0.419 | 0.447 | 0.472 | 0.495 | 0.516 | 0.553 | 0.598 | 0.655 | 0.731 | 0.780 |
| | 1% | .01 | 0.302 | 0.365 | 0.413 | 0.453 | 0.486 | 0.515 | 0.541 | 0.563 | 0.584 | 0.602 | 0.634 | 0.673 | 0.722 | 0.786 | 0.825 |
| | pow | .50 | 0.187 | 0.224 | 0.250 | 0.276 | 0.301 | 0.317 | 0.338 | 0.351 | 0.371 | 0.382 | 0.409 | 0.451 | 0.499 | 0.579 | 0.639 |
| | pow | .80 | 0.310 | 0.354 | 0.386 | 0.413 | 0.437 | 0.456 | 0.476 | 0.491 | 0.508 | 0.521 | 0.547 | 0.581 | 0.625 | 0.690 | 0.735 |
| | 5% | .05 | 0.277 | 0.322 | 0.360 | 0.394 | 0.425 | 0.452 | 0.476 | 0.498 | 0.519 | 0.537 | 0.571 | 0.612 | 0.665 | 0.737 | 0.784 |
| | 5% | .01 | 0.393 | 0.433 | 0.467 | 0.497 | 0.524 | 0.548 | 0.569 | 0.588 | 0.505 | 0.622 | 0.651 | 0.686 | 0.731 | 0.791 | 0.829 |
| | pow | .50 | 0.271 | 0.291 | 0.312 | 0.327 | 0.347 | 0.359 | 0.378 | 0.388 | 0.406 | 0.415 | 0.439 | 0.476 | 0.519 | 0.594 | 0.649 |
| | pow | .80 | 0.393 | 0.418 | 0.440 | 0.459 | 0.478 | 0.493 | 0.509 | 0.522 | 0.536 | 0.547 | 0.570 | 0.600 | 0.640 | 0.700 | 0.742 |
| 24 | nil | .05 | 0.151 | 0.221 | 0.273 | 0.316 | 0.353 | 0.385 | 0.414 | 0.440 | 0.463 | 0.484 | 0.522 | 0.568 | 0.628 | 0.708 | 0.759 |
| | nil | .01 | 0.246 | 0.319 | 0.371 | 0.413 | 0.448 | 0.478 | 0.505 | 0.528 | 0.550 | 0.569 | 0.603 | 0.644 | 0.695 | 0.763 | 0.806 |
| | pow | .50 | 0.146 | 0.187 | 0.218 | 0.244 | 0.263 | 0.285 | 0.299 | 0.320 | 0.332 | 0.351 | 0.378 | 0.411 | 0.460 | 0.544 | 0.600 |
| | pow | .80 | 0.261 | 0.312 | 0.347 | 0.376 | 0.398 | 0.420 | 0.438 | 0.456 | 0.471 | 0.486 | 0.513 | 0.545 | 0.591 | 0.659 | 0.705 |
| | 1% | .05 | 0.180 | 0.241 | 0.289 | 0.329 | 0.364 | 0.395 | 0.422 | 0.447 | 0.470 | 0.490 | 0.527 | 0.572 | 0.631 | 0.709 | 0.760 |
| | 1% | .01 | 0.284 | 0.343 | 0.389 | 0.427 | 0.459 | 0.488 | 0.513 | 0.536 | 0.556 | 0.575 | 0.607 | 0.647 | 0.698 | 0.765 | 0.807 |
| | pow | .50 | 0.175 | 0.210 | 0.233 | 0.258 | 0.274 | 0.296 | 0.308 | 0.328 | 0.347 | 0.358 | 0.384 | 0.416 | 0.473 | 0.546 | 0.609 |
| | pow | .80 | 0.292 | 0.333 | 0.363 | 0.389 | 0.410 | 0.430 | 0.447 | 0.464 | 0.481 | 0.493 | 0.519 | 0.550 | 0.597 | 0.661 | 0.709 |
| | 5% | .05 | 0.264 | 0.306 | 0.342 | 0.375 | 0.403 | 0.430 | 0.453 | 0.475 | 0.495 | 0.513 | 0.546 | 0.588 | 0.642 | 0.716 | 0.765 |
| | 5% | .01 | 0.376 | 0.413 | 0.445 | 0.474 | 0.500 | 0.523 | 0.544 | 0.563 | 0.581 | 0.597 | 0.626 | 0.662 | 0.708 | 0.770 | 0.811 |
| | pow | .50 | 0.259 | 0.276 | 0.296 | 0.310 | 0.328 | 0.340 | 0.357 | 0.367 | 0.384 | 0.392 | 0.415 | 0.452 | 0.495 | 0.562 | 0.621 |
| | pow | .80 | 0.377 | 0.399 | 0.420 | 0.437 | 0.455 | 0.469 | 0.485 | 0.497 | 0.511 | 0.522 | 0.543 | 0.574 | 0.613 | 0.672 | 0.717 |
| 26 | nil | .05 | 0.140 | 0.206 | 0.256 | 0.297 | 0.332 | 0.363 | 0.391 | 0.417 | 0.439 | 0.460 | 0.498 | 0.544 | 0.605 | 0.687 | 0.740 |
| | nil | .01 | 0.229 | 0.298 | 0.349 | 0.389 | 0.423 | 0.453 | 0.479 | 0.503 | 0.524 | 0.543 | 0.577 | 0.619 | 0.672 | 0.743 | 0.788 |
| | pow | .50 | 0.135 | 0.174 | 0.202 | 0.222 | 0.245 | 0.259 | 0.280 | 0.291 | 0.311 | 0.320 | 0.347 | 0.388 | 0.436 | 0.512 | 0.570 |
| | pow | .80 | 0.245 | 0.293 | 0.326 | 0.352 | 0.376 | 0.395 | 0.414 | 0.429 | 0.445 | 0.458 | 0.484 | 0.519 | 0.564 | 0.631 | 0.679 |

## One-Stop PV Table

|    |     |     |       |       |       |       |       |       |       |       |       |       |       |       |       |
|----|-----|-----|-------|-------|-------|-------|-------|-------|-------|-------|-------|-------|-------|-------|-------|
|    | 1%  | .05 | 0.169 | 0.226 | 0.271 | 0.310 | 0.343 | 0.373 | 0.400 | 0.424 | 0.446 | 0.467 | 0.503 | 0.549 | 0.608 | 0.689 | 0.741 |
|    | 1%  | .01 | 0.268 | 0.323 | 0.367 | 0.404 | 0.436 | 0.464 | 0.488 | 0.511 | 0.531 | 0.550 | 0.582 | 0.623 | 0.675 | 0.744 | 0.789 |
|    | pow | .50 | 0.165 | 0.197 | 0.218 | 0.241 | 0.257 | 0.277 | 0.289 | 0.308 | 0.318 | 0.336 | 0.362 | 0.393 | 0.440 | 0.515 | 0.580 |
|    | pow | .80 | 0.277 | 0.315 | 0.343 | 0.368 | 0.388 | 0.407 | 0.423 | 0.440 | 0.453 | 0.468 | 0.493 | 0.524 | 0.568 | 0.633 | 0.683 |
|    | 5%  | .05 | 0.254 | 0.293 | 0.327 | 0.357 | 0.385 | 0.410 | 0.433 | 0.454 | 0.474 | 0.492 | 0.524 | 0.566 | 0.620 | 0.696 | 0.747 |
|    | 5%  | .01 | 0.360 | 0.395 | 0.426 | 0.453 | 0.478 | 0.501 | 0.522 | 0.540 | 0.558 | 0.574 | 0.603 | 0.639 | 0.686 | 0.751 | 0.794 |
|    | pow | .50 | 0.248 | 0.264 | 0.282 | 0.295 | 0.312 | 0.322 | 0.339 | 0.348 | 0.364 | 0.372 | 0.395 | 0.421 | 0.463 | 0.540 | 0.593 |
|    | pow | .80 | 0.362 | 0.383 | 0.402 | 0.418 | 0.435 | 0.448 | 0.463 | 0.475 | 0.488 | 0.498 | 0.520 | 0.547 | 0.586 | 0.648 | 0.692 |
| 28 | nil | .05 | 0.130 | 0.193 | 0.240 | 0.279 | 0.314 | 0.344 | 0.371 | 0.396 | 0.418 | 0.439 | 0.476 | 0.522 | 0.583 | 0.667 | 0.722 |
|    | nil | .01 | 0.214 | 0.280 | 0.329 | 0.368 | 0.401 | 0.430 | 0.456 | 0.480 | 0.501 | 0.520 | 0.554 | 0.596 | 0.650 | 0.723 | 0.771 |
|    | pow | .50 | 0.126 | 0.162 | 0.189 | 0.208 | 0.229 | 0.243 | 0.263 | 0.273 | 0.292 | 0.301 | 0.327 | 0.357 | 0.405 | 0.491 | 0.541 |
|    | pow | .80 | 0.230 | 0.276 | 0.308 | 0.333 | 0.355 | 0.373 | 0.392 | 0.407 | 0.422 | 0.435 | 0.460 | 0.492 | 0.536 | 0.607 | 0.653 |
|    | 1%  | .05 | 0.160 | 0.213 | 0.256 | 4.293 | 0.325 | 0.354 | 0.380 | 0.404 | 0.426 | 0.446 | 0.481 | 0.527 | 0.586 | 0.669 | 0.724 |
|    | 1%  | .01 | 0.254 | 0.306 | 0.348 | 0.383 | 0.414 | 0.442 | 0.466 | 0.488 | 0.508 | 0.527 | 0.559 | 0.600 | 0.653 | 0.725 | 0.772 |
|    | pow | .50 | 0.156 | 0.185 | 0.206 | 0.227 | 0.241 | 0.261 | 0.272 | 0.291 | 0.300 | 0.318 | 0.333 | 0.373 | 0.419 | 0.493 | 0.551 |
|    | pow | .80 | 0.263 | 0.299 | 0.325 | 0.349 | 0.368 | 0.387 | 0.402 | 0.418 | 0.431 | 0.446 | 0.467 | 0.500 | 0.544 | 0.609 | 0.658 |
|    | 5%  | .05 | 0.245 | 0.281 | 0.313 | 0.342 | 0.368 | 0.392 | 0.415 | 0.435 | 0.454 | 0.472 | 0.504 | 0.545 | 0.600 | 0.677 | 0.729 |
|    | 5%  | .01 | 0.346 | 0.379 | 0.409 | 0.435 | 0.459 | 0.481 | 0.501 | 0.520 | 0.537 | 0.553 | 0.581 | 0.618 | 0.666 | 0.733 | 0.777 |
|    | pow | .50 | 0.239 | 0.253 | 0.270 | 0.281 | 0.298 | 0.307 | 0.323 | 0.331 | 0.339 | 0.354 | 0.376 | 0.402 | 0.443 | 0.511 | 0.565 |
|    | pow | .80 | 0.349 | 0.368 | 0.386 | 0.401 | 0.417 | 0.430 | 0.444 | 0.455 | 0.465 | 0.478 | 0.498 | 0.525 | 0.563 | 0.623 | 0.668 |
| 30 | nil | .05 | 0.122 | 0.181 | 0.226 | 0.264 | 0.297 | 0.326 | 0.353 | 0.377 | 0.399 | 0.419 | 0.455 | 0.502 | 0.563 | 0.648 | 0.705 |
|    | nil | .01 | 0.201 | 0.264 | 0.311 | 0.349 | 0.381 | 0.410 | 0.435 | 0.458 | 0.479 | 0.498 | 0.532 | 0.574 | 0.629 | 0.705 | 0.754 |
|    | pow | .50 | 0.118 | 0.152 | 0.178 | 0.195 | 0.216 | 0.228 | 0.247 | 0.258 | 0.276 | 0.284 | 0.309 | 0.339 | 0.385 | 0.461 | 0.522 |
|    | pow | .80 | 0.217 | 0.261 | 0.292 | 0.315 | 0.337 | 0.354 | 0.372 | 0.386 | 0.402 | 0.414 | 0.438 | 0.469 | 0.513 | 0.581 | 0.631 |
|    | 1%  | .05 | 0.152 | 0.202 | 0.243 | 0.278 | 0.309 | 0.337 | 0.362 | 0.385 | 0.407 | 0.426 | 0.462 | 0.507 | 0.566 | 0.650 | 0.706 |
|    | 1%  | .01 | 0.241 | 0.291 | 0.331 | 0.365 | 0.395 | 0.422 | 0.446 | 0.467 | 0.487 | 0.506 | 0.538 | 0.579 | 0.633 | 0.707 | 0.755 |
|    | pow | .50 | 0.149 | 0.176 | 0.194 | 0.215 | 0.228 | 0.247 | 0.257 | 0.275 | 0.284 | 0.292 | 0.316 | 0.344 | 0.389 | 0.464 | 0.523 |
|    | pow | .80 | 0.251 | 0.285 | 0.310 | 0.332 | 0.350 | 0.368 | 0.383 | 0.399 | 0.411 | 0.422 | 0.446 | 0.476 | 0.518 | 0.584 | 0.633 |
|    | 5%  | .05 | 0.236 | 0.271 | 0.300 | 0.328 | 0.353 | 0.377 | 0.398 | 0.418 | 0.437 | 0.454 | 0.485 | 0.526 | 0.581 | 0.659 | 0.713 |

(*Continued*)

212  One-Stop PV Table

| | | | | | | | | dfHyp | | | | | | | |
|---|---|---|---|---|---|---|---|---|---|---|---|---|---|---|---|
| | | | 1 | 2 | 3 | 4 | 5 | 6 | 7 | 8 | 9 | 10 | 12 | 15 | 20 | 30 | 40 |
| 40 | 5% | .01 | 0.334 | 0.365 | 0.393 | 0.418 | 0.442 | 0.463 | 0.483 | 0.501 | 0.518 | 0.533 | 0.562 | 0.598 | 0.647 | 0.715 | 0.761 |
| | pow | .50 | 0.231 | 0.243 | 0.259 | 0.269 | 0.285 | 0.294 | 0.309 | 0.317 | 0.324 | 0.339 | 0.359 | 0.385 | 0.425 | 0.483 | 0.539 |
| | pow | .80 | 0.338 | 0.355 | 0.372 | 0.386 | 0.401 | 0.413 | 0.426 | 0.437 | 0.447 | 0.459 | 0.479 | 0.505 | 0.542 | 0.599 | 0.645 |
| | nil | .05 | 0.092 | 0.139 | 0.176 | 0.207 | 0.234 | 0.259 | 0.282 | 0.304 | 0.323 | 0.342 | 0.375 | 0.419 | 0.479 | 0.567 | 0.628 |
| | nil | .01 | 0.155 | 0.206 | 0.244 | 0.277 | 0.305 | 0.331 | 0.353 | 0.374 | 0.394 | 0.412 | 0.444 | 0.486 | 0.542 | 0.623 | 0.679 |
| | pow | .50 | 0.090 | 0.116 | 0.132 | 0.150 | 0.160 | 0.176 | 0.184 | 0.200 | 0.207 | 0.213 | 0.234 | 0.259 | 0.300 | 0.362 | 0.413 |
| | pow | .80 | 0.170 | 0.205 | 0.229 | 0.249 | 0.265 | 0.282 | 0.295 | 0.309 | 0.320 | 0.330 | 0.351 | 0.378 | 0.418 | 0.480 | 0.528 |
| | 1% | .05 | 0.123 | 0.162 | 0.194 | 0.222 | 0.248 | 0.272 | 0.294 | 0.314 | 0.333 | 0.350 | 0.383 | 0.425 | 0.484 | 0.570 | 0.631 |
| | 1% | .01 | 0.196 | 0.234 | 0.267 | 0.296 | 0.321 | 0.345 | 0.366 | 0.386 | 0.404 | 0.421 | 0.452 | 0.493 | 0.547 | 0.626 | 0.681 |
| | pow | .50 | 0.121 | 0.140 | 0.154 | 0.169 | 0.180 | 0.188 | 0.203 | 0.210 | 0.216 | 0.231 | 0.251 | 0.275 | 0.305 | 0.365 | 0.415 |
| | pow | .80 | 0.206 | 0.231 | 0.251 | 0.268 | 0.283 | 0.296 | 0.310 | 0.320 | 0.331 | 0.343 | 0.363 | 0.389 | 0.425 | 0.485 | 0.532 |
| | 5% | .05 | 0.207 | 0.231 | 0.255 | 0.277 | 0.298 | 0.317 | 0.335 | 0.352 | 0.369 | 0.384 | 0.413 | 0.450 | 0.503 | 0.583 | 0.640 |
| | 5% | .01 | 0.288 | 0.312 | 0.334 | 0.355 | 0.375 | 0.393 | 0.410 | 0.426 | 0.442 | 0.456 | 0.483 | 0.518 | 0.566 | 0.639 | 0.690 |
| | pow | .50 | 0.200 | 0.211 | 0.219 | 0.231 | 0.237 | 0.243 | 0.256 | 0.262 | 0.266 | 0.279 | 0.288 | 0.309 | 0.344 | 0.398 | 0.444 |
| | pow | .80 | 0.294 | 0.306 | 0.317 | 0.329 | 0.339 | 0.348 | 0.359 | 0.367 | 0.375 | 0.386 | 0.400 | 0.421 | 0.455 | 0.509 | 0.551 |
| 50 | nil | .05 | 0.075 | 0.113 | 0.143 | 0.170 | 0.194 | 0.215 | 0.235 | 0.254 | 0.272 | 0.288 | 0.319 | 0.359 | 0.416 | 0.503 | 0.566 |
| | nil | .01 | 0.125 | 0.168 | 0.201 | 0.229 | 0.254 | 0.277 | 0.297 | 0.316 | 0.334 | 0.350 | 0.381 | 0.420 | 0.475 | 0.557 | 0.616 |
| | pow | .50 | 0.073 | 0.094 | 0.107 | 0.121 | 0.130 | 0.143 | 0.150 | 0.156 | 0.169 | 0.174 | 0.192 | 0.214 | 0.240 | 0.296 | 0.334 |
| | pow | .80 | 0.140 | 0.169 | 0.189 | 0.206 | 0.220 | 0.234 | 0.244 | 0.254 | 0.266 | 0.275 | 0.293 | 0.317 | 0.350 | 0.407 | 0.450 |
| | 1% | .05 | 0.106 | 0.136 | 0.163 | 0.187 | 0.209 | 0.229 | 0.248 | 0.266 | 0.282 | 0.298 | 0.328 | 0.367 | 0.422 | 0.507 | 0.569 |
| | 1% | .01 | 0.167 | 0.198 | 0.225 | 0.250 | 0.272 | 0.292 | 0.311 | 0.329 | 0.346 | 0.362 | 0.390 | 0.428 | 0.481 | 0.561 | 0.619 |
| | pow | .50 | 0.103 | 0.118 | 0.128 | 0.136 | 0.149 | 0.156 | 0.168 | 0.174 | 0.179 | 0.192 | 0.200 | 0.220 | 0.245 | 0.299 | 0.348 |
| | pow | .80 | 0.177 | 0.196 | 0.212 | 0.225 | 0.238 | 0.249 | 0.261 | 0.270 | 0.278 | 0.289 | 0.304 | 0.326 | 0.358 | 0.412 | 0.457 |
| | 5% | .05 | 0.185 | 0.206 | 0.225 | 0.243 | 0.261 | 0.278 | 0.293 | 0.308 | 0.322 | 0.336 | 0.361 | 0.396 | 0.446 | 0.523 | 0.582 |
| | 5% | .01 | 0.257 | 0.277 | 0.295 | 0.313 | 0.330 | 0.346 | 0.361 | 0.375 | 0.389 | 0.402 | 0.426 | 0.459 | 0.505 | 0.578 | 0.631 |
| | pow | .50 | 0.182 | 0.187 | 0.197 | 0.202 | 0.207 | 0.211 | 0.222 | 0.226 | 0.229 | 0.240 | 0.247 | 0.264 | 0.286 | 0.334 | 0.378 |
| | pow | .80 | 0.263 | 0.272 | 0.282 | 0.290 | 0.298 | 0.305 | 0.314 | 0.321 | 0.327 | 0.336 | 0.347 | 0.366 | 0.392 | 0.440 | 0.481 |

One-Stop PV Table   213

|    |     |     |       |       |       |       |       |       |       |       |       |       |       |       |       |       |       |
|----|-----|-----|-------|-------|-------|-------|-------|-------|-------|-------|-------|-------|-------|-------|-------|-------|-------|
| 60 | nil | .05 | 0.062 | 0.095 | 0.121 | 0.144 | 0.165 | 0.184 | 0.202 | 0.218 | 0.234 | 0.249 | 0.277 | 0.314 | 0.368 | 0.452 | 0.515 |
|    | nil | .01 | 0.105 | 0.142 | 0.171 | 0.196 | 0.218 | 0.238 | 0.256 | 0.273 | 0.290 | 0.305 | 0.333 | 0.370 | 0.423 | 0.503 | 0.563 |
|    | pow | .50 | 0.061 | 0.079 | 0.090 | 0.102 | 0.109 | 0.115 | 0.126 | 0.131 | 0.135 | 0.147 | 0.155 | 0.173 | 0.205 | 0.245 | 0.280 |
|    | pow | .80 | 0.118 | 0.144 | 0.161 | 0.176 | 0.187 | 0.198 | 0.209 | 0.217 | 0.225 | 0.235 | 0.249 | 0.270 | 0.302 | 0.350 | 0.390 |
|    | 1%  | .05 | 0.094 | 0.119 | 0.141 | 0.162 | 0.181 | 0.198 | 0.215 | 0.231 | 0.246 | 0.260 | 0.287 | 0.323 | 0.375 | 0.456 | 0.519 |
|    | 1%  | .01 | 0.147 | 0.173 | 0.196 | 0.217 | 0.237 | 0.255 | 0.272 | 0.288 | 0.303 | 0.317 | 0.344 | 0.379 | 0.430 | 0.508 | 0.567 |
|    | pow | .50 | 0.091 | 0.103 | 0.111 | 0.117 | 0.128 | 0.133 | 0.138 | 0.149 | 0.153 | 0.157 | 0.171 | 0.189 | 0.211 | 0.260 | 0.294 |
|    | pow | .80 | 0.156 | 0.172 | 0.185 | 0.196 | 0.207 | 0.216 | 0.224 | 0.234 | 0.241 | 0.248 | 0.263 | 0.283 | 0.310 | 0.360 | 0.399 |
|    | 5%  | .05 | 0.172 | 0.188 | 0.204 | 0.220 | 0.235 | 0.249 | 0.263 | 0.276 | 0.288 | 0.301 | 0.324 | 0.355 | 0.402 | 0.476 | 0.534 |
|    | 5%  | .01 | 0.235 | 0.251 | 0.267 | 0.282 | 0.297 | 0.311 | 0.324 | 0.337 | 0.349 | 0.361 | 0.383 | 0.413 | 0.458 | 0.528 | 0.582 |
|    | pow | .50 | 0.166 | 0.174 | 0.178 | 0.182 | 0.190 | 0.194 | 0.197 | 0.200 | 0.210 | 0.213 | 0.218 | 0.233 | 0.251 | 0.295 | 0.326 |
|    | pow | .80 | 0.241 | 0.249 | 0.256 | 0.262 | 0.270 | 0.276 | 0.281 | 0.287 | 0.295 | 0.299 | 0.309 | 0.325 | 0.348 | 0.391 | 0.425 |
| 70 | nil | .05 | 0.054 | 0.082 | 0.105 | 0.125 | 0.143 | 0.160 | 0.176 | 0.191 | 0.206 | 0.219 | 0.245 | 0.279 | 0.329 | 0.410 | 0.472 |
|    | nil | .01 | 0.091 | 0.123 | 0.149 | 0.171 | 0.190 | 0.208 | 0.225 | 0.241 | 0.256 | 0.270 | 0.296 | 0.331 | 0.381 | 0.459 | 0.519 |
|    | pow | .50 | 0.053 | 0.068 | 0.077 | 0.088 | 0.094 | 0.099 | 0.109 | 0.114 | 0.117 | 0.128 | 0.134 | 0.150 | 0.170 | 0.205 | 0.249 |
|    | pow | .80 | 0.103 | 0.125 | 0.140 | 0.153 | 0.163 | 0.172 | 0.182 | 0.190 | 0.197 | 0.206 | 0.218 | 0.236 | 0.262 | 0.306 | 0.346 |
|    | 1%  | .05 | 0.086 | 0.106 | 0.125 | 0.143 | 0.160 | 0.176 | 0.190 | 0.205 | 0.218 | 0.231 | 0.255 | 0.289 | 0.337 | 0.415 | 0.476 |
|    | 1%  | .01 | 0.132 | 0.154 | 0.174 | 0.193 | 0.210 | 0.226 | 0.242 | 0.256 | 0.270 | 0.283 | 0.308 | 0.341 | 0.389 | 0.465 | 0.523 |
|    | pow | .50 | 0.083 | 0.092 | 0.099 | 0.108 | 0.113 | 0.117 | 0.121 | 0.131 | 0.134 | 0.137 | 0.150 | 0.166 | 0.185 | 0.220 | 0.251 |
|    | pow | .80 | 0.142 | 0.154 | 0.164 | 0.175 | 0.183 | 0.191 | 0.198 | 0.207 | 0.213 | 0.219 | 0.232 | 0.250 | 0.274 | 0.316 | 0.351 |
|    | 5%  | .05 | 0.160 | 0.175 | 0.189 | 0.202 | 0.215 | 0.228 | 0.240 | 0.251 | 0.263 | 0.274 | 0.295 | 0.324 | 0.367 | 0.437 | 0.494 |
|    | 5%  | .01 | 0.218 | 0.232 | 0.246 | 0.259 | 0.272 | 0.285 | 0.296 | 0.308 | 0.319 | 0.330 | 0.350 | 0.378 | 0.419 | 0.487 | 0.540 |
|    | pow | .50 | 0.157 | 0.160 | 0.163 | 0.171 | 0.174 | 0.176 | 0.179 | 0.187 | 0.190 | 0.192 | 0.203 | 0.209 | 0.225 | 0.255 | 0.293 |
|    | pow | .80 | 0.224 | 0.230 | 0.236 | 0.242 | 0.247 | 0.252 | 0.257 | 0.263 | 0.268 | 0.272 | 0.283 | 0.294 | 0.314 | 0.350 | 0.384 |
| 80 | nil | .05 | 0.047 | 0.072 | 0.093 | 0.111 | 0.127 | 0.142 | 0.157 | 0.170 | 0.183 | 0.196 | 0.219 | 0.251 | 0.298 | 0.375 | 0.435 |
|    | nil | .01 | 0.080 | 0.109 | 0.131 | O.151 | 0.169 | 0.185 | 0.201 | 0.215 | 0.229 | 0.242 | 0.266 | 0.299 | 0.346 | 0.421 | 0.480 |
|    | pow | .50 | 0.046 | 0.060 | 0.068 | 0.077 | 0.083 | 0.087 | 0.096 | 0.100 | 0.103 | 0.106 | 0.118 | 0.133 | O.151 | 0.183 | 0.213 |
|    | pow | .80 | 0.091 | 0.111 | 0.124 | 0.136 | 0.145 | 0.153 | 0.162 | 0.168 | 0.175 | 0.180 | 0.193 | 0.210 | 0.233 | 0.273 | 0.307 |
|    | 1%  | .05 | 0.079 | 0.096 | 0.113 | 0.129 | 0.144 | O.158 | 0.171 | 0.184 | 0.196 | 0.208 | 0.230 | 0.261 | 0.307 | 0.381 | 0.440 |

*(Continued)*

214  *One-Stop* PV *Table*

| | | | | | | | | | *dfHyp* | | | | | | |
|---|---|---|---|---|---|---|---|---|---|---|---|---|---|---|---|
| | | 1 | 2 | 3 | 4 | 5 | 6 | 7 | 8 | 9 | 10 | 12 | 15 | 20 | 30 | 40 |
| | 1% .01 | 0.121 | 0.140 | 0.158 | 0.174 | 0.190 | 0.204 | 0.218 | 0.231 | 0.244 | 0.256 | 0.278 | 0.309 | 0.355 | 0.428 | 0.485 |
| | pow .50 | 0.078 | 0.083 | 0.089 | 0.097 | 0.101 | 0.105 | 0.108 | 0.117 | 0.120 | 0.122 | 0.134 | 0.140 | 0.165 | 0.197 | 0.226 |
| | pow .80 | 0.130 | 0.139 | 0.149 | 0.158 | 0.165 | 0.172 | 0.178 | 0.186 | 0.191 | 0.196 | 0.208 | 0.221 | 0.246 | 0.284 | 0.316 |
| | 5% .05 | 0.152 | 0.164 | 0.176 | 0.189 | 0.200 | 0.212 | 0.222 | 0.232 | 0.243 | 0.253 | 0.272 | 0.298 | 0.338 | 0.405 | 0.460 |
| | 5% .01 | 0.204 | 0.217 | 0.229 | 0.241 | 0.253 | 0.264 | 0.274 | 0.285 | 0.295 | 0.305 | 0.323 | 0.349 | 0.388 | 0.453 | 0.505 |
| | pow .50 | 0.147 | 0.153 | 0.156 | 0.158 | 0.161 | 0.162 | 0.170 | 0.172 | 0.174 | 0.176 | 0.186 | 0.190 | 0.204 | 0.231 | 0.257 |
| | pow .80 | 0.211 | 0.216 | 0.221 | 0.225 | 0.229 | 0.233 | 0.239 | 0.243 | 0.247 | 0.250 | 0.260 | 0.270 | 0.287 | 0.319 | 0.347 |
| 90 | nil .05 | 0.042 | 0.064 | 0.083 | 0.099 | 0.114 | 0.128 | 0.141 | 0.154 | 0.166 | 0.177 | 0.199 | 0.228 | 0.272 | 0.345 | 0.404 |
| | nil .01 | 0.071 | 0.097 | 0.118 | 0.136 | 0.152 | 0.167 | 0.181 | 0.194 | 0.207 | 0.219 | 0.242 | 0.272 | 0.317 | 0.390 | 0.447 |
| | pow .50 | 0.041 | 0.053 | 0.060 | 0.069 | 0.074 | 0.078 | 0.086 | 0.089 | 0.092 | 0.094 | 0.106 | 0.119 | 0.135 | 0.165 | 0.193 |
| | pow .80 | 0.081 | 0.099 | 0.111 | 0.122 | 0.130 | 0.137 | 0.145 | 0.151 | 0.157 | 0.162 | 0.174 | 0.189 | 0.210 | 0.246 | 0.277 |
| | 1% .05 | 0.072 | 0.089 | 0.104 | 0.118 | 0.131 | 0.144 | 0.156 | 0.168 | 0.179 | 0.190 | 0.210 | 0.239 | 0.281 | 0.352 | 0.409 |
| | 1% .01 | 0.111 | 0.129 | 0.144 | 0.159 | 0.173 | 0.186 | 0.199 | 0.211 | 0.223 | 0.234 | 0.255 | 0.284 | 0.326 | 0.397 | 0.453 |
| | pow .50 | 0.071 | 0.077 | 0.081 | 0.088 | 0.092 | 0.095 | 0.103 | 0.106 | 0.108 | 0.110 | 0.121 | 0.126 | 0.150 | 0.179 | 0.195 |
| | pow .80 | 0.119 | 0.128 | 0.136 | 0.144 | 0.151 | 0.156 | 0.164 | 0.169 | 0.174 | 0.178 | 0.189 | 0.201 | 0.223 | 0.258 | 0.284 |
| | 5% .05 | 0.144 | 0.156 | 0.167 | 0.178 | 0.188 | 0.198 | 0.208 | 0.217 | 0.227 | 0.235 | 0.253 | 0.277 | 0.315 | 0.378 | 0.431 |
| | 5% .01 | 0.193 | 0.205 | 0.216 | 0.226 | 0.237 | 0.247 | 0.257 | 0.266 | 0.275 | 0.284 | 0.301 | 0.326 | 0.362 | 0.424 | 0.474 |
| | pow .50 | 0.142 | 0.144 | 0.146 | 0.148 | 0.154 | 0.156 | 0.158 | 0.160 | 0.161 | 0.169 | 0.171 | 0.182 | 0.195 | 0.221 | 0.236 |
| | pow .80 | 0.200 | 0.204 | 0.208 | 0.212 | 0.216 | 0.220 | 0.223 | 0.227 | 0.230 | 0.235 | 0.241 | 0.253 | 0.269 | 0.298 | 0.320 |
| 100 | nil .05 | 0.038 | 0.058 | 0.075 | 0.090 | 0.103 | 0.116 | 0.128 | 0.140 | 0.151 | 0.161 | 0.182 | 0.209 | 0.251 | 0.320 | 0.377 |
| | nil .01 | 0.064 | 0.088 | 0.107 | 0.123 | 0.138 | 0.152 | 0.165 | 0.177 | 0.189 | 0.200 | 0.221 | 0.250 | 0.292 | 0.362 | 0.418 |
| | pow .50 | 0.037 | 0.048 | 0.055 | 0.062 | 0.067 | 0.070 | 0.077 | 0.081 | 0.083 | 0.085 | 0.096 | 0.108 | 0.123 | 0.150 | 0.166 |
| | pow .80 | 0.074 | 0.090 | 0.101 | 0.110 | 0.118 | 0.124 | 0.132 | 0.137 | 0.142 | 0.147 | 0.158 | 0.172 | 0.191 | 0.225 | 0.250 |
| | 1% .05 | 0.067 | 0.082 | 0.096 | 0.108 | 0.121 | 0.132 | 0.143 | 0.154 | 0.164 | 0.174 | 0.193 | 0.220 | 0.260 | 0.327 | 0.383 |
| | 1% .01 | 0.104 | 0.119 | 0.133 | 0.147 | 0.160 | 0.172 | 0.183 | 0.194 | 0.205 | 0.215 | 0.235 | 0.262 | 0.303 | 0.370 | 0.424 |
| | pow .50 | 0.066 | 0.071 | 0.075 | 0.081 | 0.085 | 0.087 | 0.094 | 0.097 | 0.099 | 0.101 | 0.111 | 0.115 | 0.128 | 0.154 | 0.179 |

One-Stop PV Table 215

|  |  |  |  |  |  |  |  |  |  |  |  |  |  |  |  |  |
|---|---|---|---|---|---|---|---|---|---|---|---|---|---|---|---|---|
|  | pow | .80 | 0.111 | 0.119 | 0.126 | 0.133 | 0.139 | 0.144 | 0.150 | 0.155 | 0.159 | 0.163 | 0.173 | 0.184 | 0.202 | 0.233 | 0.260 |
|  | 5% | .05 | 0.139 | 0.150 | 0.158 | 0.168 | 0.178 | 0.187 | 0.196 | 0.204 | 0.213 | 0.221 | 0.237 | 0.260 | 0.295 | 0.355 | 0.406 |
|  | 5% | .01 | 0.184 | 0.195 | 0.205 | 0.214 | 0.224 | 0.233 | 0.242 | 0.250 | 0.259 | 0.267 | 0.283 | 0.306 | 0.340 | 0.399 | 0.448 |
|  | pow | .50 | 0.135 | 0.137 | 0.142 | 0.144 | 0.145 | 0.147 | 0.148 | 0.155 | 0.156 | 0.157 | 0.160 | 0.169 | 0.181 | 0.204 | 0.218 |
|  | pow | .80 | 0.190 | 0.194 | 0.198 | 0.201 | 0.204 | 0.208 | 0.211 | 0.215 | 0.218 | 0.221 | 0.226 | 0.236 | 0.250 | 0.277 | 0.297 |
| 120 | nil | .05 | 0.032 | 0.049 | 0.063 | 0.075 | 0.087 | 0.098 | 0.108 | 0.118 | 0.128 | 0.137 | 0.155 | 0.179 | 0.216 | 0.279 | 0.332 |
|  | nil | .01 | 0.054 | 0.074 | 0.090 | 0.104 | 0.117 | 0.129 | 0.140 | 0.151 | 0.161 | 0.171 | 0.189 | 0.215 | 0.253 | 0.317 | 0.370 |
|  | pow | .50 | 0.031 | 0.040 | 0.046 | 0.049 | 0.056 | 0.059 | 0.061 | 0.068 | 0.070 | 0.072 | 0.080 | 0.084 | 0.096 | 0.119 | 0.142 |
|  | pow | .80 | 0.062 | 0.076 | 0.085 | 0.092 | 0.099 | 0.105 | 0.110 | 0.116 | 0.120 | 0.124 | 0.133 | 0.143 | 0.159 | 0.188 | 0.212 |
|  | 1% | .05 | 0.061 | 0.073 | 0.084 | 0.095 | 0.105 | 0.115 | 0.124 | 0.133 | 0.142 | 0.151 | 0.167 | 0.191 | 0.226 | 0.287 | 0.339 |
|  | 1% | .01 | 0.092 | 0.105 | 0.117 | 0.128 | 0.139 | 0.149 | 0.159 | 0.169 | 0.178 | 0.187 | 0.204 | 0.228 | 0.264 | 0.326 | 0.377 |
|  | pow | .50 | 0.059 | 0.063 | 0.068 | 0.071 | 0.073 | 0.075 | 0.081 | 0.083 | 0.085 | 0.086 | 0.095 | 0.098 | 0.109 | 0.132 | 0.154 |
|  | pow | .80 | 0.098 | 0.105 | 0.111 | 0.116 | 0.121 | 0.125 | 0.130 | 0.134 | 0.137 | 0.141 | 0.149 | 0.158 | 0.173 | 0.199 | 0.223 |
|  | 5% | .05 | 0.130 | 0.138 | 0.147 | 0.154 | 0.162 | 0.170 | 0.171 | 0.185 | 0.192 | 0.199 | 0.213 | 0.233 | 0.264 | 0.318 | 0.365 |
|  | 5% | .01 | 0.170 | 0.179 | 0.187 | 0.195 | 0.203 | 0.211 | 0.219 | 0.226 | 0.233 | 0.241 | 0.254 | 0.274 | 0.305 | 0.358 | 0.404 |
|  | pow | .50 | 0.126 | 0.128 | 0.129 | 0.134 | 0.135 | 0.136 | 0.137 | 0.138 | 0.144 | 0.145 | 0.147 | 0.155 | 0.159 | 0.178 | 0.198 |
|  | pow | .80 | 0.175 | 0.178 | 0.181 | 0.184 | 0.187 | 0.189 | 0.192 | 0.194 | 0.198 | 0.200 | 0.204 | 0.213 | 0.222 | 0.244 | 0.265 |
| 150 | nil | .05 | 0.025 | 0.039 | 0.051 | 0.061 | 0.070 | 0.079 | 0.088 | 0.096 | 0.104 | 0.112 | 0.127 | 0.148 | 0.179 | 0.235 | 0.282 |
|  | nil | .01 | 0.043 | 0.060 | 0.073 | 0.084 | 0.095 | 0.105 | 0.114 | 0.123 | 0.132 | 0.140 | 0.156 | 0.178 | 0.211 | 0.268 | 0.315 |
|  | pow | .50 | 0.025 | 0.032 | 0.037 | 0.040 | 0.045 | 0.047 | 0.049 | 0.054 | 0.056 | 0.058 | 0.065 | 0.068 | 0.078 | 0.097 | 0.108 |
|  | pow | .80 | 0.050 | 0.061 | 0.069 | 0.075 | 0.080 | 0.085 | 0.089 | 0.094 | 0.097 | 0.101 | 0.108 | 0.116 | 0.129 | 0.153 | 0.170 |
|  | 1% | .05 | 0.054 | 0.063 | 0.072 | 0.080 | 0.089 | 0.096 | 0.104 | 0.112 | 0.119 | 0.126 | 0.140 | 0.160 | 0.190 | 0.243 | 0.289 |
|  | 1% | .01 | 0.080 | 0.090 | 0.099 | 0.108 | 0.117 | 0.126 | 0.134 | 0.142 | 0.149 | 0.157 | 0.171 | 0.192 | 0.223 | 0.277 | 0.323 |
|  | pow | .50 | 0.052 | 0.056 | 0.058 | 0.060 | 0.062 | 0.066 | 0.068 | 0.069 | 0.071 | 0.071 | 0.078 | 0.081 | 0.090 | 0.108 | 0.119 |
|  | pow | .80 | 0.086 | 0.090 | 0.094 | 0.098 | 0.102 | 0.106 | 0.109 | 0.112 | 0.115 | 0.117 | 0.124 | 0.131 | 0.143 | 0.165 | 0.181 |
|  | 5% | .05 | 0.120 | 0.126 | 0.132 | 0.139 | 0.146 | 0.152 | 0.158 | 0.165 | 0.170 | 0.176 | 0.187 | 0.204 | 0.230 | 0.277 | 0.319 |
|  | 5% | .01 | 0.155 | 0.162 | 0.168 | 0.175 | 0.181 | 0.188 | 0.194 | 0.200 | 0.206 | 0.212 | 0.224 | 0.240 | 0.267 | 0.313 | 0.354 |
|  | pow | .50 | 0.116 | 0.120 | 0.121 | 0.122 | 0.123 | 0.123 | 0.124 | 0.125 | 0.130 | 0.130 | 0.132 | 0.139 | 0.141 | 0.157 | 0.167 |
|  | pow | .80 | 0.160 | 0.163 | 0.164 | 0.166 | 0.168 | 0.169 | 0.171 | 0.173 | 0.176 | 0.178 | 0.181 | 0.188 | 0.194 | 0.212 | 0.226 |

*(Continued)*

# One-Stop PV Table

| | | | | | | | | dfHyp | | | | | | | |
|---|---|---|---|---|---|---|---|---|---|---|---|---|---|---|---|
| | | 1 | 2 | 3 | 4 | 5 | 6 | 7 | 8 | 9 | 10 | 12 | 15 | 20 | 30 | 40 |
| 200 | nil .05 | 0.019 | 0.030 | 0.038 | 0.046 | 0.053 | 0.060 | 0.067 | 0.073 | 0.080 | 0.086 | 0.097 | 0.114 | 0.139 | 0.185 | 0.225 |
| | nil .01 | 0.033 | 0.045 | 0.055 | 0.064 | 0.072 | 0.080 | 0.087 | 0.094 | 0.101 | 0.108 | 0.120 | 0.138 | 0.165 | 0.212 | 0.253 |
| | pow .50 | 0.019 | 0.024 | 0.028 | 0.030 | 0.034 | 0.036 | 0.037 | 0.041 | 0.042 | 0.043 | 0.045 | 0.051 | 0.059 | 0.068 | 0.083 |
| | pow .80 | 0.038 | 0.046 | 0.052 | 0.057 | 0.061 | 0.064 | 0.068 | 0.071 | 0.074 | 0.076 | 0.081 | 0.088 | 0.098 | 0.114 | 0.130 |
| | 1% .05 | 0.046 | 0.053 | 0.060 | 0.065 | 0.072 | 0.078 | 0.084 | 0.089 | 0.095 | 0.100 | 0.111 | 0.127 | 0.151 | 0.195 | 0.234 |
| | 1% .01 | 0.067 | 0.074 | 0.081 | 0.088 | 0.095 | 0.101 | 0.107 | 0.113 | 0.119 | 0.125 | 0.136 | 0.153 | 0.178 | 0.223 | 0.262 |
| | pow .50 | 0.045 | 0.047 | 0.048 | 0.051 | 0.052 | 0.053 | 0.054 | 0.055 | 0.059 | 0.060 | 0.062 | 0.067 | 0.070 | 0.084 | 0.092 |
| | pow .80 | 0.071 | 0.075 | 0.078 | 0.081 | 0.083 | 0.085 | 0.088 | 0.090 | 0.093 | 0.095 | 0.098 | 0.105 | 0.112 | 0.128 | 0.141 |
| | 5% .05 | 0.109 | 0.114 | 0.118 | 0.123 | 0.128 | 0.133 | 0.138 | 0.142 | 0.147 | 0.151 | 0.160 | 0.173 | 0.194 | 0.232 | 0.267 |
| | 5% .01 | 0.138 | 0.143 | 0.148 | 0.153 | 0.158 | 0.163 | 0.168 | 0.172 | 0.177 | 0.182 | 0.191 | 0.204 | 0.225 | 0.263 | 0.298 |
| | pow .50 | 0.106 | 0.106 | 0.109 | 0.110 | 0.110 | 0.111 | 0.111 | 0.112 | 0.112 | 0.112 | 0.117 | 0.118 | 0.124 | 0.132 | 0.139 |
| | pow .80 | 0.142 | 0.143 | 0.145 | 0.146 | 0.148 | 0.149 | 0.150 | 0.151 | 0.152 | 0.153 | 0.157 | 0.160 | 0.167 | 0.177 | 0.187 |
| 300 | nil .05 | 0.013 | 0.020 | 0.026 | 0.031 | 0.036 | 0.041 | 0.045 | 0.050 | 0.054 | 0.058 | 0.067 | 0.078 | 0.097 | 0.130 | 0.160 |
| | nil .01 | 0.022 | 0.030 | 0.037 | 0.043 | 0.049 | 0.054 | 0.059 | 0.064 | 0.069 | 0.073 | 0.082 | 0.095 | 0.114 | 0.150 | 0.181 |
| | pow .50 | 0.013 | 0.016 | 0.018 | 0.020 | 0.023 | 0.024 | 0.025 | 0.025 | 0.028 | 0.029 | 0.030 | 0.035 | 0.040 | 0.046 | 0.052 |
| | pow .80 | 0.025 | 0.031 | 0.035 | 0.038 | 0.041 | 0.044 | 0.046 | 0.047 | 0.050 | 0.052 | 0.055 | 0.060 | 0.066 | 0.077 | 0.086 |
| | 1% .05 | 0.037 | 0.042 | 0.046 | 0.050 | 0.054 | O.OSB | 0.062 | 0.066 | 0.070 | 0.074 | 0.081 | 0.092 | 0.109 | 0.141 | 0.170 |
| | 1% .01 | 0.053 | 0.058 | 0.062 | 0.067 | 0.071 | 0.075 | 0.079 | 0.084 | 0.088 | 0.092 | 0.100 | 0.111 | 0.129 | 0.162 | 0.192 |
| | pow .50 | 0.036 | 0.037 | 0.038 | 0.040 | 0.041 | 0.042 | 0.042 | 0.043 | 0.043 | 0.046 | 0.047 | 0.048 | 0.053 | 0.058 | 0.064 |
| | pow .80 | 0.056 | O.OSB | 0.060 | 0.062 | 0.063 | 0.064 | 0.066 | 0.067 | 0.068 | 0.070 | 0.073 | 0.076 | 0.081 | 0.091 | 0.099 |
| | 5% .05 | 0.097 | 0.100 | 0.103 | 0.107 | 0.109 | 0.112 | 0.115 | 0.119 | 0.122 | 0.125 | 0.131 | 0.140 | 0.154 | 0.182 | 0.208 |
| | 5% .01 | 0.119 | 0.122 | 0.126 | 0.129 | 0.132 | 0.136 | 0.139 | 0.142 | 0.145 | 0.149 | 0.155 | 0.164 | 0.179 | 0.207 | 0.233 |
| | pow .50 | 0.094 | 0.095 | 0.095 | 0.095 | 0.098 | 0.098 | 0.098 | 0.098 | 0.099 | 0.099 | 0.099 | 0.103 | 0.104 | 0.108 | 0.113 |
| | pow .80 | 0.122 | 0.123 | 0.123 | 0.124 | 0.126 | 0.126 | 0.127 | 0.127 | 0.128 | 0.129 | 0.130 | 0.133 | 0.136 | 0.142 | 0.149 |
| 400 | nil .05 | 0.010 | 0.015 | 0.019 | 0.023 | 0.027 | 0.031 | 0.034 | 0.038 | 0.041 | 0.044 | 0.051 | 0.060 | 0.074 | 0.100 | 0.124 |
| | nil .01 | 0.016 | 0.023 | 0.028 | 0.033 | 0.037 | 0.041 | 0.045 | 0.049 | 0.052 | 0.056 | 0.063 | 0.072 | 0.088 | 0.116 | 0.141 |
| | pow .50 | 0.009 | 0.012 | 0.014 | 0.015 | 0.017 | 0.018 | 0.019 | 0.019 | 0.021 | 0.022 | 0.023 | 0.026 | 0.030 | 0.035 | 0.039 |

One-Stop PV Table 217

|  |  | .80 |  |  |  |  |  |  |  |  |  |  |  |  |  |
|---|---|---|---|---|---|---|---|---|---|---|---|---|---|---|---|
|  | pow | .80 | 0.019 | 0.024 | 0.026 | 0.029 | 0.031 | 0.033 | 0.034 | 0.036 | 0.038 | 0.039 | 0.041 | 0.045 | 0.050 | 0.058 | 0.065 |
|  | 1% | .05 | 0.033 | 0.036 | 0.039 | 0.042 | 0.045 | 0.048 | 0.051 | 0.054 | 0.057 | 0.060 | 0.065 | 0.073 | 0.087 | 0.112 | 0.135 |
|  | 1% | .01 | 0.045 | 0.049 | 0.052 | 0.055 | 0.059 | 0.062 | 0.065 | 0.068 | 0.071 | 0.074 | 0.080 | 0.089 | 0.103 | 0.129 | 0.153 |
|  | pow | .50 | 0.032 | 0.033 | 0.033 | 0.034 | 0.034 | 0.036 | 0.036 | 0.036 | 0.037 | 0.037 | 0.040 | 0.040 | 0.044 | 0.049 | 0.053 |
|  | pow | .80 | 0.048 | 0.049 | 0.050 | 0.051 | 0.052 | 0.054 | 0.054 | 0.OSS | 0.056 | 0.057 | 0.059 | 0.061 | 0.066 | 0.073 | 0.079 |
|  | 5% | .05 | 0.089 | 0.092 | 0.094 | 0.097 | 0.099 | 0.101 | 0.104 | 0.106 | 0.109 | 0.111 | 0.115 | 0.122 | 0.133 | 0.155 | 0.175 |
|  | 5% | .01 | 0.108 | 0.111 | 0.113 | 0.116 | 0.118 | 0.121 | 0.123 | 0.126 | 0.128 | 0.130 | 0.135 | 0.142 | 0.154 | 0.175 | 0.196 |
|  | pow | .50 | 0.088 | 0.088 | 0.089 | 0.089 | 0.089 | 0.089 | 0.089 | 0.089 | 0.089 | 0.092 | 0.092 | 0.092 | 0.093 | 0.096 | 0.100 |
|  | pow | .80 | 0.111 | 0.111 | 0.112 | 0.112 | 0.113 | 0.113 | 0.113 | 0.114 | 0.114 | 0.116 | 0.116 | 0.118 | 0.119 | 0.124 | 0.128 |
| 500 | nil | .05 | 0.008 | 0.012 | 0.015 | 0.014 | 0.022 | 0.025 | 0.028 | 0.030 | 0.033 | 0.036 | 0.041 | 0.048 | 0.060 | 0.081 | 0.102 |
|  | nil | .01 | 0.013 | 0.018 | 0.022 | 0.026 | 0.030 | 0.033 | 0.036 | 0.039 | 0.042 | 0.045 | 0.051 | 0.059 | 0.071 | 0.094 | 0.115 |
|  | pow | .50 | 0.008 | 0.010 | 0.011 | 0.012 | 0.014 | 0.014 | 0.015 | 0.015 | 0.017 | 0.018 | 0.018 | 0.021 | 0.022 | 0.028 | 0.032 |
|  | pow | .80 | 0.015 | 0.019 | 0.021 | 0.023 | 0.025 | 0.026 | 0.028 | 0.029 | 0.030 | 0.031 | 0.033 | 0.036 | 0.039 | 0.047 | 0.052 |
|  | 1% | .05 | 0.030 | 0.032 | 0.035 | 0.037 | 0.040 | 0.042 | 0.044 | 0.046 | 0.049 | 0.OSI | 0.056 | 0.062 | 0.073 | 0.093 | 0.113 |
|  | 1% | .01 | 0.040 | 0.043 | 0.046 | 0.048 | 0.051 | 0.053 | 0.056 | 0.058 | 0.061 | 0.063 | 0.068 | 0.075 | 0.086 | 0.108 | 0.128 |
|  | pow | .50 | 0.029 | 0.030 | 0.030 | 0.030 | 0.031 | 0.031 | 0.032 | 0.033 | 0.033 | 0.033 | 0.033 | 0.036 | 0.037 | 0.040 | 0.043 |
|  | pow | .80 | 0.043 | 0.044 | 0.044 | 0.045 | 0.046 | 0.046 | 0.048 | 0.048 | 0.049 | 0.049 | 0.051 | 0.053 | 0.055 | 0.061 | 0.065 |
|  | 5% | .05 | 0.085 | 0.087 | 0.089 | 0.091 | 0.092 | 0.094 | 0.096 | 0.098 | 0.100 | 0.102 | 0.105 | 0.111 | 0.120 | 0.137 | 0.154 |
|  | 5% | .01 | 0.101 | 0.103 | 0.105 | 0.107 | 0.109 | 0.111 | 0.113 | 0.115 | 0.117 | 0.119 | 0.123 | 0.128 | 0.138 | 0.155 | 0.173 |
|  | pow | .50 | 0.083 | 0.083 | 0.083 | 0.083 | 0.083 | 0.083 | 0.085 | 0.085 | 0.085 | 0.085 | 0.086 | 0.086 | 0.088 | 0.089 | 0.092 |
|  | pow | .80 | 0.103 | 0.103 | 0.104 | 0.104 | 0.104 | 0.104 | 0.106 | 0.106 | 0.106 | 0.106 | 0.107 | 0.108 | 0.110 | 0.112 | 0.116 |
| 600 | nil | .05 | 0.006 | 0.010 | 0.013 | 0.016 | 0.018 | 0.021 | 0.023 | 0.025 | 0.028 | 0.030 | 0.034 | 0.040 | 0.050 | 0.069 | 0.086 |
|  | nil | .01 | 0.011 | 0.OIS | 0.019 | 0.022 | 0.025 | 0.028 | 0.030 | 0.033 | 0.035 | 0.038 | 0.042 | 0.049 | 0.060 | 0.080 | 0.098 |
|  | pow | .50 | 0.006 | 0.008 | 0.009 | 0.010 | 0.011 | 0.012 | 0.012 | 0.013 | 0.014 | 0.015 | 0.015 | 0.017 | 0.018 | 0.023 | 0.026 |
|  | pow | .80 | 0.013 | 0.016 | 0.018 | 0.019 | 0.021 | 0.022 | 0.023 | 0.024 | 0.025 | 0.026 | 0.028 | 0.030 | 0.033 | 0.039 | 0.043 |
|  | 1% | .05 | 0.027 | 0.029 | 0.032 | 0.034 | 0.036 | 0.038 | 0.040 | 0.041 | 0.043 | 0.045 | 0.049 | 0.055 | 0.063 | 0.081 | 0.097 |
|  | 1% | .01 | 0.037 | 0.039 | 0.041 | 0.044 | 0.046 | 0.048 | 0.050 | 0.052 | 0.054 | 0.056 | 0.060 | 0.066 | 0.075 | 0.093 | 0.110 |
|  | pow | .50 | 0.026 | 0.027 | 0.028 | 0.028 | 0.028 | 0.028 | 0.029 | 0.030 | 0.030 | 0.030 | 0.031 | 0.031 | 0.033 | 0.036 | 0.039 |
|  | pow | .80 | 0.039 | 0.040 | 0.040 | 0.041 | 0.041 | 0.042 | 0.042 | 0.043 | 0.044 | 0.044 | 0.045 | 0.046 | 0.049 | 0.053 | 0.057 |
|  | 5% | .05 | 0.081 | 0.083 | 0.084 | 0.086 | 0.087 | 0.089 | 0.091 | 0.092 | 0.094 | 0.096 | 0.099 | 0.103 | 0.111 | 0.125 | 0.140 |
|  | 5% | .01 | 0.096 | 0.098 | 0.100 | 0.101 | 0.103 | 0.104 | 0.106 | 0.108 | 0.109 | 0.111 | 0.114 | 0.119 | 0.127 | 0.142 | 0.156 |

*(Continued)*

218  One-Stop PV Table

|  |  |  | 1 | 2 | 3 | 4 | 5 | 6 | 7 | 8 | 9 | 10 | 12 | 15 | 20 | 30 | 40 |
|---|---|---|---|---|---|---|---|---|---|---|---|---|---|---|---|---|---|
|  | pow | .50 | 0.080 | 0.081 | 0.081 | 0.081 | 0.081 | 0.081 | 0.081 | 0.081 | 0.081 | 0.081 | 0.081 | 0.083 | 0.083 | 0.096 | 0.086 |
|  | pow | .80 | 0.098 | 0.098 | 0.099 | 0,099 | 0.099 | 0.099 | 0.099 | 0.100 | 0.100 | 0.100 | 0.100 | 0.102 | 0.103 | 0.106 | 0.107 |
| 1000 | nil | .05 | 0.004 | 0.006 | 0.008 | 0.009 | 0.011 | 0.012 | 0.014 | 0.015 | 0.017 | 0.018 | 0.021 | 0.024 | 0.031 | 0.042 | 0.053 |
|  | nil | .01 | 0.007 | 0.009 | 0.011 | 0.013 | 0.015 | 0.017 | 0.018 | 0.020 | 0.021 | 0.023 | 0.026 | 0.030 | 0.037 | 0.049 | 0.061 |
|  | pow | .50 | 0.004 | 0.005 | 0.006 | 0.006 | 0.007 | 0.007 | 0.007 | 0.008 | 0.009 | 0.009 | 0.009 | 0.010 | 0.011 | 0.014 | 0.016 |
|  | pow | .80 | 0.008 | 0.010 | 0.011 | 0.012 | 0.013 | 0.013 | 0.014 | 0.014 | 0.015 | 0.016 | 0.017 | 0.018 | 0.020 | 0.023 | 0.026 |
|  | 1% | .05 | 0.023 | 0.024 | 0.025 | 0.026 | 0.028 | 0.029 | 0.030 | 0.031 | 0.032 | 0.033 | 0.035 | 0.039 | 0.044 | 0.055 | 0.065 |
|  | 1% | .01 | 0.030 | 0.031 | 0.032 | 0.033 | 0.034 | 0.036 | 0.037 | 0.038 | 0.039 | 0.040 | 0.043 | 0.046 | 0.052 | 0.063 | 0.074 |
|  | pow | .50 | 0.022 | 0.023 | 0.023 | 0.023 | 0.023 | 0.023 | 0.023 | 0.024 | 0.024 | 0.024 | 0.024 | 0.024 | 0.026 | 0.028 | 0.028 |
|  | pow | .80 | 0.031 | 0.031 | 0.031 | 0.032 | 0.032 | 0.032 | 0.032 | 0.033 | 0.033 | 0.033 | 0.034 | 0.034 | 0.036 | 0.038 | 0.040 |
|  | 5% | .05 | 0.073 | 0.074 | 0.075 | 0.076 | 0.077 | 0.078 | 0.079 | 0.080 | 0.081 | 0.082 | 0.084 | 0.087 | 0.091 | 0.100 | 0.109 |
|  | 5% | .01 | 0.085 | 0.086 | 0.087 | 0.087 | 0.088 | 0.089 | 0.090 | 0.091 | 0.092 | 0.093 | 0.095 | 0.098 | 0.103 | 0.112 | 0.121 |
|  | pow | .50 | 0.073 | 0.073 | 0.073 | 0.073 | 0.073 | 0.073 | 0.073 | 0.073 | 0.073 | 0.073 | 0.073 | 0.074 | 0.074 | 0.075 | 0.076 |
|  | pow | .80 | 0.086 | 0.086 | 0.086 | 0.086 | 0.086 | 0.086 | 0.086 | 0.086 | 0.087 | 0.087 | 0.087 | 0.088 | 0.088 | 0.089 | 0.090 |
| 10000 | nil | .05 | 0.000 | 0.001 | 0.001 | 0.001 | 0.001 | 0.001 | 0.001 | 0.002 | 0.002 | 0.002 | 0.002 | 0.002 | 0.003 | 0.004 | 0.006 |
|  | nil | .01 | 0.001 | 0.001 | 0.001 | 0.001 | 0.002 | 0.002 | 0.002 | 0.002 | 0.002 | 0.002 | 0.003 | 0.003 | 0.004 | 0.005 | 0.006 |
|  | pow | .50 | 0.000 | 0.000 | 0.001 | 0.001 | 0.001 | 0.001 | 0.001 | 0.001 | 0.001 | 0.001 | 0.001 | 0.001 | 0.001 | 0.001 | 0.001 |
|  | pow | .80 | 0.001 | 0.001 | 0.001 | 0.001 | 0.001 | 0.001 | 0.001 | 0.001 | 0.002 | 0.002 | 0.002 | 0.002 | 0.002 | 0.002 | 0.003 |
|  | 1% | .05 | 0.013 | 0.014 | 0.014 | 0.014 | 0.014 | 0.014 | 0.014 | 0.014 | 0.014 | 0.014 | 0.015 | 0.015 | 0.015 | 0.016 | 0.011 |
|  | 1% | .01 | 0.015 | 0.015 | 0.015 | 0.015 | 0.015 | 0.016 | 0.016 | 0.016 | 0.016 | 0.016 | 0.016 | 0.017 | 0.017 | 0.018 | 0.019 |
|  | pow | .50 | 0.013 | 0.013 | 0.013 | 0.013 | 0.013 | 0.013 | 0.013 | 0.013 | 0.013 | 0.013 | 0.013 | 0.013 | 0.013 | 0.013 | 0.013 |
|  | pow | .80 | 0.015 | 0.015 | 0.015 | 0.015 | 0.015 | 0.015 | 0.015 | 0.015 | 0.015 | 0.015 | 0.015 | 0.015 | 0.015 | 0.015 | 0.015 |
|  | 5% | .05 | 0.057 | 0.057 | 0.057 | 0.057 | 0.057 | 0.057 | 0.057 | 0.057 | 0.057 | 0.058 | 0.058 | 0.058 | 0.059 | 0.059 | 0.060 |
|  | 5% | .01 | 0.060 | 0.060 | 0.060 | 0.060 | 0.060 | 0.060 | 0.061 | 0.061 | 0.061 | 0.061 | 0.061 | 0.061 | 0.062 | 0.063 | 0.064 |
|  | pow | .50 | 0.057 | 0.057 | 0.057 | 0.057 | 0.057 | 0.057 | 0.057 | 0.057 | 0.057 | 0.057 | 0.057 | 0.057 | 0.057 | 0.057 | 0.057 |
|  | pow | .80 | 0.060 | 0.060 | 0.060 | 0.060 | 0.060 | 0.060 | 0.060 | 0.060 | 0.060 | 0.060 | 0.060 | 0.060 | 0.060 | 0.060 | 0.060 |

*dfHyp*

## R Code Used to Generate One-Stop *PV* Table

```
# ONE-STOP PV TABLE
# User defined functions
noncenfn<-function(df, pv) df * pv / (1 - pv)
noncenfn2<-function(df, FF) df * FF
critFn<-function(df1, df2, alph, effect) qf((1-alph), df1, df2, noncenfn(df2, effect))
findF<-function(FF) 1-pf(critF, dfhyp, dferr, FF * dfhyp)-powtarget
PV<-function(FF) FF * dfhyp /(FF * dfhyp + dferr)

options(digits=3)

# Global variables
dfhyps<-c(1, 2, 3, 4, 5, 6, 7, 8, 9, 10, 12, 15, 20, 30, 40, 60, 120)
dferrs<-c(3, 4, 5, 6, 8, 10, 12, 14, 16, 18, 20, 22, 24, 26, 28, 30, 40, 50, 60,
70, 80, 90, 100, 120, 150, 200, 300, 400, 500,600, 1000, 10000)

onestop<-matrix(nrow=12, ncol=length(dfhyps))
rownames(onestop)<-c("nil .05", "nil .01", "pow .5", "pow .8", "1%
.05", "1% .01", "pow .5", "pow .8", "5% .05", "5% .01", "pow .5",
"pow .8")
colnames(onestop)<-dfhyps

for (dferr in dferrs) {
   cat("\ndfErr = ", dferr, "\t\t\t\t\tdfHyp\n")

   for (i in 1:length(dfhyps)) {
     dfhyp=dfhyps[i]

     # Nil hypotheses
     effect=0
     critF<-critFn(dfhyp, dferr, 0.05, effect)
     onestop[1, i]=PV(critF)
     onestop[2, i]=PV(critFn(dfhyp, dferr, 0.01, effect))

     powtarget=0.5
     Fresult<-uniroot(findF,c(0,650))
     onestop[3, i]=PV(Fresult$root)
     powtarget=0.8
     Fresult<-uniroot(findF,c(0,650))
     onestop[4, i]=PV(Fresult$root)

     # 1% minimum effects
     effect=0.01
```

```
        critF<-critFn(dfhyp, dferr, 0.05, effect)
        onestop[5, i]=PV(critF)
        onestop[6, i]=PV(critFn(dfhyp, dferr, 0.01, effect))

        powtarget=0.5
        Fresult<-uniroot(findF,c(0,650))
        onestop[7, i]=PV(Fresult$root)
        powtarget=0.8
        Fresult<-uniroot(findF,c(0,650))
        onestop[8, i]=PV(Fresult$root)

        # 5% minimum effects
        effect=0.05
        critF<-critFn(dfhyp, dferr, 0.05, effect)
        onestop[9, i]=PV(critF)
        onestop[10,i]=PV(critFn(dfhyp, dferr, 0.01, effect))

        powtarget=0.5
        Fresult<-uniroot(findF,c(0,650))
        onestop[11, i]=PV(Fresult$root)
        powtarget=0.8
        Fresult<-uniroot(findF,c(0,650))
        onestop[12,i]=PV(Fresult$root)
    }
    print(onestop)
}
```

# Appendix D

$df_{err}$ Needed for *Power* of .80 for Nil and Minimum-Effect Hypothesis Tests

| PV | d | 1 | 2 | 3 | 4 | 5 | 6 | 7 | 8 | 9 | 10 | 12 | 15 | 20 | 30 | 40 | 60 | 120 |
|---|---|---|---|---|---|---|---|---|---|---|---|---|---|---|---|---|---|---|
| .02 | 0.29 | 385 | 473 | 533 | 579 | 627 | 662 | 694 | 722 | 762 | 787 | 832 | 909 | 993 | 1176 | 1313 | 1413 | 2010 |
|  |  | **3225** | **3242** | **3301** | **3266** | **3334** | **3349** | **3364** | **3429** | **3442** | **3454** | **3479** | **3570** | **3621** | **3900** | **4042** | **4379** | **5260** |
| .03 | 0.35 | 255 | 313 | 353 | 384 | 416 | 439 | 460 | 479 | 505 | 522 | 552 | 603 | 660 | 782 | 874 | 1008 | 1341 |
|  |  | **1058** | **1086** | **1104** | **1122** | **1139** | **1176** | **1185** | **1199** | **1212** | **1254** | **1271** | **1303** | **1377** | **1518** | **1615** | **1833** | **2299** |
| .04 | 0.41 | 190 | 233 | 263 | 286 | 310 | 328 | 343 | 358 | 377 | 390 | 413 | 451 | 494 | 585 | 654 | 774 | 1031 |
|  |  | **573** | **590** | **607** | **623** | **650** | **658** | **670** | **683** | **694** | **716** | **736** | **779** | **836** | **920** | **993** | **1151** | **1458** |
| .05 | 0.46 | 151 | 186 | 209 | 228 | 247 | 261 | 273 | 285 | 300 | 310 | 329 | 359 | 402 | 466 | 522 | 618 | 825 |
|  |  | **373** | **389** | **405** | **422** | **444** | **445** | **457** | **472** | **483** | **492** | **509** | **541** | **586** | **652** | **728** | **843** | **1075** |
| .06 | 0.51 | 125 | 154 | 173 | 189 | 204 | 216 | 227 | 236 | 249 | 257 | 273 | 298 | 333 | 388 | 434 | 514 | 687 |
|  |  | **269** | **285** | **299** | **313** | **323** | **334** | **343** | **357** | **365** | **373** | **387** | **414** | **450** | **506** | **568** | **654** | **854** |
| .07 | 0.55 | 106 | 131 | 148 | 161 | 174 | 184 | 193 | 204 | 212 | 220 | 233 | 255 | 285 | 331 | 371 | 440 | 601 |
|  |  | **208** | **223** | **245** | **246** | **255** | **267** | **275** | **283** | **290** | **297** | **315** | **338** | **362** | **419** | **460** | **533** | **718** |
| .08 | 0.59 | 92 | 114 | 128 | 140 | 152 | 160 | 168 | 178 | 185 | 191 | 203 | 222 | 248 | 289 | 324 | 384 | 525 |
|  |  | **166** | **180** | **192** | **202** | **211** | **219** | **226** | **236** | **243** | **249** | **265** | **279** | **307** | **357** | **393** | **457** | **606** |
| .09 | 0.60 | 81 | 100 | 113 | 124 | 134 | 142 | 149 | 157 | 164 | 169 | 179 | 196 | 220 | 256 | 287 | 341 | 466 |
|  |  | **139** | **151** | **161** | **170** | **180** | **187** | **193** | **199** | **208** | **214** | **224** | **241** | **266** | **303** | **343** | **400** | **532** |
| .10 | 0.67 | 73 | 90 | 101 | 110 | 120 | 127 | 133 | 141 | 146 | 152 | 161 | 176 | 197 | 230 | 258 | 312 | 419 |
|  |  | **117** | **130** | **139** | **147** | **154** | **162** | **168** | **174** | **179** | **187** | **196** | **212** | **234** | **268** | **297** | **355** | **473** |
| .11 | 0.70 | 66 | 81 | 91 | 101 | 108 | 115 | 120 | 127 | 132 | 137 | 148 | 159 | 178 | 208 | 238 | 283 | 388 |
|  |  | **101** | **113** | **122** | **129** | **136** | **143** | **149** | **154** | **159** | **166** | **174** | **189** | **205** | **240** | **266** | **318** | **426** |
| .12 | 0.74 | 60 | 74 | 83 | 92 | 99 | 104 | 110 | 116 | 121 | 125 | 135 | 145 | 163 | 190 | 218 | 259 | 355 |
|  |  | **89** | **99** | **108** | **115** | **121** | **127** | **133** | **138** | **142** | **147** | **157** | **170** | **185** | **217** | **241** | **289** | **388** |
| .13 | 0.77 | 55 | 68 | 76 | 84 | 90 | 96 | 101 | 106 | 111 | 115 | 124 | 133 | 150 | 178 | 200 | 238 | 327 |
|  |  | **80** | **89** | **97** | **104** | **109** | **114** | **121** | **125** | **129** | **133** | **142** | **152** | **168** | **197** | **220** | **264** | **355** |
| .14 | 0.81 | 50 | 62 | 70 | 78 | 83 | 88 | 94 | 98 | 102 | 106 | 114 | 123 | 138 | 165 | 185 | 220 | 302 |
|  |  | **72** | **80** | **87** | **94** | **99** | **104** | **110** | **114** | **118** | **121** | **130** | **139** | **154** | **181** | **202** | **243** | **327** |

df for Nil and Minimum-Effect Hypothesis Tests 223

| | | | | | | | | | | | | | | | | | | |
|---|---|---|---|---|---|---|---|---|---|---|---|---|---|---|---|---|---|---|
| .15 | | 47 | 58 | 65 | 72 | 77 | 82 | 87 | 91 | 95 | 98 | 106 | 115 | 129 | 153 | 172 | 205 | 286 |
| | 0.84 | 65 | 73 | 80 | 86 | 91 | 95 | 101 | 105 | 108 | 112 | 120 | 128 | 142 | 168 | 187 | 225 | 303 |
| .16 | | 43 | 54 | 61 | 67 | 72 | 76 | 81 | 85 | 88 | 92 | 99 | 107 | 120 | 143 | 161 | 192 | 268 |
| | 0.87 | 59 | 67 | 73 | 79 | 84 | 88 | 93 | 97 | 100 | 103 | 111 | 119 | 132 | 156 | 174 | 209 | 283 |
| .17 | | 40 | 50 | 57 | 63 | 68 | 72 | 76 | 80 | 83 | 86 | 93 | 101 | 112 | 134 | 151 | 183 | 251 |
| | 0.91 | 54 | 61 | 68 | 73 | 77 | 81 | 85 | 90 | 93 | 96 | 103 | 110 | 123 | 145 | 162 | 195 | 269 |
| .18 | | 38 | 47 | 53 | 59 | 63 | 67 | 71 | 75 | 78 | 81 | 87 | 96 | 106 | 126 | 142 | 172 | 236 |
| | 0.94 | 49 | 57 | 63 | 68 | 72 | 76 | 79 | 83 | 86 | 89 | 96 | 103 | 115 | 136 | 152 | 183 | 252 |
| .19 | | 36 | 44 | 50 | 55 | 59 | 63 | 67 | 70 | 73 | 77 | 82 | 90 | 101 | 119 | 136 | 163 | 227 |
| | 0.97 | 45 | 53 | 58 | 63 | 67 | 71 | 74 | 78 | 81 | 84 | 90 | 97 | 108 | 127 | 144 | 172 | 238 |
| .20 | | 34 | 42 | 47 | 52 | 56 | 60 | 64 | 67 | 69 | 73 | 77 | 85 | 96 | 112 | 129 | 154 | 214 |
| | 1.00 | 42 | 49 | 55 | 59 | 63 | 67 | 69 | 73 | 76 | 79 | 84 | 91 | 101 | 120 | 137 | 162 | 234 |
| .22 | | 30 | 37 | 42 | 47 | 51 | 54 | 57 | 60 | 62 | 65 | 70 | 76 | 86 | 102 | 116 | 139 | 194 |
| | 1.06 | 37 | 43 | 48 | 52 | 56 | 59 | 62 | 65 | 68 | 70 | 75 | 81 | 91 | 107 | 123 | 146 | 204 |
| .24 | | 27 | 34 | 39 | 42 | 46 | 49 | 52 | 54 | 57 | 59 | 63 | 69 | 78 | 93 | 105 | 128 | 178 |
| | 1.12 | 32 | 38 | 43 | 47 | 50 | 53 | 56 | 59 | 61 | 63 | 68 | 74 | 83 | 97 | 111 | 134 | 185 |
| .26 | | 25 | 31 | 35 | 38 | 42 | 44 | 47 | 49 | 52 | 54 | 58 | 63 | 71 | 85 | 96 | 117 | 163 |
| | 1.19 | 29 | 34 | 38 | 42 | 45 | 48 | 51 | 53 | 55 | 57 | 61 | 67 | 75 | 90 | 101 | 122 | 169 |
| .28 | | 22 | 28 | 32 | 35 | 38 | 41 | 43 | 45 | 48 | 49 | 53 | 58 | 65 | 78 | 90 | 107 | 152 |
| | 1.25 | 26 | 31 | 35 | 38 | 41 | 43 | 46 | 48 | 50 | 53 | 56 | 61 | 69 | 82 | 92 | 111 | 156 |
| .30 | | 21 | 26 | 30 | 32 | 35 | 37 | 40 | 42 | 44 | 45 | 49 | 53 | 61 | 72 | 83 | 100 | 142 |
| | 1.31 | 24 | 28 | 32 | 35 | 37 | 40 | 42 | 44 | 46 | 48 | 51 | 56 | 63 | 75 | 86 | 103 | 144 |
| .32 | | 19 | 24 | 27 | 30 | 33 | 35 | 37 | 39 | 40 | 42 | 45 | 50 | 56 | 68 | 76 | 93 | 131 |
| | 1.37 | 21 | 26 | 29 | 32 | 34 | 37 | 39 | 40 | 42 | 44 | 47 | 52 | 58 | 69 | 79 | 96 | 135 |
| .34 | | 18 | 22 | 25 | 27 | 30 | 32 | 34 | 36 | 38 | 39 | 42 | 46 | 52 | 63 | 72 | 87 | 123 |
| | 1.44 | 20 | 24 | 27 | 30 | 32 | 34 | 36 | 37 | 39 | 41 | 44 | 48 | 54 | 64 | 73 | 89 | 125 |
| .36 | | 16 | 20 | 23 | 26 | 28 | 30 | 32 | 33 | 35 | 37 | 39 | 43 | 49 | 59 | 67 | 81 | 115 |
| | 1.50 | 18 | 22 | 25 | 27 | 30 | 32 | 33 | 35 | 36 | 38 | 41 | 44 | 50 | 60 | 68 | 83 | 117 |
| .38 | | 15 | 19 | 22 | 24 | 26 | 28 | 30 | 31 | 33 | 34 | 37 | 40 | 45 | 55 | 62 | 76 | 108 |
| | 1.57 | 17 | 20 | 23 | 25 | 27 | 29 | 31 | 32 | 34 | 35 | 38 | 41 | 47 | 56 | 64 | 78 | 110 |

*(Continued)*

## dfHyp

| PV | d | 1 | 2 | 3 | 4 | 5 | 6 | 7 | 8 | 9 | 10 | 12 | 15 | 20 | 30 | 40 | 60 | 120 |
|---|---|---|---|---|---|---|---|---|---|---|---|---|---|---|---|---|---|---|
| .40 | 1.63 | 14 | 18 | 20 | 23 | 24 | 26 | 28 | 29 | 31 | 32 | 35 | 38 | 43 | 52 | 59 | 72 | 101 |
|     |      | **15** | **19** | **21** | **23** | **25** | **27** | **29** | **40** | **32** | **33** | **35** | **39** | **44** | **53** | **60** | **73** | **103** |
| .42 | 1.70 | 13 | 17 | 19 | 20 | 20 | 24 | 26 | 27 | 29 | 30 | 32 | 36 | 40 | 48 | 55 | 67 | 96 |
|     |      | **14** | **18** | **20** | **22** | **24** | **25** | **27** | **28** | **30** | **31** | **33** | **36** | **41** | **49** | **56** | **69** | **98** |
| .44 | 1.77 | 12 | 15 | 18 | 20 | 21 | 23 | 24 | 26 | 27 | 28 | 30 | 33 | 38 | 45 | 52 | 64 | 91 |
|     |      | **13** | **16** | **19** | **20** | **22** | **24** | **25** | **27** | **28** | **29** | **31** | **34** | **39** | **47** | **53** | **65** | **92** |
| .46 | 1.85 | 12 | 14 | 17 | 19 | 20 | 22 | 23 | 24 | 25 | 26 | 29 | 31 | 35 | 43 | 49 | 60 | 85 |
|     |      | **12** | **15** | **17** | **19** | **21** | **22** | **24** | **25** | **26** | **27** | **29** | **32** | **37** | **44** | **50** | **61** | **87** |
| .48 | 1.92 | 11 | 14 | 16 | 18 | 19 | 20 | 22 | 23 | 24 | 25 | 27 | 30 | 34 | 40 | 46 | 57 | 81 |
|     |      | **12** | **14** | **16** | **18** | **19** | **21** | **22** | **23** | **25** | **26** | **28** | **30** | **34** | **41** | **47** | **58** | **82** |
| .50 | 2.00 | 10 | 13 | 15 | 16 | 18 | 19 | 20 | 21 | 22 | 24 | 25 | 28 | 32 | 38 | 44 | 54 | 76 |
|     |      | **11** | **13** | **15** | **17** | **18** | **20** | **21** | **22** | **23** | **24** | **26** | **29** | **32** | **39** | **45** | **54** | **77** |
| .60 | 2.45 | 8  | 10 | 11 | 12 | 13 | 14 | 15 | 16 | 17 | 18 | 19 | 21 | 24 | 29 | 33 | 41 | 58 |
|     |      | **8**  | **10** | **11** | **13** | **14** | **14** | **16** | **16** | **17** | **18** | **19** | **21** | **24** | **29** | **34** | **41** | **59** |
| .70 | 3.06 | 6  | 7  | 8  | 9  | 10 | 11 | 11 | 12 | 13 | 13 | 15 | 16 | 18 | 22 | 25 | 30 | 44 |
|     |      | **6**  | **7**  | **8**  | **9**  | **10** | **11** | **12** | **12** | **13** | **13** | **14** | **16** | **18** | **22** | **25** | **31** | **44** |

Note: The $df_{err}$ for testing the nil hypothesis ($\alpha = .05$) are shown in plain text. The $df_{err}$ for testing the hypothesis that treatments account for 1% of the variance in the DV or less ($\alpha = .05$) are shown in boldface text below.

## df for Nil and Minimum-Effect Hypothesis Tests

### R Text for generating the $df_{err}$ needed to test the traditional Null Hypothesis with power = .80, α = .05

```
# dferr and N-needed for traditional null hypothesis to achieve power = 0.8
# User defined functions
noncenfn<-function(df, pv) df * pv / (1 - pv)
noncenfn2<-function(df, FF) df * FF
critFn<-function(df1, df2, alph, effect) qf((1-alph), df1, df2, noncenfn(df2, effect))
finddferr<-function(dferr) 1-pf(critFn(dfhyp, dferr, 0.05, mineffect), dfhyp, dferr, noncenfn(dferr, PV))-powtarget

# Global variables
dfhyps<-c(1, 2, 3, 4, 5, 6, 7, 8, 9, 10, 12, 15, 20, 30, 40, 60, 120)
PVs<-c(1:20, seq(22, 50, by = 2), 60, 70)

mineffect=0
powtarget=.8
Nneeded<-vector(length = length(dfhyps))

cat("PV d\t\t\t\t\tdfHyp\n")
cat(" ", dfhyps, "\n")

for (i in PVs) {
   PV=i/100
   cat(PV, round(2 * sqrt(PV) / sqrt(1 - PV), digits = 2))

   for (j in 1:length(dfhyps)) {
      dfhyp=dfhyps[j]
      k<-uniroot(finddferr, c(1,5000))
      Nneeded[j]=ceiling(k$root) # dferr needed
      # Nneeded[j]=ceiling(k$root + dfhyp + 1) # UNCOMMENT FOR Nneeded
      }
   print(Nneeded)
}
```

### R Text for generating the $df_{err}$ needed to test the Hypothesis that Treatments account for 1% or less of the variance with power =.80, α =.05

```
# dferr and N-needed for the hypothesis that treatments account for 1% of
the variance or less to achieve power = 0.8
# User defined functions
noncenfn<-function(df, pv) df * pv / (1 - pv)
```

```
noncenfn2<-function(df, FF) df * FF
critFn<-function(df1, df2, alph, effect) qf((1-alph), df1, df2, noncenfn(df2,
  effect))
finddferr<-function(dferr) 1-pf(critFn(dfhyp, dferr, 0.05, mineffect), dfhyp,
  dferr, noncenfn(dferr, PV))-powtarget

# Global variables
dfhyps<-c(1, 2, 3, 4, 5, 6, 7, 8, 9, 10, 12, 15, 20, 30, 40, 60, 120)
PVs<-c(2:20, seq(22, 50, by = 2), 60, 70)

mineffect=.01
powtarget=.8
Nneeded<-vector(length = length(dfhyps))

cat("PV d\t\t\t\t\tdfHyp\n")
cat(" ", dfhyps, "\n")

for (i in PVs) {
   PV=i/100
   cat(PV, round(2 * sqrt(PV) / sqrt(1 - PV), digits = 2))

   for (j in 1:length(dfhyps)) {
      dfhyp=dfhyps[j]
      k<-uniroot(finddferr, c(1,7000))
      Nneeded[j]=ceiling(k$root) # dferr needed
      # Nneeded[j]=ceiling(k$root + dfhyp + 1) # UNCOMMENT
      FOR Nneeded
      }
```

# References

Aguinis, H. (1995). Statistical power problems with moderated multiple regression in management research. *Journal of Management*, 21, 1141–1158.

Aguinis, H. (2004). *Regression analysis for categorical moderators*. New York: Guilford.

Aguinis, H., Beaty, J. C., Boik, R. J. & Pierce, C. A. (2005). Effect size and power in assessing moderating effects of categorical variables using multiple regression. *Journal of Applied Psychology*, 90, 94–107.

Aguinis, H. & Gottfredson, R. K. (2010). Best-practice recommendations for estimating interactions using moderated multiple regression. *Journal of Organizational Behavior*, 31, 776–786.

Aguinis, H., Gottfredson, R. K. & Culpepper, S. A. (2013). Best-practice recommendations for estimating cross-level interaction effects using multilevel modeling. *Journal of Management*, 39(6), 1490–1528.

Algina, J. & Keselman, H. J. (1997). Detecting repeated measures effects with univariate and multivariate statistics. *Psychological Methods*, 2, 208–218.

Amrhein, V., Greenland, S. & McShane, B. (2019). Scientists rise up against statistical significance. *Nature*, 567, 305–307.

Banks, G. C., Rogelberg, S. G., Woznyj, H. M., Landis, R. S. & Rupp, D. E. (2016). Editorial: Evidence on questionable research practices: The good, the bad and the ugly. *Journal of Business and Psychology*, 31, 323–338.

Beaty, J. C., Cleveland, J. N. & Murphy, K. R. (2001). The relationship between personality and contextual performance in "strong" versus "weak" situations. *Human Performance*, 14, 125–148.

Bedeian, A. G., Taylor, S. G. & Miller, A. N. (2010). Management science on the credibility bubble: Cardinal sins and various misdemeanors. *Academy of Management Learning & Education*, 9, 715–725.

Bettis, R. A., Ethiraj, S., Gambardella, A., Helfat, C. & Mitchell, W. (2016). Creating repeatable cumulative knowledge in strategic management: A call for a broad and deep conversation among authors, referees, and editors. *Strategic Management Journal*, 37, 257–261.

Bikel, R. (2007). *Multilevel analysis for applied research: It's just regression*. New York: Guilford Press.

Bliese, P. D. & Hanges, P. J. (2004). Being both too liberal and too conservative: The perils of treating grouped data as though they were independent. *Organizational Research Methods* 7(4), 400–417.

Bosco, F. A., Aguinis, H., Singh, K., Field, J. G. & Pierce, C. A. (2015). Correlational effect size benchmarks. *Journal of Applied Psychology*, *100*, 431–449.

Bradley, D. R. & Russell, R. R. (1998). Some cautions regarding statistical power in split-plot designs. *Behavioral Research Methods: Instruments and Computers*, *30*, 462–477.

Bunce, D. & West. M. A. (1995). Self-perceptions and perceptions of group climate as predictors of individual innovation at work. *Applied Psychology: An International Review*, *44*, 199–215.

Carroll, J. B. (1993). *Human cognitive abilities: A survey of factor-analytic studies*. Cambridge, England: Cambridge University Press.

Cascio. W. F. & Zedeck, S. (1983). Open a new window in rational research planning: Adjust alpha to maximize statistical power. *Personnel Psychology*, *36*, 517–526.

Clapp, J. F. & Rizk, K. H. (1992). Effect of recreational exercise on midtrimester placental growth. *American Journal of Obstetrics and Gynecology*, *167*, 1518–1521.

Cohen. J. (1962). The statistical power of abnormal-social psychological research. *Journal of Abnormal and Social Psychology*, *65*, 145–153.

Cohen. J. (1988). *Statistical power analysis for the behavioral sciences* (2nd cd.). Hillsdale, NJ: Lawrence Erlbaum Associates.

Cohen. J. (1994). The earth is round (p < .05). *American Psychologist*, *49*, 997–1003.

Cohen. J. & Cohen, P. (1983). *Applied multiple regression/correlation analysis for the behavioral sciences*. Hillsdale, NJ: Lawrence Erlbaum Associates.

Cohen. J., Cohen, P., West, S. G. & Aiken, L. S. (2003). *Applied multiple regression/correlation analysis for the behavioral sciences* (3rd ed.). Mahwah, NJ: Lawrence Erlbaum Associates.

Cortina, J. M. & Dunlap, W. P. (1997). On the logic and purpose of significance testing. *Psychological Methods*, *2*, 161–173.

Cowles, M. (1989). *Statistics in psychology: An historical perspective*. Hillsdale, NJ: Lawrence Erlbaum Associates.

Cowles, M. & Davis, C. (1982). On the origins of the .05 level of statistical significance. *American Psychologist*, *37*, 553–558.

Dumas-Mallet, E., Button, K. S., Boraud, T., Gonon, F. & Mufanò, M. R. (2017). Low statistical power in biomedical science: A review of three human research domains. *Royal Society Open Science*, *4*(2), 160254.

Fanelli, D. (2009). How many scientists fabricate and falsify research? A systematic review and meta-analysis of survey data. *PLoS One*, *4*, e5738.

Ferguson, C. J. (2009). An effect size primer: A guide for researchers and clinicians. *Professional Psychology: Research and Practice*, *40*, 532–538.

Fick, P. L. (1995). Accepting the null hypothesis. *Memory and Cognition*, *23*, 132–138.

Fidler, F., Thomason, N., Cumming, G., Finch, S. & Leeman, J. (2004). Editors can lead researchers to confidence intervals but can't make them think. *Psychological Science*, *15*, 119–126.

Glass, G. V., McGaw, B. & Smith, M. L. (1981). *Meta-analysis in social research*. Berkely, CA: Sage.

Greenhouse, S. W. & Geisser. S. (1959). On method in the analysis of profile data. *Psychometrika*, *24*, 95–112.

Greenwald. A. G. (1993). Consequences of prejudice against the null hypothesis. In G. Keren & C. Lewis (Eds.). *A handbook for data analysis in the behavioral sciences: Methodological issues* (pp. 419–448). Hillsdale, NJ: Lawrence Erlbaum Associates.

Grissom, R. J. (1994). Probability of the superior outcome of one treatment over another. *Journal of Applied Psychology*, *79*, 314–316.

Guilford, J. P. & Fruchter, B. (1978). *Fundamental statistics in psychology and education* (6th ed.). New York: McGraw-Hill.

Gutenberg, R. L., Arvey, R. D., Osburn, H. G. & Jenneret, P. R. (1983). Moderating effects of decision-making/information processing job dimensions on test validities. *Journal of Applied Psychology*, 68, 602–608.

Haase, R. R., Waechter, D. M. & Solomon, G. S. (1982). How significant is a significant difference? Average effect size of research in counseling psychology. *Journal of Counseling Psychology*, 29, 58–65.

Hagen, R. L. (1997). In praise of the null hypothesis statistical test. *American Psychologist*, 52, 15–24.

Hedges, L. (1981). Distribution theory for Glass's estimator of effect size and related estimators. *Journal of Educational Statistics*, 6, 107–128.

Hedges, L. V. (1987). How hard is hard science, how soft is soft science? *American Psychologist*, 42, 443–455.

Hedges, L. V. & Olkin, I. (1985). *Statistical method for meta-analysis*. New York: Academic Press.

Himel, H. N., Liberati. A., Laird, R. D. & Chalmers, T. C. (1986). Adjuvant chemotherapy for breast cancer: A pooled estimate based on published randomized control trials. *Journal of the American Medical Association*, 256, 1148–1159.

Hoenig, J. M. & Heisey, D. M. (2001). The abuse of power: The pervasive of power calculations for data analysis. *The American Statistician*, 55, 19–24.

Horton, R. L. (1978). *The general linear model: Data analysis in the social and behavioral sciences*. New York: McGraw-Hill.

Hox, J. J., Moerbeek, M. & van de Schoot, R. (2018). *Multilevel analysis: Techniques and applications*. New York: Routledge.

Hunter. J. E. & Hunter, R. F. (1984). Validity and utility of alternative predictors of job performance. *Psychological Bulletin*, 96, 72–98.

Hunter, J. E. & Hirsh, H. R. (1987). Applications of meta-analysis. In C. L. Cooper & I. T. Robertson (Eds.). *International review of industrial and organizational psychology* (pp. 321–357). Chichester, England: Wiley.

Hunter, J. E. & Schmidt, F. L. (1990). *Methods of meta-analysis: Correcting error and bias in research findings*. Newbury Park, CA: Sage.

Ioannidis, J. P. (2005). Why most published research findings are false. *PLOS Medical, 2*, e124.

John, L. K., Loewenstein, G. & Prelec, D. (2012). Measuring the prevalence of questionable research practices with incentives for truth-telling. *Psychological Science*, 23, 524–532.

Kelly, K. (2008). Sample size planning for the squared multiple correlation coefficient: Accuracy in parameter estimation via narrow confidence intervals. *Multivariate Behavioral Research*, 43, 524–555.

Kelly, K. & Lai, K. (2011). Accuracy in parameter estimation for the root mean square error of approximation: Sample size planning for narrow confidence intervals. *Multivariate Behavioral Research*, 46, 1–32.

Kelly, K. & Preacher, K. J. (2012). On effect size. *Psychological Methods*, 17, 137–152.

Kelley, K. & Rausch, J. R. (2006). Sample size planning for the standardized mean difference: Accuracy in parameter estimation via narrow confidence intervals. *Psychological Methods*, 11, 363–385.

Keselman, H. J., Miller, C. W. & Holland, B. (2011). Many tests of significance: New methods for controlling type I errors. *Psychological Methods*, 16, 420–431.

# References

Kirk, R. (1995). *Experimental design: Procedures for the behavioral sciences* (3rd ed.). CA: Brooks/Cole.

Kirk, R. (1996). Practical significance: A concept whose time has come. *Educational and Psychological Measurements, 56*, 746–759.

Kraemer, H. C. & Thiemann, S. (1987). *How many subjects?* Newbury Park. CA: Sage.

Kreft, I. (1996). Are multilevel techniques necessary? An overview including simulation studies. Unpublished manuscript, California State University, Los Angeles.

Labovitz, S. (1968). Criteria for selecting a significance level: A note on the sacredness of .05. *American Sociologist, 3*, 220–222.

Lai, K. & Kelly, K. (2012). Accuracy in parameter estimation for targeted effects in structural equation modeling: Sample size planning for narrow confidence intervals. *Psychological Methods, 16*, 127–148.

Landy, F. J., Farr, J. L. & Jacobs, R. R. (1982). Utility concepts in performance measurement. *Organizational Behavior and Human Performance. 30*, 15–40.

Levine, M. & Epsom, M. H. H. (2001). Post-hoc power analysis: An idea whose time has passed? *Pharacotherapy, 21*, 405–409.

Lipsey, M. W. (1990). *Design sensitivity*. Newbury Park. CA: Sage.

Lipsey, M. W. & Wilson, D. B. (1993). The efficacy of psychological, educational, and behavioral treatment. *American Psychologist, 48*, 1181–1209.

Liu, H. & Yuan, K. (2021). New measures of effect size in moderation analysis. *Psychological Methods, 26*, 680–700.

Maas, C. J. M. & Hox, J. J. (2005). Sufficient sample sizes for multilevel modeling. *Methodology, 1*, 86–92.

Maxwell. S. E. & Delaney, H. D. (1990). *Designing experiments and analyzing data*. Belmont. CA: Wadsworth.

Maxwell. S. E., Kelly, K. & Rausch, J. R. (2008). Sample size planning for statistical power and accuracy in parameter estimation. *Annual Review of Psychology, 59*, 537–563.

McDaniel. M. A. (1988). Does pre-employment drug use predict on-the-job suitability? *Personnel Psychology, 41*, 717–729.

McGraw, K. O. & Wong, S. P. (1992). A common language effect size statistic. *Psychological Bulletin, 111*, 361–365.

McNutt, M. (2014). Reproducibility. *Science, 343*, 229.

Meehl, P. (1978). Theoretical risks and tabular asterisks: Sir Karl, Sir Ronald, and the slow progress of psychology. *Journal of Consulting and Clinical Psychology, 46*, 806–834.

Meinck, S. & Vandenplas, C. (2012). Sample size requirements in HLM: An empirical study. IERI Monograph Series, Special Issue 1. Educational Testing Service, Princeton: NJ.

Mone, M. A., Mueller, G. C. & Mauland, W. (1996). The perceptions and usage of statistical power in applied psychology and management research. *Personnel Psychology, 49*, 103–120.

Morrison. D. E. & Henkel, R. E. (1970). *The significance test controversy: A reader*. Chicago: Aldine.

Murphy, K. R. (1990). If the null hypothesis is impossible, why test it? *American Psychologist, 45*, 403–404.

Murphy, K. R. (2021). In praise of Table 1: The importance of making better use of descriptive statistics. *Industrial and Organizational Psychology: Perspectives on Science and Practice, 14*, 461–477.

Murphy, K. R. & Aguinis, H. (2019). HARKing: How badly can cherry picking and question trolling produce bias in published results? *Journal of Business and Psychology, 34*, 1–17.

Murphy, K. R. & Myors, B. (1999). Testing the hypothesis that treatments have negligible effects: Minimum-effect tests in the general linear model. *Journal of Applied Psychology, 84,* 234–248.

Murphy, K. R. & Russell, C. J. (2017). Mend it or end it: Redirecting the search for interactions in the organizational sciences. *Organizational Research Methods, 20,* 549–573.

Nagel, S. S. & Neff, M. (1977). Determining an optimal level of statistical significance. *Evaluation Studies Review Annual, 2,* 146–158.

Newman, D. A., Jacobs, R. R. & Bartram, D. (2007). Choosing the best method for local validation: Relative accuracy of a local study versus Bayes analysis. *Journal of Applied Psychology, 92,* 1394–1413.

O'Boyle, E., Banks, G. C., Carter, K., Walter, S. & Yuan, Z. (2019). A 20-year review of outcome reporting bias in moderated multiple regression. *Journal of Business and Psychology, 34,* 19–37.

O'Brien, R. G. & Castelloe, J. M. (2007). Sample size analysis for traditional hypothesis testing: concepts and issues. In A. Dmitrienko, C. Chuang-Stein & R. D'Agostino (Ed.). *Pharmaceutical statistics using SAS: A practical guide* (pp. 237–271). Cary, NC: SAS.

Open Science Collaboration (2015). Estimating the reproducibility of psychological science. *Science 349,* aac4716. DOI: 10.1126/science.aac4716

Osburn, H. G., Callender, J. C., Greener, J. M. & Ashworth, S. (1983). Statistical power of tests of the situational specificity hypothesis in validity generalization studies: A cautionary note. *Journal of Applied Psychology, 68,* 115–122.

Overall, J. E. (1996) How many repeated measurements are useful? *Journal of Clinical Psychology, 52,* 243–252.

Pashler, H. & Wagenmakers, E. J. (2012). Editors' introduction to the special section on replicability in psychological science: A crisis of confidence? *Perspectives on Psychological Science, 7,* 528–530.

Patnaik, P. B. (1949). The non-central t- and F-distributions and their applications. *Biometrika, 36,* 202–232.

Pearson, E. S. & Hartley, H. O. (1951). Charts of a power-function for analysis of variance tests, derived from the non-central F-distribution. *Biometrika, 38,* 112–130.

Preacher, K. J. & Kelley, K. (2011). Effect size measures for mediation models: Quantitative strategies for communicating indirect effects. *Psychological Methods, 16,* 93–115.

Raudenbush, S. W. & Bryk, A. S. (2002). *Hierarchical linear models: Applications and data analysis methods* (2nd ed.). Thousand Oaks, CA: Sage.

Rights, J. D. & Sterba, S. K. (2019). Quantifying explained variance in multilevel models: An integrative framework for defining R-squared measures. *Psychological Methods, 24,* 309–338..

Rights, J. D. & Sterba. S. K. (2020) New recommendations on the use of R-squared differences in multilevel model comparisons. *Multivariate Behavioral Research, 55*(4), 568–599.

Rosenthal, R. (1991). *Meta-analytic procedures for social research.* Newbury Park. CA: Sage.

Rosenthal. R. (1993). Cumulating evidence. In G. Keren & C. Lewis (Eds.). *A handbook for data analysis in the behavioral sciences: Methodological issues* (pp. 519–559). Hillsdale, NJ: Lawrence Erlbaum Associates.

Rouanet, H. (1996). Bayesian methods for assessing the importance of effects. *Psychological Bulletin, 119,* 149–158.

Ryan, T. A. (1962). The experiment as the unit for computing rates of error. *Psychological Bulletin, 59,* 301–305.

Sackett, P. R., Harris, M. M., and Orr, J. M. (1986). On seeking moderator variables in the meta-analysis of correlational data: A Monte Carlo investigation of statistical power and resistance to Type I error. *Journal of Applied Psychology, 71,* 302–310.

Scherbaum, C. A. & Ferreter, J. M. (2009). Estimating statistical power and required sample sizes for organizational research using multilevel modeling. *Organizational Research Methods, 12,* 347–367..

Schmidt, F. L. (1992). What do the data really mean? Research findings, meta-analysis and cumulative knowledge in psychology. *American Psychologist, 47,* 1173–1181.

Schmidt, F. L. (1996). Statistical significance testing and cumulative knowledge in psychology: Implications for training of researchers. *Psychological Methods, 1,* 115–129.

Schmidt, F. L. & Hunter, J. E. (1998). The validity and utility of selection methods in personnel psychology: Practical and theoretical implications of 85 years of research findings. *Psychological Bulletin, 124,* 262–274.

Schmidt, F. L., Hunter, J. E., McKenzie. R. C. & Muldrow. T. W. (1979). Impact of valid selection procedures on work-force productivity. *Journal of Applied Psychology, 71,* 432–439.

Schmidt. F. L., Mack, M. J. & Hunter. J. E. (1984). Selection utility in the occupation of U.S. park ranger for three modes of test use. *Journal of Applied Psychology, 69,* 490–497.

Sedlmeier, P. & Gigerenzer, G. (1989). Do studies of statistical power have an effect on the power of studies? *Psychological Bulletin, 105,* 309–316.

Serlin. R. A. & Lapsley, D. K. (1985). Rationality in psychological research: The good-enough principle. *American Psychologist, 40,* 73–83.

Serlin, R. A. & Lapsley. D. K. (1993). Rational appraisal of psychological research and the good-enough principle. In G. Keren & C. Lewis (Eds.). *A handbook for data analysis in the behauioral sciences: Methodological issues* (pp. 199–228). Hillsdale, NJ: Lawrence Erlbaum Associates.

Siegel. S. (1956). *Nonparametric statistics for the behavioral sciences.* New York: McGraw-Hill.

Snijders, T. A. B. (2005). Power and sample size in multilevel linear models'. In B. S. Everitt and D. C. Howell (Eds.). *Encyclopedia of statistics in behavioral science.* Volume 3 (pp. 1570–1573). Chicester, UK: Wiley.

Snijders, T. A. B. & Bosker, R. J. (2012). *Multilevel analysis: An introduction to basic and advanced multilevel modeling* (2nd ed.). London: Sage.

Stevens, J. P. (1980). Power of multivariate analysis of variance tests. *Psychological Bulletin, 86,* 728–737.

Stevens, J. (1988). *Applied multivariate statistics for the social sciences.* Hillsdale, NJ: Lawrence Erlbaum Associates.

Stevens, J. P. (2002). *Applied multivariate statistics for the social sciences.* (4th ed.). Hillsdale, NJ: Lawrence Erlbaum Associates.

Tatsuoka. M. (1993a). Effect size. In G. Keren & C. Lewis (Eds.). *A handbook for data analysis in the behavioral sciences: Methodological issues* (pp. 461–479). Hillsdale.NJ: Lawrence Erlbaum Associates.

Tatsuoka, M. (1993b). Elements of the general linear model. In G. Keren & C. Lewis (Eds.). *A handbook for data analysis in the behavioral sciences: Statistical issues* (pp. 3–42). Hillsdale, NJ: Lawrence Erlbaum Associates.

Tukey, J. W. (1949). Comparing individual means in the analysis of variance. *Biometrics, 5,* 99–114.

Valentine, J. C. & Cooper, H. (2003). *Effect size substantive interpretation guidelines: Issues in the interpretation of effect sizes.* Washington, DC: What Works Clearinghouse.

Van Iddekinge, C. H., Aguinis, H., LeBreton, J. M., Mackey, J. D. & DeOrtentiis, P. S. (2021). Assessing and interpreting interaction effects: A reply to Vancouver, Carlson, Dhanani, and Colton (2021). *Journal of Applied Psychology*, 106, 476–488.

Vonesh, E. F. & Schork, M. A. (1986). Sample sizes in the multivariate analysis of repeated measurements. *Biometrics*, 42, 601–610.

Wasserstein, R. L. & Lazar, N. A. (2016). The ASA's statement on p-values: Context, process, and purpose. *The American Statistician*, 70(2), 129–133.

Wasserstein, R. L., Schirm, A. L. & Lazar, N. A. (2019). Moving to a world beyond "$p < 0.05$". *The American Statistician*, 73(sup1), 1–19.

Wilcox. R. R. (1992). Why can methods for comparing means have relatively low power, and what can you do to correct the problem? *Current Directions in Psychological Science*, 1, c101–c105.

Wilkinson. L. & Task Force on Statistical Inference (1999). Statistical methods in psychology journals: Guidelines and explanations. *American Psychologist*, 54, 594–604.

Wolach, A. H. (1983). *BASIC analysis of variance programs for microcomputers*. Belmont, CA: Brooks/Cole.

Wolach, A. H. & McHale, M. A. (1987). F-Ratios for fixed, mixed, and random model ANOVAs. *Behavior Research Methods, Instruments, & Computers*, 19, 409–412.

Yates, F. (1933). The analysis of replicated experiments when the field results are incomplete. *Empire Journal of Experimental Agriculture*, 1, 129–142.

Yuen, K. K. (1974). The two-sample trimmed t for unequal population variances. *Biometrika*, 61, 165–170.

Zimmerman. D. W. & Zumbo, B. D. (1993). The relative power of parametric and nonparametric statistical methods. In G. Keren & C. Lewis (Eds.). *A handbook for data analysis in the behavioral sciences: Methodological issues* (pp. 481–518). Hillsdale. NJ: Lawrence Erlbaum Associates.

Zwick, R. & Marascuilo, L. A. (1984). Selection of pairwise comparison procedures for parametric and nonparametric analysis of variance models. *Psychological Bulletin*, 95, 148–155.

# Index

**NOTE**: Please note that the numbers in bold represents 'tables' and numbers in italics represents 'figures'. Also note that the numbers followed by 'n' represents notes respectively.

accuracy in parameter estimation (AIPE) 25–6
*alpha* error 5
alternative hypothesis 52; domain of 55
alternative multilevel models: comparing and making sense of 157
Analysis of variance (ANOVA) 111–13, 119; results, Rat Running Study **112**

*beta* error 6
between-groups design 138
between *vs* within-subjects designs: sources of variance in **139**

carry-over effects 137
Chi-squared values 165–6
Chi-square test 164
choosing significance levels: balancing risks in 85–7
Cochrane collaboration 65
cognitive ability 16
common language effect size (CLE) 31
conditional probability 6–7; *sensitivity* 6; *specificity* 6
confidence intervals: for $PV$ and $d$ 35
confidence intervals 23–4
correlation and regression analysis: low power in tests for moderators, implications of 100–101; multiple regression 93–7; testing for moderators, power in 97–9; working with large samples, perils of 90–3

data collection design 149
deductive methods 77–9

descriptive statistics 1
deviance 156
different questions imply different levels of power 121–2

effect size 30–1, 62–3; conventions **43**, 78–9; focus on 183; measures 34
effect size information from $F$ Ratios, retrieving 113
Epsilon 142
ES levels 23

factorial experiment *120*
$F$-equivalents 40, 96; and power, calculation 77–81
fixed/mixed/random models: denominators for **134**
$F$ statistic 34; significance of 63
$F$ value 41–2

Group X Occasion factor 142

HARKing 181
Hawthorn effects 56
hierarchical regression 38–9; models 95–6; results reported in Bunce and West **95**
highest mean 115
*Human Performance* 173
hypothesis tests 23–4

inductive methods 74–7
inferential statistics 1
Interpreting Research 180
*Intraclass correlation* 155

large samples 182–3
least significant difference (LSD) procedure 114–15; power for 115
linear models 32–3, 43; complex 51
lowest mean 115

maximizing power 171–2
Mean squares (MS) 112
"minimum-effect" hypotheses 59, 107, 167; are meaningful 60; definition 64–7; power of 67; reasonable chance of being true 61–2; testing 68, 94; tests of 63; Type I error 71; worked example of 69
minimum effect size 50
moderated multiple regression **159**
"moderated" relationship 150
moderator effects small 98–9
more realistic decision table **58**
multifactor ANOVA designs: calculating PV from F and df in 124–5; complete ANOVA Table **129**; estimating power in 122–3; factorial Analysis of Variance 119–26; factorial ANOVA from Means and Standard Deviations 126–31; fixed/mixed/random models 133–4; general design principles 131–2; PV from F, estimating 123–4; reconstructing from descriptive statistics 127–8; sources of variance, shell of **127**; study design *126*; study results *127*
multilevel analysis: preliminaries 157; symbols 154; terms 154
multiple regression models 93–5
multivariate analysis of variance (MANOVA) 147

N, ambiguity of 166
"negligible" effect 46
nil hypothesis 4, 52, 59, 167; is impossibly precise 55
nil is a point hypothesis 54–5
nil treatment effects: in World of Abstractions 55–6
noncentral F 36–7; distribution to assess power 37–8
noncentrality parameter 29, 169n6
nonparametric tests 44
non-treatment/control group 2
null, accepting the 173–6
null hypothesis 4, 25–7; critics of 185
null umbrella 60

number of observations required to achieve power **137**
number of subjects required to achieve power **138**

omega squared 130
one-sample test 104
One-Stop F Calculator 94
One-Stop F Table 45, 68, 91, 94
"one-tailed" statistical test 3
one-tailed test 109
Open Science Consortium (2015) study 178
ordinary least squares (OLS) 154

partial eta squared 130
percentage of variance (PV) 31; values **81**
*Perspectives on Psychological Science* 177
Placental Volumes Report **106**
Planning Research 179–80
polar bear traps 57–8
post-hoc power analyses 23
power, estimation 105–6
power analyses 113; as alternatives to traditional null hypothesis tests 59–63; analytic and tabular methods of 44–5; applying 22–3; appropriate decision criteria, determining 85–9; avoid post-hoc power analyses 89; benefits/costs/implications for hypothesis testing 179; calculating power 79–81; conclusion, implications of 56–8; costs associated with 183–4; direct benefits of 179–81; effect size 30–3, 74–9; essentials of *12*; four applications of 79; F Statistic 30–3; from F to power analysis 44; general linear model 30–3; an illustration of 156–8; implications of 170, 184–5; indirect benefits of 182–3; intraclass correlation 155–6; large/medium/small effects 42–3; minimum-effect hypotheses, tests of 52, 172–8; minimum-effect hypothesis, testing 63–7; for moderators 99–100; multilevel equation 152–5; for multilevel studies 149; nil hypothesis testing 52–4; noncentral F 36–8; nonparametric and robust statistics 43–4; one-stop tables 67–71, 75–6; power tables 45; quantitative ability study 166–8; R Code/Shiny Web App 75–6; simple and general model 28;

236  Index

simple and general software for 47–9; traditional null hypothesis, tests 170–2; regression 159–60; sample size 81–4, 168–9; sensitivity of studies, determining 84–5; transforming from F to PV 41–2; translating common statistics and ES measures into F 38–41; Type I error 71–2; using 20–1, 73; using changes in model fit 164–8; using One-Stop F table 46–7; web-based app for 49–51
power and replication crisis 18
power estimation 92–3, 96–7, 107–8, 129–30
power levels 23
pre-employment drug use report, predictive validity of **91**
pre-test and post-test comparison **110**

quantitative ability study **167**

random intercepts 155
randomized Block Study: sources of variance in **136**
range hypothesis 60–1
R code 47–9, 67, 69, 82, 146, 161–2
regression coefficient 158
rejecting hypothesis, probability of **72**
repeated-measures and multivariate analyses: complexities in estimating power in 141–2; independent groups vs repeated measures 137–41; mixed designs 142–4; multivariate analysis of variance 147–8; randomized Block ANOVA 135–7; split-plot designs 146–7; within-subject vs between-subject factors, power for 145–6
repeated-measures design 139
replication crisis 177–8
Ryan's procedure 115–17; order means used in **115**

sample size 22–3, 83–4, **117**
sample size estimation 97, 108
sampling distributions 9–10
sampling error 2
sensible alpha 88–9
sensitive procedures 182–3
Shiny Web app 47, 67, 69, 82, 94, 116, 118n1, 146

simple linear regression 169n2
singly noncentral F distribution 36
smoking reduction treatments, re-analysis of 109
soft science 180
split-plot factorial design *143*, **144**; estimating power 145; with two repeated-measures factors **147**
squared multiple correlation coefficient 90
statistical power 177–8; steps to determine **13**
statistical power analysis 172
statistical significance 12, 14, 19, 22, 23, 73, 80, 85, 179
statistical tests, power of: decision criteria on power 7–9, 16–17; errors in *56*; general linear model 17; hypothesis tests vs. confidence intervals 23–6; outcomes of *5*, 5–7; power, desired level of 22; power analysis, mechanics of 9–14, 17, 20–3; sensitivity effects 7–9, 14; size effects 7–9, 14–16, **15**, 21–2; structure of 2–5
statistic delta 16

t distribution vs normal distribution 106–7
testing minimum-effect hypotheses: balancing errors in 176–7
time spent reading and achievement: possible patterns of relationships 151
traditional nil hypothesis 54, 65
traditional null hypothesis 107
traditional vs Minimum-effect tests 91–2, 106–7
t Statistic: degrees of freedom **104**
t-Tests and the one-way analysis of variance: analysis of variance 111–13; designing 117; independent groups 105–8; one-tailed vs two-tailed tests 108–9; repeated measures/dependent t-Test 109–11
two-sample test 104
two-tailed test 109
Type I error 5, 27n1, 87–8, 170; in minimum-effect tests revisited 174–5
Type II error 6, 170
Type I vs Type II errors 87

utility theory 66

Made in the USA
Monee, IL
03 May 2026